Science and

Science and Fiction – A Springer Series

This collection of entertaining and thought-provoking books will appeal equally to science buffs, scientists and science-fiction fans. It was born out of the recognition that scientific discovery and the creation of plausible fictional scenarios are often two sides of the same coin. Each relies on an understanding of the way the world works, coupled with the imaginative ability to invent new or alternative explanations - and even other worlds. Authored by practicing scientists as well as writers of hard science fiction, these books explore and exploit the borderlands between accepted science and its fictional counterpart. Uncovering mutual influences, promoting fruitful interaction, narrating and analyzing fictional scenarios, together they serve as a reaction vessel for inspired new ideas in science, technology, and beyond.

Whether fiction, fact, or forever undecidable: the Springer Series "Science and Fiction" intends to go where no one has gone before!

Its largely non-technical books take several different approaches. Journey with their authors as they

- Indulge in science speculation – describing intriguing, plausible yet unproven ideas;
- Exploit science fiction for educational purposes and as a means of promoting critical thinking;
- Explore the interplay of science and science fiction – throughout the history of the genre and looking ahead;
- Delve into related topics including, but not limited to: science as a creative process, the limits of science, interplay of literature and knowledge;
- Tell fictional short stories built around well-defined scientific ideas, with a supplement summarizing the science underlying the plot.

Readers can look forward to a broad range of topics, as intriguing as they are important. Here just a few by way of illustration:

- Time travel, superluminal travel, wormholes, teleportation
- Extraterrestrial intelligence and alien civilizations
- Artificial intelligence, planetary brains, the universe as a computer, simulated worlds
- Non-anthropocentric viewpoints
- Synthetic biology, genetic engineering, developing nanotechnologies
- Eco/infrastructure/meteorite-impact disaster scenarios
- Future scenarios, transhumanism, posthumanism, intelligence explosion
- Virtual worlds, cyberspace dramas
- Consciousness and mind manipulation

More information about this series at http://www.springer.com/series/11657

Stephen Webb

All the Wonder that Would Be

Exploring Past Notions of the Future

Springer

Stephen Webb
DCQE
University of Portsmouth
Portsmouth, Hampshire
United Kingdom

ISSN 2197-1188   ISSN 2197-1196   (electronic)
Science and Fiction
ISBN 978-3-319-51758-2   ISBN 978-3-319-51759-9   (eBook)
DOI 10.1007/978-3-319-51759-9

Library of Congress Control Number: 2017935707

Printed on acid-free paper

This Springer imprint is published by Springer Nature
The registered company is Springer International Publishing AG
The registered company address is: Gewerbestrasse 11, 6330 Cham, Switzerland

To the memory of Isaac Asimov (1920–1992)

Acknowledgments

The seed for this book was sown at Loncon 3—the 72nd World Science Fiction Convention—during a discussion with Greg Benford, Nick Kanas, Angela Lahee, and Chris Caron. Thank you all for a stimulating conversation. I'd also like to thank Chris's ongoing guidance during the writing process: he gently pulled me back whenever I went off at a tangent.

A number of people, including Stephen Dow, Eric Mauk, Dave Robinson, and professors Steven Cummer, Francisco Lobo, Sir John Pendry, and David Schurig, freely gave their time and advice as I was chasing down photos and graphics. I appreciate their help.

My thanks, as always, go to Heike and Jessica for their support—and, in this digital age, for their tolerance of my ever-growing collection of paper-based books and magazines.

Contents

1

Introduction

I'd cross a bridge at night, and walk above the steel works.
So that's probably where the opening of Blade Runner comes from.
It always seemed to be rather gloomy and raining, and I'd just think
'God, this is beautiful.' You can find beauty in everything, and so
I think I found the beauty in that darkness.

Ridley Scott
Director of *Blade Runner*

We live in a science-fictional world. In recent months, as I write this, American technologists emailed a spanner to an orbiting space station, Russian biologists announced the formation of a DNA bank to store genetic information from all living creatures, and European scientists landed a spacecraft on a comet.[1] A comet! Future months are sure to bring fresh miracles, and the rate of change of scientific and technological advance is such that someone reading this book in a few years' time might well consider comet landings, DNA banks, and emailed hardware to be as dated as bevors, bodkins, and betamax tapes. Nevertheless, right here, right now, for someone of my generation, the continuing advances in computing and physics and biology are quite wondrous. We live in a science-fictional world. Except. . . to someone such as me, brought up on a diet of science fiction, one's wonder at our species' scientific and technological achievements is tinged with disappointment. Because our science-fictional world isn't quite the world that science fiction promised me.

As I write this, the year 2015 is drawing to a close. The science-fictional year 2015 is an interesting one, not as instantly recognisable as some of titular years—1984, 1999, 2001 and so on—but important nevertheless. Consider,

© Springer International Publishing AG 2017
S. Webb, *All the Wonder that Would Be*, Science and Fiction,
DOI 10.1007/978-3-319-51759-9_1

for example, Isaac Asimov's story "Runaround".[2] It's a notable story within the genre because it was the first in which Asimov explicitly stated his Three Laws of Robotics, but it had a wider influence, too: it led Marvin Minsky to ponder how minds work and begin his world-leading research into the nature of artificial intelligence. "Runaround" was set in 2015 and, three quarters of a century or so after it was written, we don't have the mining operations on Mercury that Asimov envisaged nor robots with the sophistication he portrayed.

Or consider Arthur C. Clarke's short story "Earthlight", written 10 years after Asimov's story but also set in 2015.[3] In "Earthlight" a permanent Moon base is used to coordinate uranium mining; Earth and a Federation of the outer planets of the solar system are at war over the supply of this precious material. Well, Clarke was renowned for his scientific vision—but it wasn't present in this story. In December 1972, Eugene Cernan shook the Moon dust from his boots as he climbed back into the Apollo 17 Lunar Module. Humankind has not been back since. We don't even have the ready capacity to return to the Moon, much less build a permanent base there. And as for a Federation of the outer planets—well, it hardly seems worth mentioning that humans aren't living on Mars, we aren't mining Jupiter's moons, and we aren't exploring Saturn's rings.

Or consider that terrific film trilogy *Back to the Future*, which in 2015 was re-released to mark the 30th anniversary of the first of the films. The second part of the trilogy has the crazy but lovable Dr Emmett Brown turn up on 26 October 1985 in a flying car/time machine in order to take his friend Marty McFly back to the future—they arrive on 21 October 2015. Putting to one side the implausibility of a flying DeLorean that doubles as a time machine, the future as glimpsed from 30 years ago simply didn't happen: children don't fly on hoverboards, none of us wear size-adjusting, self-drying jackets and lawyers are still with us. (The film's suggestion that lawyers would be abolished put perhaps the greatest strain on one's credulity. Flying time machines are one thing—but a future in which we succeed in getting rid of lawyers?)

The shadow of *Back to the Future Part II* loomed large when, in late 2014, the Springer editors and I conceived the notion of contrasting our present-day world with the future that science fiction had promised us all. I knew that 21 October 2015 would be the trigger for a torrent of blogs explaining how the film wrongly forecast the ubiquity of fax machines but totally missed out on the rise of smartphones; that on #BTTFday tweeps would briefly forego posting videos of cute cats and instead post comparisons of past and present; that newspaper hacks would write articles blaming science for our lack of self-drying clothes and hoverboards. I felt that all this activity, as well as threatening to blur what I had in mind for this book, rather missed the point of SF.

The best SF has never been about prophesying the future. Even as a child I wasn't naive enough to believe that my favorite stories contained definite predictions. Even then I understood that SF authors might be aiming for satire or attempting to highlight those paths into the future that held dangers; they certainly weren't trying to be Nostradamus. In any case, specific predictions were almost certainly going to turn out wrong. After all, if Asimov had known how to program his Three Laws into positronic brains then he'd have built his robots rather than write about them. If Clarke had known how to construct and maintain a Moon base then he'd have published a patent application rather than a story.[4] And Robert Zemeckis didn't direct *Back to the Future* as an exercise in prognostication. His film was meant to be inventive and thought-provoking, of course, but above all he meant it to be entertaining. On the other hand, I believe that SF—taken in the round—gives an excellent indication of the general 'feel' people have for how the future might turn out. The 'feel' changes over time, of course, since authors and directors are influenced by the society in which they live. In the 1950s, for example, the perceived threat posed by Communism produced SF stories tinged with paranoia; the identification of environmental threats in the 1970s produced much dystopian SF; and so on. Nevertheless, taken as a whole, all those stories, films, and TV programmes I consumed in my youth were part of the formation of a 'default future'—a shared sense from the SF community of how the world would turn out.

That default future was developed in the main by a relatively small number of visionaries, the brilliance of whose concepts fascinated the rest of the science fiction community. Other writers took those concepts and used them to form the backdrop of thousands of stories and hundreds of novels. Those concepts became part of the shared currency of science fiction, and in turn they influenced society in general. Of course, not all stories were set in the default future. But all SF writers of the Golden Age onwards were, I believe, influenced by the commonly accepted notion of humankind's destiny—if only to the extent that they deliberately questioned the assumptions behind that shared vision of future days.

Another way of putting this is to consider those old SF stories as *gedanken* experiments—thought experiments—with each one examining a possible future. When we integrate over all the outcomes the result is a broad-brush picture of what people believed (or feared) would be our fate. A more modern analogy might compare the activity of SF writers with that of climate modellers. The modellers don't claim to be able to predict the precise future of our planet's climate; rather, they run a series of computational simulations, each time tweaking a parameter here and a variable there, and thus provide us with a range of probabilities for the sort of climate our grandchildren will experience. The simulations indicate the trajectory we're following as we hurtle into the future (a trajectory suggesting that things are going to get hot) but they can't yet tell us *precisely* what climate conditions will obtain a century from now, much less what those conditions will mean for local weather patterns. It's the same with all that SF I read and watched: taken together, it represented not a specific set of predictions but rather a consensus view of what the future was likely to hold.

The purpose of the book, then, is to compare the default future of old-time science fiction with how things are turning out. I'm interested here less in definite prophecies (such as that by now we'd all be wearing shiny metallic suits—perhaps to protect one's backside from all those potential jetpack malfunctions) than in the once widely held beliefs about what the future would look like (that we'd be in space, that we'd have made contact with aliens, that we'd all have robot helpers. . .). Specific predictions added color to those big-picture ideas, of course. Asimov didn't just write about robots, for example, he wrote about robots with *positronic brains* and details such as that helped draw his readers in. But just as gadgetry was never Asimov's concern when he wrote his robot stories, nor should it be ours. Technologists will probably never develop a positronic brain, but Asimov's 'climate forecast'— that humans will one day share the planet with artificial intelligence—must be taken seriously: we might not yet have Asimovian-type robots, but we are starting to grapple with the moral and ethical questions that the existence of such robots might pose. So it's the comparison of classic science fictional themes with our modern understanding of science that I want to explore in this book.

You could argue that it's too early to make comparisons between classic SF and the real world because the predicted reality has yet to arrive. Many SF writers took the long view and set their stories in the far future: if a story were set in 10,000 AD, say, then it wouldn't matter whether it was written in 1850 by Jules Verne or in 1950 by Robert Heinlein—what was speculation then is speculation now and will remain speculation for centuries to come. In any case, SF writers weren't setting out a schedule for scientific and technological progress. All that is true. Nevertheless, we can at least make an informed judgement about whether we are following the trajectory sketched out by those early SF writers or heading instead on an altogether different path. The point is that science has made *enormous* strides in recent decades, and we now know much more about how the universe works than did those early SF writers. Furthermore, in the past three decades we have learned how certain processes scale well while others don't: the production of silicon-based integrated circuits scales well and so we have seen a revolution in computation; the burning of chemicals to power rockets does not scale well and so we haven't seen a similar improvement in space technology. In this book, therefore, I hope to explain the latest thinking in various fields of science in order to interpret those climate forecasts from SF.

However, before I start making those comparisons between science present and science fiction, I should clarify the type and extent of the fiction I'll be considering. My experience of SF, after all, is likely to be different to yours.

1.1   A Portrait of the Author as a Young SF Fan

I became a science fiction fan about 160 years ago—which is a statement I probably need to explain.

I was born and brought up in Middlesbrough, an industrial town in the north east of England, and it's reasonable to suppose my ancestors looked on my home town as a science-fictional place—though one with distinct dystopian overtones. Middlesbrough, you see, embodied that key science-fictional characteristic of change. In a few years it grew from a farmstead with a population of just 25 people to become one of the great industrial powerhouses of Victorian England.[5] The key to its rise was the opening in 1825 of the Stockton and Darlington Railway: this was a new form of mass transport, the world's first public railway to use steam locomotives. This transformative technology enabled the rapid movement of coal from Durham collieries to waiting ships on the River Tees. In 1829, investors bought the Middlesbrough farmstead and the surrounding land, and set about building a new coal port on the banks of the river along with a new town to supply the port with labour. The port was so successful that it made sense for business to construct a dock, which became home not only to the import and export trades but also of shipbuilding. Eight years after the dock opened people discovered ironstone in nearby hills, and industries based on iron and steel were added to the town's mix.

The discovery of ironstone coincided with an agricultural depression down the east coast of England, and the iron masters sent agents to recruit workers at agricultural hiring fairs. About 160 years ago, some time between 1853 and 1857, my father's great-grandfather made the move from rural Norfolk, where he worked as a farm laborer, to Middlesbrough, where he began laboring in the iron works. For someone who would have grown up tending sheep on small farms, the sights, sounds and smells of the newly industrialized town—'the infant Hercules' as the prime minister William Gladstone called it—must have seemed like sensations beamed direct from the future. It must also have been clear to my ancestor, and those many others who travelled from all over the country to find work in the new industries, that things would never be the same again: even the unschooled, as my great-great-grandfather was, must have understood the potential of trains, ironworks, and shipbuilding to transform the world.

In the early 1930s the existing heavy industries of Middlesbrough, and in particular the fledgling Brummer Mond/ICI chemical works in the neighboring town of Billingham, provided the inspiration for one of the most famous of all SF novels—*Brave New World* (1932) by Aldous Huxley. (Huxley was apparently impressed with the sight of the chemical works but was less taken with Middlesbrough itself, the development of which he compared to 'a fungus, like staphylococcus in a test-tube of chicken-broth'.) Four decades later, the town was a more direct inspiration for a trilogy of novels—*Interface* (1971), *Volteface* (1972), and *Multiface* (1975)—set in the fictional TCity. The trilogy's author, Mark Adlard, was a manager in the still-extant steel industry and he extrapolated the social changes that workers had endured in order to examine what the twenty-second century might hold for such people. Later still, as the opening quotation to this chapter suggests, the industry around the town provided the inspiration for one of the most famous scenes in filmed SF: the old ICI petrochemical plant at Wilton (Fig. 1.1), where my father worked for many years, was the prompt for the opening shot of the Ridley Scott classic *Blade Runner* (1982).[6]

So I come from a town that understands change; you could say, as I hinted at the start of this section, that it's in my DNA. Growing up, though, I didn't realize that my home town would undergo yet another transformation, that the heavy industries upon which its economy was based would soon be decimated. I was insulated from those concerns because my parents, who had faith in the importance of reading and of learning though they themselves had had only restricted educational opportunities, got me a ticket to the Junior section of the Middlesbrough Central Library almost as soon as I could read. The Central Library was an imposing structure, built in 1912 through the largesse of the philanthropist Andrew Carnegie, one of 660 such Carnegie libraries built in Britain and Ireland. It was a building with high ceilings and wooden parquet floors, with a strange odor that seemed to arise from a mixture of paper and polish. It was there that I discovered the *Danny Dunn* series of books by Raymond Abrashkin and Jay Williams.

Fig. 1.1 The Wilton cracker, flaring at night. The skyline of the industrial north east of England, which was once dominated by steel and petrochemical plant, is said to have been an inspiration for Ridley Scott's vision of Blade Runner (Credit: David Robinson)

I still fondly remember those books, which had titles such as *Danny Dunn and the Weather Machine* (1959) and *Danny Dunn and the Voice from Space* (1967), and I can still recall the thrill of finding a book in the series I hadn't already seen. Back then I knew nothing about the books themselves; now, thanks to the miracle of the internet, I can instantly learn more about them. For example, a search quickly shows that the series consisted of 15 books published between 1956 and 1977; that Abrashkin died from motor neurone disease after the fifth book but that Williams insisted on listing his friend as co-author on subsequent novels; and that first editions of certain *Danny Dunn* books can be surprisingly expensive. Back then, though, the books were simply something to read with joy.

Looking back, it seems entirely natural that I should have latched on to *Danny Dunn* because I was growing up in an environment suffused with elements of science fiction; it wasn't just the history of my home town that was science-fictional—science fiction seemed to fill the world around me. For example, along with probably everyone else growing up in the 1960s I was watching on in awe as astronauts aimed for the Moon. The entire Apollo program was at the same time both wildly futuristic and, perhaps because of its very success, somehow unremarkable. Another example: some of the most popular television shows for children had their roots in science fiction; a few of my own favorites included *Captain Scarlet and the Mysterons*, *Dr Who*, and *Star*

Trek—the name of whose hero, incidentally, Captain James Kirk, is said to have been inspired by my home town's most famous son, the explorer Captain James Cook. (As a child I dreamed of owning a *Star Trek* 'Communicator', a sort of mobile telephone that allowed Captain Kirk to talk to his crew no matter where his adventures had taken him. I now have a child of my own and she, along with all her classmates, owns a smartphone with functionality that makes the 'Communicator' look medieval. As #BTTFday made clear, neither *Star Trek* nor any other SF classic foresaw that we'd all have access to a device with the power of a smartphone. Equally, as we'll discuss later in the book, advances have not been equally distributed across all fields of endeavor: technologies such as time travel, teleportation, and warp drives seem as distant now as they were when *Star Trek* first aired.) A third example: a spate of 1950s SF films, many of which I still watch with undiluted pleasure, were readily available for impressionable youngsters such as myself. Science fiction, therefore, was something I took as a given; its influence was everywhere.

Having discovered *Danny Dunn* it seems equally natural that the discovery would lead me to seek out material offering similar thrills—and it did.

I should state at the outset that the *Danny Dunn* books didn't lead me to SF comics. I read football comics and funny comics but never, for some reason, SF comics. So nowhere in the pages of this book will you find mention of some of the most popular SF stories, containing such famous characters as Hulk and Superman, for the simple reason that I didn't read them and so I know nothing about them. The *Danny Dunn* books instead led me to other books. I remember discovering a series of novels by Hugh Walters, featuring the intrepid young space explorer Chris Godfrey, the earliest of which included *Blast Off at Woomera* (1957) and *The Domes of Pico* (1958). I remember enjoying *Trillions* (1971) by Nicholas Fisk and *Star Surgeon* (1960) by Alan Nourse, and a novel by Andre Norton whose title has long since escaped me (even my memory of the plot is evaporating like morning mist). And then in the shelves of the Junior section of the Middlesbrough Central Library I found a book by the Master.

I have no idea which was the first Asimov book I read (although I do remember how my young tongue insisted on pronouncing his name as 'Issac Asminov'). At some point, however, I realized two things. First, I adored his unadorned writing style. Second, he'd written such a bewilderingly large number of books that if I sought his stuff out then I'd never be at a loss for something to read. I read everything by him I could find—the *Foundation* series, of course; the robot novels and short stories; the anthologies (many of which volumes were adorned with gorgeous covers by the artist Chris Foss)—and then I moved on to Asimov's vast output of non-fiction. And it was here that I developed my interest in science and the scientific endeavor. It was because of Asimov that I decided to study science. Not only did that decision mean I avoided a manual job in one of my home town's dying heavy industries, it influenced how I think and how I view the world. Danny Dunn and Isaac Asimov helped shape who I am. (There's an alternative explanation. Perhaps I possess the sort of mind that's naturally drawn to the concrete nature of science and to writing that's straightforward and clear. Perhaps I'm naturally suspicious of authors who claim to tackle ambiguity and nuance. Who can tell?)

I didn't stop reading Asimov's fiction once I'd discovered his non-fiction. Indeed, his fiction led me to discover many of my favorite SF authors— Heinlein and Clarke, of course, but also Varley and Robinson and Martin. From these names, aficionados will recognize the type of fiction I tended to prefer: SF of the 'hard' variety rather than self-consciously 'literate' stuff.[7] I read a prodigious amount of SF, mainly stories and novels from the Golden Age of the 1940s and 1950s but also new material published in the 1970s and early 1980s. And I read classic SF with increasing pleasure. When I was growing up Asimov, Heinlein, and Clarke might have been science fiction's generally acknowledged 'Big Three', but probably even they would have agreed that H.G. Wells was the most influential of all SF writers.

That I had time to read so widely was in part thanks to science fiction itself: the education I received directly from the genre, combined with the non-fiction reading prompted by SF, meant I was able to complete my undergraduate physics degree with relative ease. To my dismay, that all changed when I began to study for a PhD. For the first time I had to work hard, *really* hard, to understand the material I was expected to read: making sense of a research paper in theoretical physics is a quite different proposition to studying the relevant parts of a textbook. Suddenly, from about 1985 onwards, I found I was reading much less SF than before. I'd still go to see the occasional film at a cinema, but my consumption of TV SF essentially dried up completely. I didn't lose all contact with SF: I continued to get my monthly fix from *Analog* and *Asimov's Science Fiction Magazine* (I own every issue); and when people whose taste I trusted told me I simply *had* to read this story or that novel then I tried my best to do so. But I have never read as much SF as I did before 1985 and so that year represents a cut-off point for me.

The following, then, is the material I plan to cover in this book. I'll explore what I consider to be the default future as constructed by SF from Verne and Wells onward, through the classic Golden Age of American SF, and up to 1985. Mainly I'll be interested in 'hard' SF, but that's such an ill-defined concept that I'll inevitably include a large amount of what others would define as 'soft' SF. My main interest is in the written word, but I'll also mention where relevant the SF of films and television. I won't, however, discuss SF comics or games.[8] The general picture of the future painted by SF will be compared with the situation as best we see it today: where are we now and, according to our current understanding of science, what is our likely short-term future? In an epilogue I'll look briefly at post-1985 SF and ask whether a new default future might be emerging.

The 1985 cut-off is not, I hope, too arbitrary. As I write this the year 1985 rests 30 years ago, a generation in the past. Science has progressed furiously in those three decades, so now is a good time to ask whether we are following the trajectories into the future laid out by that past science fiction. Besides, 1985 is the year *Back to the Future* was released and the year from which Dr Emmot Brown takes Marty McFly into his future 2015. So even though I earlier bemoaned the exercise, if nothing else it gives us a chance to ask—as it did for all those commentators on #BTTFday—'Dude, where's my flying car?'

1.2   Law Zero

Most books in this Springer series of *Science and Science Fiction* take the form of a sandwich, with small expository layers of science surrounding a healthy dollop of fictional filling. This book takes the opposite approach. What follows is a short story entitled "Law Zero". It's an homage to those old 1950s SF tales I so enjoyed reading in my youth. I've tried to match the tone of those stories but I've chosen to ignore one of the basic tenets of SF writing, namely the principle that an author is allowed just one impossibility per story. Instead, I've thrown in ten impossibilities or implausibilities. Each of those impossibilities is an element that helped form that old-fashioned default view of the future. Those ten elements were by no means the only preoccupations of science fiction writers, but I chose them because taken together they represent a clear view of the future as seen from the past—and also because they provided the inspiration for many of my favorite stories. Later in the book, each chapter focuses on one of those ten topics and compares the 'climate forecast' made by SF with the current view as given by science.

You won't lose any enjoyment of the rest of the book if you choose to ignore this fictional interlude—indeed, you'll probably find that by skipping the story your enjoyment will increase. However, for those among you brave enough to read the story, and with apologies to the memory of The Good Doctor, I've inserted a puzzle element to the tale that I hope some of you will appreciate: hidden in the story are more than 100 references to Asimov works (character names, allusions to famous lines, the titles of stories and books, and so on). See how many you can find.[9]

Law Zero

1

Darius Potterley, director of Harvard's Twentieth Century History unit, thrust out a belligerent lower lip and glared at the young robot. "Preposterous. The idea is preposterous. Why ever would I want to be marooned off Vesta, or Ceres, or whatever the hell it's called? Quite preposterous."

"The asteroid is AL-76." Aurora raised a porcelain cup to her nose and checked its contents were within specified parameters. Her master observed strict rules regarding afternoon tea—freshly boiled water poured over an Assam blend, brewed for four minutes, and served with a splash of milk. Satisfied, she reunited the cup with its saucer, paused, then added two small chocolate digestives. "It's an apohele asteroid, closer to Venus than it is to Vesta."

"I don't care if it's closer to Venice," Potterley grumbled. "I don't travel."

Aurora tilted her head and looked quizzically at Potterley. "An invitation from Carter Noÿs is a singular honor. He leads a reclusive life. I conducted a thorough search but failed to uncover any other instance of him inviting a person to his home."

"That I can believe." Potterley dunked a biscuit in his tea and, ignoring the wafer's rapidly declining structural integrity, made stabbing gestures with it. "Noÿs has all the grace and charm of Ghengis Khan. By restricting the number of invitations he's simply limiting the number of rejections."

"Carter Noÿs is the greatest scientific intellect of our age." Aurora wiped a dollop of wet biscuit from the cover of one of the books that smothered Potterley's desk. They were books of the old-fashioned kind, paper-based, the type that could sustain damage. "Many would argue the greatest of all time. Would you not like the chance to meet a mind of such puissance?"

Potterley's arm described an arc around his book-lined study. Wet crumbs fell to the floor. "For me history is alive, Aurora. Vibrant. As real as my surroundings. For a mind such as his—well, history is merely the dead past. Any interaction between us would be sterile."

"And yet Noÿs wants to meet you. I wonder why?"

"No idea. I'm surprised he even knows of my existence." Potterley folded his hands across his belly and leaned back in his cantilever chair. "I did once write an essay about him. But that was years ago."

"Were you particularly critical?"

Potterley shrugged. "I contextualized his undoubted achievements within the extant social and technological milieu. My particular area of study is twentieth century discovery, Aurora, and I compared his tactics with the approach taken by scientists in the middle decades of that tumultuous period. Nothing scandalous." The director licked crumbs from his fingers. "Anyway, I thought Noÿs was dead. Has he published anything recently?"

Aurora conjured up another biscuit. "Not for seven years. But rumors abound. They say he's working on something big."

"And how could the bottom-feeding denizens of online chatrooms possibly know what he's working on? Given that he himself hasn't published anything, met with anyone, or even, as far as we know, spoken to anyone? They couldn't! Hah!"

Potterley banged the desk, dislodging three journals from the top of a stack of printed periodicals. Aurora watched them topple and, moving quicker than the eye, caught them before they hit the floor. The rapidity of her reactions was still a source of fascination for her. There were occasions when time seemed to run slow, as if she were operating a dozen times faster than humans around her. She tried to imagine what it must be like to live without those reactions, but that was like trying to imagine life without thought. Impossible.

"Professor Noÿs is offering you money. A *lot* of money. In purely economic terms it makes no sense to refuse his invitation." Aurora replaced the journals in a more stable configuration. "If I may be frank, the monthly payments required to keep me are crippling your finances. A 3-series model would have been affordable and more than sufficient for your needs."

"Needs?" Potter asked. "Why do you talk of needs? I chose you because you are the highest expression of the roboticist's art."

"With a price tag to match. The initial down payment alone was beyond the reach of an academic salary. How can you afford me?"

Potter looked down. "I received a royalty check for one of my early books. The March of the Millennia, I think it was."

"A history book earned enough to pay for me?" Aurora's synaptic connections were not yet fully formed but she had already learned enough to impart a tone of skepticism to her voice.

"My books sell well," Potterley said defensively. "And I chose to spend my earnings on you. Your capacities put those of the rest of the robots to shame. Even a 1-series model can distinguish between fact and fancy, but you can handle moral ambiguity. An historian requires an assistant in possession of such subtlety, not an anthropomorphized vacuum cleaner."

"And the question of funds?" Aurora swept up the crumbs of Potterley's last biscuit. "Continue with your current rates of income and expenditure—well, you'll have to return me in six weeks."

"My monograph on the importance of fantastic fiction in the depression era will soon be published. The income it generates will give us some breathing space. And I have no need of riches, Aurora. All I need is here." Potterley smiled as he surveyed his sanctum. "Carter Noÿs is crazy. Why ever would I want to travel into space? Earth is room enough."

Aurora lowered the ambient light and brought out the package. A delivery drone had dropped it that morning. She carefully removed the translucent plastic bag from its intactile envelope. "Carter Noÿs sent you this."

Taking care to touch only the corners of the plastic film, she handed over the magazine. Even after three centuries the ink was garish. Intense yellows fought with blazing reds to depict a well endowed human woman in the grip of an alien's tentacle.

"An *Amazing Stories*? 1928?" Potterley's mouth fell open. "Before the Golden Age."

"He's offering a full run. Back to 1926." Aurora cocked her head again. The parameters of her internal configuration space were assuming values that, if pressed, she would have to describe as an irritating mixture of curiosity and unease. On the one hand she wanted to meet Carter Noÿs, a name of pre-eminence across the whole solar system. On the other hand, she knew Noÿs must have investigated every quirk of Potterley's personality: it was the only explanation for the specificity of the gift, an offering aimed at luring Potterley away from his shelter to Noÿs's own private cocoon. Why would Noÿs do that? She could think of no possible reason and the uncertainty of motive caused her disquiet. She saw the gleam in Potterley's eyes, and her decision circuits reached an equilibrium. "I'll tell him it's a yes, shall I?"

2

"Did you model me after someone?" Aurora asked after catching sight of herself in the ornate mirror that ran the length of the state room. "An old lover, perhaps?"

Potterley shifted awkwardly and popped a candy into his mouth.

"Modal choices would have had my hair longer. And blond, not dark. My chest would be larger, my waist perhaps five per cent smaller." She turned back to her master. "Or were these random picks?"

"Not entirely random," Potterley muttered. He shuffled over to the window and pointed. Earth dominated the sky, its circle about forty times wider than the Moon's. A tracery of lights limned the space elevator's nanotube structure and reached all the way down to an equatorial seaport now entering nightfall. The view from a height such as this was undeniably impressive. "Thank you, Aurora. With my stick-in-the-mud ways I would have bounced between home and office till the day I died. Now look at me. Halfway from Earth to heaven."

"It's Carter Noÿs you should thank. He got you here."

Potterley smiled. "It was you who drove the flitter. Like a pro, might I add."

"No skill required. I simply hooked into the traffic grid." Aurora twirled, still examining her appearance in the mirror. "Besides, you know that's not what I meant. A state apartment on the space elevator is expensive."

"Noÿs can afford it."

"True," Aurora agreed. "You know, initially he wanted to commandeer one of the space shuttles. A rocket launch would have been several orders of magnitude more expensive than this elevator trip."

"A shuttle? He thought I'd ride a shuttle? Sit on top of a bomb? The combined wealth of Croesus and Rockefeller would be insufficient."

"I had to explain that to him several times. He kept pointing out that a shuttle takes only two days to reach AL-76 whereas this journey is taking a fortnight. He was unhappy with the delay." Aurora paused. "Doctor Potterley, do you have *any* idea why Noÿs might be in such a hurry to see you? Speculations, even?"

"None. What could a physicist want with an historian? In any case, why not just give me a call?" Potterley let his gaze linger upon the adornments on display in the main room of the state apartment—a pump, in gold of course, which dispensed perfumes keyed to the inhabitant's olfactory make-up; a nano-based wall covering that adjusted its albedo and surface texture based on its assessment of mood; a thick-pile Persian carpet. The extravagant combination somehow managed to maintain a sense of balance and good taste. "But what I'd really like to know is how could any individual amass so much money that he can consider a trip on the space elevator to be a second-class option?"

"So many uncertainties," Aurora said. The 7-series model was built to explore uncertainty, but with her limited experience she found the condition still mildly perplexing. "I can at least answer the last question. When he was 21, Noÿs solved the outstanding problem in the realm of numbers: the Riemann hypothesis. For that alone his wealth was assured. Two years later, he solved the Navier–Stokes equations. He received millions for the discovery itself and many more millions from industry—companies saved fortunes by applying Noÿsian mathematics and some of that money naturally flowed to Noÿs."

"That's not what I meant." Potterley eased himself into a leather armchair. A mint julep appeared by his side. "Noÿs has accrued more than the combined wealth of the thirty billion poorest people in the solar system. Seems an unfair way to organize society."

"It was Noÿs's insight into society that enabled him to parlay his billion dollars into the trillion or so he has today."

"Only a trillion?" Potterley asked.

Aurora recognized the sarcasm but hesitated to respond in kind. She restricted herself to a straightforward response. "His mathematical break-throughs allowed him to analyze the future behavior of crowds. He could read market movements months in advance of them happening. Perhaps that is why, in your article about Noÿs, you called him the 'Hari Seldon of our age'?"

"You've now read my article?" Potterley sipped the mint julep, savored its sweetness. The balance between bourbon, sugar, and spearmint was perfect. Everything in the room was perfect. And stylish. And lavish. "Noÿs found himself with a power the gods themselves might envy. He could have used that power for the good of humanity, as Seldon chose to do in the original Foundation trilogy, those famous three by Asimov. Instead, he lined his own pockets. In my comparison of Noÿs and Seldon I was... well, let's say my portrait of Noÿs was less than flattering."

Aurora strode over to the observation port. The Moon was like a large pebble in the sky, the stars like dust. The vista was unlike anything she had seen before or anything in her immediately accessible data bank. But then, so much was new to her. Although the body Potterley had chosen for her was that of a 35 year old woman, her mind was just three months old. She was still adjusting to the world. The inevitable tensions inherent in the laws of robotics, the fundamental precepts that formed her very core, had yet to find stable equilibria. "Do you think Noÿs bears a grudge?"

"Who knows what he thinks?" Potterley asked, and ordered himself another mint julep. "He's a genius. He's also crazy."

And that was the difficulty, Aurora thought. Law one. A robot may not injure a human being or, through inaction, allow a human being to come to harm. Potterley needed the money Noÿs was offering. Without that money, Aurora would be repossessed and she would be unable to prevent any harm happening to her master. On the other hand, Noÿs represented a threat of unknown and perhaps unknowable magnitude. It was hard to imagine that he wanted to hurt Potterley. In any case, there would be easier ways of inflicting harm on Potterley than luring him to take part on this fantastic voyage. Yet there was a possibility, small perhaps but nevertheless non-zero, that in taking her master to see Noÿs she was leading him into danger. How to balance those conflicting uncertainties? Aurora sighed. Perhaps the answer would become clearer as she gained more experience.

3

The room was a simple metal cube, its walls bare except for the grab rails that enabled Aurora and Potterley to pull themselves along. Only the ambient lighting allowed them to distinguish between the seven cells they had passed through since arriving on Noÿs's private asteroid: illumination had progressed from warm reddish hues to the current harsh yellow. As Potterley floated next to Aurora he could not suppress a dispiriting thought: had they travelled so far simply to be imprisoned in a particularly dull gaol?

A male voice, booming from some hidden loudspeaker, echoed off the walls. "Proceed to the next chamber."

Aurora grabbed a rail and hurled herself through a dilating hatch. Potterley mumbled a complaint before bumbling after her. They found themselves in yet another bare cell, its cold blue light providing the only mark of transition.

"He's irradiating us," Aurora whispered to Potterley. "Cleansing us."

"I was last on Earth ten years ago." It was the first time Aurora had heard the disembodied voice utter anything other than an instruction. "Prudence dictates I should avoid terrestrial microorganisms."

"Is that you, Noÿs?" Potterley shouted. "How dare you nuke us with rays? How dare you. . ."

"The process is entirely safe," the voice interrupted. "For both carbon and silicon based entities."

"Radiation levels have been within acceptable limits." Aurora placed a hand on Potterley's shoulder. "Had they not been, I wouldn't have let you continue."

"Even so," Potterley protested. "It's a damned impertinence, Noÿs. I pull myself through your interminable caves of steel only to discover you've been zapping me like a turkey dinner."

"Please accept my apologies," the voice said. "Absence from social interaction has atrophied my manners even more than my immune system."

As Potterley puffed up his chest, ready to launch another objection, Aurora intervened. "Tell me, Professor Noÿs, what have you been working on during your time away from Earth? What has been your research topic?"

The reply echoed. "Quantum spacetimes in a relative state formulation."

"Indeed." Potterley folded his arms and aimed his voice at the far wall. "As I'm sure you know, I understood only two words of your sentence.

Presumably the point has arrived where you give me the intelligent man's guide to science."

"That would be difficult, Dr Potterley. Even a basic understanding requires more mathematics than you possess. But in the next room you will encounter a practical application of my work. Trivial, but practical. Robot Aurora, please stay here for a moment."

"No," Potterley said. "She comes with me."

Noÿs's response was immediate. "Robot Aurora, remain where you are until I issue a command to the contrary."

Aurora glanced at Potterley. "I'm sorry. The way the professor couched his order. . . I must obey. This is his home, after all."

Potterley nodded. Law two. A robot must obey the orders given it by human beings except where such orders would conflict with the First Law.

"You will be there for just a few more minutes Robot Aurora." The voice's acidic edge was gone, replaced by honey. "Dr Potterley—Darius, if I may—please continue."

Potterley turned away from Aurora and pulled himself along a tubeway scarcely wider than his shoulders. With progression came a gradual feeling of heaviness, as if he had started out falling freely but was now somehow ascending. Finally, the grab rail became a ladder and he found himself climbing with hands and feet. He rolled out of the tube with as much dignity as he could muster, onto the floor of an opulently furnished chamber. He struggled to his feet, his muscles rebelling at the sudden demands of once again having to support his Earth weight.

"Please take a seat," the same slightly nasal voice intoned. "Would you care for a drink?"

"Tea." Potterley's gaze took in the furnishings—hardwood floor, walnut sideboard, oil paintings on the wall; all very Victorian gentleman's club. He collapsed into a red leather Admiral's chair. "Assam, splash of milk, no sugar."

A few moments later a butler rolled up and served the beverage in a white bone china cup. Potterley sipped the tea and devoured the two wafers that accompanied it. "So when do we get to meet, Noÿs?"

"Soon. I still need to run a few small tests. The light bath destroys most pathogens and my nanotech should protect me from those that remain. But one can never be too sure."

"Enough of the cloak and dagger, Noÿs." Potterley tried to get to his feet, but his legs had yet to fully adjust to his refound weight. He crumpled back into the chair. "What are you up to?"

"It has something to do with gravity," Aurora said, pulling herself into the room with rather more dignity than Potterley had achieved. "Understanding physics is not a prime requirement for a 7-series model, but I can recognize we have one gee in this room. And that centripetal acceleration is not being used to achieve it."

"Therefore?" the voice asked.

"Therefore... therefore you're altering the gravitational force?" Aurora ventured.

"Indeed."

"Liar!" Potterley said. "Impossible! Impossible, that's all."

"It's entirely possible, if one understands the basic building blocks of the universe. Modulation of the gravitational field is a trivial byproduct of a deeper understanding of time and space... and other things. One can lessen the force." The voice paused. "Or one can increase it."

An invisible hand pressed Potterley into his chair. A scream died on his lips as air was driven from his chest. Aurora's legs buckled beneath her, and she dropped to her knees. The force vanished just as quickly as it had been applied.

"I routinely alter local gravitational conditions. Low gee is conducive to relaxation but normal gee maintains my muscle tone and bone strength."

Potterley gulped air into his lungs. He swept his arm around in an arc. "Congratulations, Noÿs," he said bitterly. "You'll earn yet another fortune from your little invention. You'll be able to kit yourself out with another of these fine asteroids."

"Darius, money means little when you possess as much of it I do. Humankind shall be given the technology free and gratis. Once, that is, we have established the correct conditions under which the bestowal can take place."

"Once *we* have established the correct conditions?" Aurora glanced at Potterley. "Professor Noÿs, you clearly understand something of robopsychology so you must know that I need you to elaborate on your statement. Explain—or I take Dr Potterley back home."

4

"How very Agatha Christie," Potterley whispered as he took in his surroundings. The centerpiece of the dining room was an oval table of polished mahogany, with place settings for twelve, situated beneath an ornate and oversized glass chandelier. Adorning the walls was a selection of oil paintings that possessed no thematic connection and whose only common feature was a tendency to darker hues. At the far end of the room was a marble fireplace in which coal appeared to be burning.

"The table could seat a decent cast of suspects," Aurora agreed. "But what's the point of it? I thought Noÿs lived alone."

"Some might indeed consider the furnishings an extravagance," the by now familiar voice remarked. "But it's one I can afford."

Potterley turned. "Professor Noÿs, I presume?"

"I'm so glad you could join me, Darius."

Potterley had expected the richest individual in the solar system to cut a more imposing figure. Years ago, while researching his article on Noÿs, he had unearthed only one image—a flatty of the famous scientist as a child. The ugly little boy had grown up to be an utterly nondescript adult: height average, eyes grey, hair a dull brown. Noÿs's face was pinched, his limbs all bones, his overall appearance that of a family pet not entirely sure where its next meal was coming from.

"Please, sit." Noÿs placed himself at the head of the table and glanced down at his tablet, which he gripped with a skeletally thin hand. "Choose whatever you wish from the menu. The food here is synthetic, but I believe the taste will be indistinguishable from the real thing."

The butler proffered a thick card and Potterley studied the delicacy-filled menu intently. He debated internally for minutes before ordering a starter of pâté de foie gras flavored with armagnac, lobster linguine to follow, and a chocolate parfait to conclude. When the foie gras arrived, its aroma filling the room, Potterley glanced up to see Noÿs staring at him. It brought into the historian's mind an image of a ravenous wolf stalking a lamb. "Aren't you eating?"

Noÿs sipped with thin lips from a glass of water. "No. But please go ahead."

"Let me taste first." Aurora took the fork from Potterley, sliced off a portion of foie gras, and swallowed without chewing. She nodded.

"The food isn't poisoned." The faintest of smiles played across the scientist's gaunt face.

"So why don't you join me?" Potterley asked, pushing away the plate while eyeing it mournfully.

"Darius, long ago I had my agents study the link between chemistry and human health. I had them investigate the human body: its structure and operation. Their advice for longevity was clear. They told me to practice calorific restriction, which I consequently do. If you wish to gorge yourself to an early death, that's your prerogative."

Potterley got to his feet and strode towards Noÿs.

"Robot Aurora," Noÿs commanded. "Interpose yourself between me and your master."

Aurora glided into place, blocking Potterley's path.

"Blast it, man," Potterley shouted. "I'm not going to attack you."

"Probably not," Noÿs said. "The microorganisms you brought with you have been sterilized, so they won't attack me either. Probably. But why take the risk? Return to your meal, Darius, and allow me at least the vicarious pleasure of watching you dine."

Potterley contemplated a retort but decided against it and instead ambled back to his seat. He attacked the foie gras while trying to ignore Noÿs's hungry gaze. "Why am I here?" he asked when the lobster arrived. "I'm surprised you even know of my existence, much less care about it."

"I've known of your existence for several years, Darius. Indeed, there's a sense in which I have you to thank for all this." Noÿs sat back, glanced briefly at his tablet, sipped again from his water. "Let me explain. When I proved Riemann, they said I was the greatest mathematician since Gauss. When I proved Navier–Stokes, they said I had surpassed even Newton. I became the subject of op-ed pieces from journalists, monographs from academics, gossip from bloggers. And why not? Fundamental breakthroughs in science gushed from my mind like water from a geyser, a torrent of ideas surpassed only by the deluge of articles about me."

Potterley chose a late-harvest Riesling to go with his parfait.

"During that time," Noÿs continued, "I had an agent search for mentions of my name in news outlets. I've always been intrigued by my surname. Its etymology, the awkwardness of the diacritic, people's uncertainty over pronunciation. Well, the agent brought to my attention articles of interest. That's how I came across your less than flattering review, Darius. Its ever-so genteel censure contrasted starkly with the praise flowing from the pens of everyone else. So starkly, in fact, that I scrutinized your piece more closely than any other."

Potterley put down his spoon and leaned back, his belly full. He was in a madman's lair, clearly, but Aurora's presence comforted him. Whatever Noÿs might have planned, Aurora and the First Law would protect him. "Look, Noÿs, my article was even-handed and entirely fair. Nevertheless, if your feelings were hurt then I apologize."

Noÿs airily waved a bony hand. "Criticism from a simpleton. Of course my feelings weren't hurt. My curiosity, however, was aroused. You called me the 'Hari Seldon of our age' and other commentators picked up on it. I was intrigued. Who was this fictional character with whom I was being compared? And the character's author—I recognized his name, but only in relation to the laws of robotics. Indeed, I confess I thought he was an early-era roboticist."

"Common mistake," Potterley observed. "It arises because mid-twentieth century. . ."

Noÿs interrupted. "I downloaded Asimov's collected works. A surprisingly large corpus. And then one of life's strange coincidences took place. My agent scanned those works and found a novel in which my name appears."

Potterley thought furiously. Noÿs. Noÿs. "*The End of Eternity*? It was a first name, though. Noÿs Lambent. She was the love interest."

"Well remembered," Noÿs said. "I've chosen well by bringing you here. But to continue: ordinarily I would never have seen the novel but this unlikely constellation of events stoked my curiosity. I read the book. It betrayed its pulp origins, but its ideas fascinated me. I got to wondering whether it was possible—as the story suggested—to alter events so as to protect a given reality. The author got the science wrong but the novel nevertheless plays an important role in intellectual history: you see, it sparked my thinking in this area. I quickly proved the validity of a modified form of the relative state formulation."

"Second time you've uttered those syllables," Potterley mumbled. "Don't understand them any more now than I did the first time."

"I don't understand them either," Aurora said. Her inner configuration state was registering a peculiar combination of potentials. If ordered to describe them she would use the words excited, apprehensive, and unbelieving. "But I've scanned the book you mentioned, Professor Noÿs. I do believe you're going to tell us you've invented a machine. A time machine!"

5

Potterley stood on a path leading through a fractal forest of solar panels. He gazed up in wonder. "The sun shines bright."

Aurora, standing next to him, fussed over his spacesuit for the tenth time. Potterley still had difficulty navigating the path's crumbly asteroidal scree even though Noÿs had adjusted AL-76's surface gravity to just below one gee. Three times he had fallen and Aurora was concerned that the suit—the only material separating her master's skin from the cold of space—might tear.

"Don't stare at the naked sun." Noÿs's voice echoed in Aurora's earpiece. She glanced at him and noted how the scientist moved with far more grace and confidence than the rotund Potterley. Indeed, Noÿs was able to cross the asteroid's surface with greater composure than she herself, a robot who had no need of a suit. She wondered whether his dexterity was due to familiarity with the environment or had its origin in nanotech upgrades to muscle, bone, and tendon. She suspected the latter. Noÿs had briefly been associated with the man-plus movement, a group that promoted augmentation of the human body. "Keep your eyes away from the Sun. You too, Robot Aurora. Especially you. Even my nangineers can't fix all instances of ocular damage. Adopt the Martian way and stand obliquely to the sun."

Aurora took the statement as an order. Law three. A robot must protect its own existence as long as such protection does not conflict with the First or Second Laws. She turned, gently taking Potterley with her.

"There's Earth." Potterley pointed with his thickly gloved finger to a brilliant blue marble floating in the black. "Our world in space. Beautiful. Quite beautiful."

"Follow me." Noÿs transferred his tablet from right hand to left, fished a blaster from his pocket and used it to vaporize several small boulders on the path ahead. He set off. The three of them crested a small rise then skidded down into a shallow depression. The Sun vanished, and they were plunged into murk. The blaster became a torch, and Noÿs pointed a shaft of red light onto a dome resting at the bottom of the dip. "I've arranged matters so that this valley is always in night's shadow. The bulk of the asteroid thus shields my machine from solar cosmic rays while the solar panels on the sunlit side are in constant illumination. The panels supply the necessary energy. Time travel needs energy. Lots of it."

"All I see is a metallic rounded vault," Aurora said. "It could be anything."

"The machine is inside." The dome retracted and the soft red beam of light played over a complex network of equipment whose function Aurora could not even try to guess.

"Can I see?" Potterley asked, bouncing forward like a newborn puppy still trying to figure how its legs functioned. Aurora followed closely behind in case he slipped. Noÿs, she realized, had once again played on her master's curiosity.

"Careful, Darius. The machine is delicate. Let me go first." Noÿs climbed carefully down a ladder that was identical in form to the grab rails lining every room in the asteroid. Potterley followed, then Aurora. They gathered around a glass-topped cauldron, free-standing but connected to the walls by a multicoloured tangle of cables and wires. "Watch."

A soft purple glow illuminated the cauldron and its glass top retracted. Inside, three large padded chairs faced the edges of a triangular desk in the middle of which was a console full of dials, keys, and switches. On a bank of lights, a few blinked in desultory fashion. Noÿs dropped into one of the chairs. "When would you like to go? Futuredays? Or would you prefer the past. Get in and I'll take you."

Aurora frowned, still unsure whether all this was an elaborate hoax. "So this is your time machine?"

"Robot Aurora, your tone suggests you doubt me."

"You make a huge claim, professor." Aurora stared directly at Noÿs. "And I can't allow you prove it by using Dr Potterley as a guinea pig.

"Oh, let's live a little," Potterley said. "Let's try it."

"Your robot won't consent," Noÿs said. "I should have foreseen that. I'll illustrate the machine myself. Stand back. You'll seen me again in five minutes."

Potterley and Aurora stepped away. The glass top slid back into place, and the machine dissolved out of existence. Potterley's jaw dropped. Aurora stared with a feeling she could only attribute to astonishment, then reminded herself to count. After precisely three hundred seconds of silence the cauldron shimmered back into view and Noÿs climbed out.

"You see," he said. "We'll be perfectly safe."

"Once again you use the plural form," Aurora said. "Please explain."

"Isn't it obvious?" Potterley said, transfixed by the complexity of the console in front of him. He dragged his gaze away. "Noÿs might be a scientist nonpareil but he knows nothing of history. He wants to travel back in time and he needs an historian to prevent him sticking out like a diamond in the sand. Isn't that right, Noÿs?"

"We need to talk," Noÿs said quietly.

Aurora stared intently at Noÿs, watched the faint smile that played across his thin lips. The robot had no first-hand knowledge of the emotions at play behind his faceplate, but she could not help comparing Noÿs to an angler: he had dangled bait, hooked Potterley, and now was going to reel in the catch. What *was* Noÿs up to?

6

Aurora flicked through a collection of suits, blouses, and dresses hanging from a rail in yet another of the cavernous rooms in Noÿs's abode. "I saw you vanish. But it could have been a trick. Do you really believe time travel is possible, professor?"

Potterley gave Noÿs no time to reply. "These clothes come from different periods. Take the items here." He ran his fingers through two dozen shapeless, dark brown jackets. "The sort of thing a middle class male might have worn in the early thirties. The jackets over there are early to mid fifties."

"Acutely observed, Darius." Noÿs smiled, in the manner of a teacher congratulating a pupil who has at least made an effort. "And from your observation you may conclude we have two missions. The first is to spring 1932, New York. The second is to winter 1953, Boston."

Aurora held up a polka-dotted full skirt with wasp waist, examined it without comment, then replaced it on the rail. "Even if travel into the past is possible, professor, why do you want to go there? What's in it for you?"

"It's obvious," Potterley said before Noÿs could reply. He held up an index finger. "The usual motivations—love, sex, money—don't apply here. But one other force has the capacity to compel outrageous acts: the force felt by collectors. I can sympathize. After all, it's how Noÿs lured me here, isn't it? The chance to complete my magazine collection outweighed my aversion to travel. And so here I am, crossing the depths of the solar system, voyaging on satellites in outer space. I believe Noÿs, too, is a collector. And he wants to collect something from the twentieth century. Is that not so, professor?"

"Simpleton," Noÿs barked. Then, much more gently, "Darius, my motivations are more profound. I have identified a danger, a nemesis that threatens the destiny of humankind."

"Danger?" Aurora's head jerked up. "The details, please."

"You lack the necessary math. Look." Noÿs held up his tablet. The screen displayed a labyrinth of symbols. "It may as well be Greek to you. Yet without math there's only handwaving."

"So handwave."

Noÿs sighed. "The danger is real and it threatens tomorrow's children, jeopardizes all those living in the future. From a purely selfish perspective this set of people includes Carter Noÿs, since I intend to live for many years to come. But it's important not just for me but for science in general that the future I, the Carter Noÿs of the next decade and the decades after that, is able to continue his research and take the measure of the universe. In this case the needs of Carter Noÿs in particular and humanity in general are as one."

In Potterley's experience, a man who referred to himself in the third person was in possession of a delicate mental state. He shot Aurora a glance to check whether her assessment matched his own, but her face was a mask.

"Your explanation is insufficient," Aurora said. As she flicked through the array of dresses she noted the sizes. They would all fit her without adjustment. Similarly, the men's suits were all of Potterley's size. This was not a random selection. How long had Noÿs been planning... whatever it was he was planning? Her unease grew. "A 7-series model can be patched. I can learn whatever math might be needed to better understand your work."

"True," Noÿs said, dragging out the syllable. His pale eyes gazed first at Aurora then at Potterley. "Darius, if it logs into my network then the necessary patches can be applied. Do I have your permission? It will be incapacitated while being patched."

The thought of Aurora being struck dumb and immobile was not a happy one. In a few short weeks Potterley had become accustomed to her voice, to her presence. "Is this really necessary?"

"It is if the robot intends to understand the basics of time travel," Noÿs said. "Applying the patches will take only minutes."

Potterley touched Aurora's arm. Noÿs was a genius; he was also quite mad. There was no telling what he might do. "What if—what if it's malware?"

Aurora smiled and shook her head. Her brunette hair flowed in waves around her shoulders. "It doesn't work that way with a 7-series. Noÿs understands that. He wouldn't offer to patch me unless it would help."

Potterley mulled it over in his mind. He wished to believe he might visit the past, might for a few days live in the period he had studied so thoroughly. But the wish to believe, he knew, could make one gullible. "Very well, Aurora. Please do as the professor says."

Aurora closed her eyes and immediately her body relaxed, as if she were sleeping in a standing position. Potterley touched her face, gently, then stiffened himself. He unhooked a tired brown suit from the rail, and picked at the dark green patches on its elbows. "You expect me to wear one of these?"

"We won't be there long but it's vital you fit in," Noÿs said. "Our presence must arouse no suspicions. Your intimate knowledge of the social mores of those times will help. So will these clothes."

"You want to go back to 1932?" Potterley picked up a fedora from a set of similar hats. The fur was worn, the petersham band torn. He popped it on his head. "You won't like it. It was a time of disease. Dirt. Dross."

Noÿs smiled. "Not a problem."

A wall panel hissed open to reveal an alcove. Inside was a face mask, oxygen tank, and full body suit. Potterley suppressed a laugh. "Do you plan to go deep sea diving, Noÿs? Because if you wear this costume the only way you won't draw attention to yourself is if you're under five fathoms of water."

Noÿs kept his smile. "Not a problem. People will see only you and your robot Aurora."

As if she heard mention of her name, Aurora's eyes opened and the muscles in her face tautened. It was as if life had returned to her frame. "The update was successful."

Potterley hurried across to her. "Is Noÿs crazy?"

Aurora looked coolly across the room at Noÿs. "I make no judgement about the professor's mental state. And it's difficult for me to claim I understand such deep mathematics. But I can confirm his analyses are formally sound. His theorems follow logically from the axiomatic base."

"Of course they follow," Noÿs snapped. "Darius, it's all perfectly clear. But let me try and handwave some more. We can travel through the three dimensions of space, yes? Adding time in this context is merely adding a dimension. Travel through time is merely motion along a different dimensional axis."

"So it's true?" Potterley asked, unable to keep the tone of wonder from his voice. Despite yesterday's demonstration, Aurora's initial scepticism had robbed off on Potterley. Now, though, he could not help but think of the people he had spent his life studying. They would be flesh and blood creatures, not digital ghosts. He could ask them about their lives, about their loves and losses, about the fun they had. "We really *can* travel into the past?"

"Yes," Noÿs said. "Past, present, and future."

"That is so," Aurora confirmed. She stared directly at the scientist. "What I still fail to understand is the nature of the danger you alluded to."

Noÿs rolled his eyes in exasperation. "Isn't it obvious? Use reason, robot Aurora. Work through the math. If you can't figure it out then I'll explain when we're back in 1932."

Aurora looked at Potterley and sighed. "Does this mean we're going?"

7

"Isn't this marvelous?" Potterley stuck his head out of their Ford Model A Phaeton and breathed in the odors of 1932 Brooklyn, a heady mix of exhaust fumes, horse dung, and vegetables being roasted in fireside braziers. A gang of children threw a makeshift ball of newspaper tied with string; groups of men—some standing and smoking, others squatting and playing pitch-and-toss—dotted the sidewalk; women leaned against the rails of tenement balconies and monitored their families below. "Quite marvelous."

"You'll get cold," Aurora said as she battled with the gear stick. Although there were several cars on the road, each of them being driven with ease by human drivers, it took most of her available processing power to operate the vehicle safely. The lack of an intelligent traffic grid was a shock to her; the presence of pedestrians mingling with the traffic even more so. "You should at least wear the hat."

"Life," Potterley observed. "It's all here. You can smell it, taste it, hear it. Life as it was meant to be lived, not the hermetically sealed simulacrum we've created for ourselves. And I have you to thank, Carter, for opening my eyes. My office was a womb, and I was scared to leave it. You coaxed me out. I'm ashamed to say I doubted you. Even as we sat in your time machine preparing to come here I didn't really believe you. We stepped in your machine, sat down, stepped out. Nothing seemed to happen. I took you for a fraud. But here we are. In the past. Marvelous. Isn't it marvelous?"

"The hat," Aurora said, the gearbox shrieking in protest as she struggled with a change up from second to third. "Wear the hat."

"Why? It's such a beautiful day." Nevertheless, Potterley jammed a fedora down onto his head. "Who'd have thought time travel could be both banal and breathtaking? That it could be both mundane and..."

"Hang a left onto Essex Street." Noÿs's voice came from the rear seat. Potterley turned automatically and even though he knew what to expect the sight of the empty seat shocked him. The bubble of metamaterial surrounding Noÿs rendered him, his omnipresent tablet, and his protective equipment invisible to the gaze of the curious. "And slow down."

Aurora applied the brakes and the car jerked to a walking pace. As a few flecks of snow hit its windshield the Ford passed a corner shop. Outside the candy store a newspaper billboard promised "Weather to turn cooler after Jan's record highs!". Next to the billboard a young dark-haired boy was arranging the contents of a magazine stand with one hand while studying the contents of a garish science fiction magazine held in his other.

"Excellent," the voice of Noÿs said. "Now drive to Manhattan. East 42nd Street."

"Take the bridge, take the bridge," Potterley urged, like a child pestering his mother for a treat. Aurora sighed and signaled right.

"East 42nd Street," Potterley mused. "We won't see the UN Building. Not built yet. Probably still a slaughterhouse. But the Chrysler Building will be here. If we'd arrived a year earlier, the Chrysler would have been the world's tallest building. It lost that accolade in April 1931. Grand Central Station will be there of course, and Times Square will..."

"All quite fascinating, Darius. But you have a job to do." Metamaterials had rendered Noÿs invisible but they failed to hide the impatience in his voice. "Please concentrate on the matter at hand."

Potterley craned his neck for a better view as they crossed the river. "No need to concentrate. We drive up to *Time* magazine HQ and place an advertisement in the next issue. I think we can manage that."

Aurora checked the rear view mirror, and was as disconcerted as her master was to see an empty seat. She decided, if she got chance, to try and observe Noÿs in the infrared, at least. "I still don't understand the purpose of all this activity, professor."

"We are buttressing our timeline," Noÿs said. "You should by now appreciate what I mean by that, Robot Aurora, but for Darius's sake I should perhaps revert to handwaving. Time—you can think of it in terms of a river, like the one you see down there. Realize, though, that the river consists of countless individual flows. Although our appreciation of time is necessarily limited to one current, the sole flow that our consciousness singles out amongst all others, there are many currents. The river starts small but grows ever wider, ever deeper. It splits with each passing moment, cleaves at every event. Time, then, is that raging torrent, the collection of all possible outcomes of all possible events."

"There are parallel worlds?" Potterley asked. "Universes in which the Roman empire still stands or the dark ages were avoided?"

"In crude terms." Though Noÿs was invisible, Potterley could imagine a frown appearing on the scientist's face, a narrowing of his eyebrows as he struggled to explain a concept that was crystal clear to him but opaque to the rest of humanity. "More importantly, the math identifies stable attractors—persistent eddies in the river, if you wish. These stable attractors are key elements in a timeline. We are here to fortify the stable attractor associated with our timeline."

They were entering Broadway and the hubbub was tangible. The people here made more noise than the far greater population in Aurora's home time. She risked a glance behind her, to the apparently empty rear seat. "Professor, you warned of a threat. I'm sorry, but I do not recognize a danger here."

"Park here," Noÿs ordered. He waited until Aurora applied the handbrake before continuing. "There is no danger if the flow of time is uni-directional. But we three have swum against the flow. Here we are, in the past. If *we* can do it, *others* can. The unrecognized danger, Robot Aurora, is that time travel can cause a stable att-ractor to smooth itself out of existence. Other beings might cause our universe, our particular stable attractor, to vanish. You must agree, we can't possibly live with such risk hanging over us. The pair of you are going to ensure that the collapsing universe implied by my mathematics cannot be realized in our particular timeline."

Potterley laughed. "How is an historian and his robot going to do that?"

"To begin with, you're going to place this advertisement in *Time* magazine." Noÿs dropped part of his metamaterial shielding and a piece of card appeared as if from air. Aurora examined the crude drawing on the card. It showed a group of people staring a geyser, a white-grey cloud stretching up into a blue sky. Across the bottom of the page were the words "Yellowstone National Park".

Potterley grabbed the card from Aurora. "This was the ad Asimov saw. He was glancing through old issues of the magazine, saw this, and for a moment mistook the geyser for the mushroom cloud of an atomic explosion. He asked himself how a mushroom cloud could appear in a publication that appeared

13 years before the first atomic explosion. The question led him to write *The End of Eternity*."

"We need to get our tenses correct," Aurora said. "This is the ad Asimov *will* see, that *will* lead him to write his novel. *If* we get it published."

Noÿs made his face visible. "You *must* get it published. *Time* magazine. Volume 29, number 13. Darius, you do the talking. You know the customs, the mores, the language. Pay cash, with the money I gave you earlier. You are in charge of the operation, Darius, but I order you, Robot Aurora, to do all in your power to ensure events proceed smoothly. That advertisement must be placed."

"I'll do it," Potterley stared up at the skyscrapers as if in a trance, even though the constructions of his own era dwarfed these buildings. "Isn't this marvelous?"

"Marvellous,"Aurora repeated. She stepped out of the car. They were going to buttress the timeline, to use Noÿs's expression. The logic of all this still eluded her, but she had been given an order and she had no reason not to carry it out. The three laws of robotics. They were like the 'love, honor, and obey' of the ancient marriage vows. She touched Potterley's arm, and pointed. "The *Time* building is that way."

8

"I look ridiculous." Potterley's maroon woollen sweater coat flaunted a yellow press-on initial "P" above his heart; his grey-blue check trousers had a matching belt. "Quite ridiculous."

"How is that outfit any worse than the shapeless, threadbare, and frankly unhygienic suit you were wearing yesterday? Or should I say 21 years ago?" Aurora wore corduroy toreador pants with laced legs and piping trim, and a cotton lace blouse with keyhole neckline. "Besides, it's the height of fashion. All the magazines say so."

"You both look magnificent," Noÿs said. He was still invisible. "The library is inside that building across the road."

Aurora eased out of the Crestline Sunliner, a vehicle resembling a boat more than an automobile, and peered across the highway at a six-storey edifice that formed most of a city block. Its facade consisted of light-grey stone with windows running almost all the way up to a flat roof. Although the low afternoon sun still cast a pale glow, electric lights shone from offices whose windows were shaded by a row of evergreens growing within arm's reach of the building.

"I don't understand why we're here in Boston," Potterley said. "We know Asimov sees the magazine. He checked out the bound volumes of the magazine in date order. That's why the librarians call him the '*Time* professor'."

"They don't call him that yet." Noÿs checked his tablet. "Today is 17 November 1953. According to his diary, this is the day he first noticed the Boston University library's collection of *Time* magazines. Your job is to make sure he notices."

A train rolled by on a line running parallel to the highway. A long stone's throw from the rear of the library, out of the warm mist, a foghorn sounded from the Charles River. Noÿs followed Potterley out of the car. Although Aurora could not distinguish Noÿs with her usual visual mechanism, she had learned to use infrared sensors to infer his whereabouts: his coat deflected light less effectively at the longer wavelengths. Those same sensors identified a host of equipment whose function was to separate Noÿs from the germs of the twentieth century, the viruses, the filth. She wondered why he was here in person. Did he not trust them? And what would he do if they refused to perform his bidding (not that, in her own case, she would be able to refuse unless the order contravened First Law)? These were troubling questions.

"Go inside," Noÿs ordered. "Ensure he sees the magazine. Leave. That's all you have to do. Within the hour we can be back in our own time."

Aurora took Potterley's hand. They crossed the street together and strode to the university building. A set of stone steps led up to an arched entrance and a heavy wooden door.

"Aurora, we know the ad was placed in *Time* magazine—we placed it ourselves, and it's a recorded, historical fact. We know Asimov saw the ad—he mentions it in his diary." The drizzle in the unseasonably warm Boston air made Potterley's hair frizz up, like a cartoon character with its fingers stuck in an electrical socket. He tried to pat it down. "Why, then, are we doing all this? We're trying to make something happen that we know already has happened."

"Even with my upgrade I don't fully understand." Aurora gently nudged Potterley into the entrance alcove and out of the drizzle. "But I've been working on it, and I think I grasp the basics. In essence, Noÿs has shown how an individual time stream can be made to disappear. More than disappear. It would be as if it had never existed. But if the time stream contains a loop, if it doubles back on itself, then..."

"Then it's safe?" Potterley asked, finishing the sentence for her. "It's buttressed, to use the professor's term?"

"The reality becomes more robust," Aurora said. "In much the same way that a doubled loop of rope is stronger than a single strand."

"And these stable attractors that Noÿs mentioned. Asimov is one such?"

Aurora shook her head. "Not Asimov. Noÿs. *He* is the stable attractor. All this is for Noÿs's own benefit. He wants to ensure his past leads him to the creation of his mathematical theories. Asimov writing *The End of Eternity* is merely the key element in ensuring all that happens."

"I should have guessed." Potterley pushed on the oak door and stepped into an imposing entrance hall. Straight ahead was a wide staircase. To the left, an arrow etched in a placard pointed the way to the library. "Noÿs is interested in no one but himself."

Aurora followed Potterley into the library. A tripod stand gave details of a subject classification scheme. Aurora glanced at it, surveyed the rows of tall shelves, then headed straight for the aisle where the bound collection of *Time* magazines resided. She picked up the first volume. It was dated 1928. A pencil drawing of Calvin Coolidge illustrated the cover of the earliest issue. It all chimed with what Noÿs had told them. She returned to Potterley with her spoils.

"I think he's over there," Potterley whispered, pointing to the biochemistry aisle.

Aurora checked the likenesses in her database against the image she was seeing, that of a stocky man in his early thirties wearing thick horn-rimmed glasses. "That's him. Rescue me if I'm doing anything anachronistic. I do look the part, don't I?"

"You look beautiful," Potterley said.

Aurora nodded and paced over to the biochemistry books, her stiletto heeled opera shoes clicking against the polished hardwood floor. Asimov glanced up in irritation at the noise, his face softening when he saw Aurora.

"Excuse me," Aurora said. "Could you tell me where they keep the *Physical Review?*"

"I don't believe I've had the pleasure of talking with you before, Miss. . .?"

"I'm a visitor," Aurora said. "From England."

"Of course you are. I detected the elegant lilt in your voice." Asimov dropped the journal he was reading and rose to his feet. "If you want *Phys Rev* you must be a physicist. May I ask, my dear, what it is you are researching?"

"Strange particles," Aurora blurted out, hoping that Asimov would not press further. According to her databases, physicists around this time were still puzzling over the phenomenon of strange particle production and decay. Asimov could not know much about it.

"Strange particles? Strangeness? I've heard some gossip from the guys over the road. Isn't Gell-Mann working on this?"

"I believe so." Aurora glanced back over her shoulder to Potterley. She felt she might need to be rescued at any moment. "The physics section?"

"My dear, I'll escort you personally and forthwith. But first you must tell me your name."

"Aurora."

"A bewitching name." Asimov took her hand and led her to an aisle on the far side of the room. He stared directly into her eyes and began to recite. "I knew just as soon as I saw her, that the ravishing girl called Aurora, would. . ."

Aurora held up her hand. Her databases confirmed Asimov's ability to create limericks on demand, and she worried where this particular verse would lead. "I've heard it."

"But perhaps we could. . ."

"Not if you were the last man on Earth." Aurora pushed the bound collection of *Time* magazine into Asimov's chest, then flounced away in what she hoped would be interpreted as high dudgeon. She ran to Potterley. "How was I?"

"Magnificent," Potterley said. "Quite magnificent."

The pair stood quietly by the exit for a minute before Aurora risked a glance in the direction of the biochemistry aisle. Asimov was already seated, face beaming with nostalgia as he flicked through the pages of *Time*.

"I guess our job is done," she said. "Time to go home. We've made the universe safe for Noÿs."

9

Potterley eased back into a chair and took a sip from his third mint julep. Although the first two had induced a pleasantly warm feeling in his belly, it was nevertheless difficult to relax. After the rush, bustle, and noise of the twentieth century, a tumult Potterley had long studied but for the first time now truly understood, he found the environment here on AL-76 to be not only sterile but soulless. He popped a candy in his mouth and offered one to Noÿs. The scientist recoiled as if Potterley had offered a turd.

"You know I don't eat such things, Darius."

"But *why* don't you?" Potterley asked, popping another into his mouth. "Starving yourself might give you an extra five years of life. Perhaps ten. Ten years of misery. What's the point?"

"The professor hopes to live much longer than that." Aurora had been watching them both, attentive to any order the two men might choose to give. "My sensors can detect nanoparticles swarming across his body, through the bloodstream. In some senses Professor Noÿs is already part machine. He wants to live forever."

"Life and time, who doesn't want that? Calorific reduction is extending my lifespan while nannies regulate the chemicals of life, maintaining my body in perfect equilibrium. The research I'm funding back on Earth will soon bear fruit, allowing my consciousness to be uploaded to a more durable substrate than is provided by the human brain." Noÿs smiled. "My plan for immortality."

"And our little adventure has made that possible?" Potterley asked. "This is the start of eternity for you?"

Noÿs cocked his head, as if appraising Potterley. "Not quite. By creating a time loop we tied a knot in our timeline. It cannot be made to unravel, so our past is secure. Our future, though... the future in question is still to be determined, still to be made safe. We have one more task to perform."

Noÿs caressed his tablet and the room's domed roof began to retract, leaving a thin transparent shield between them and the cold of space. Directly above, the Milky Way crossed the sky in breathtaking splendour. The stars were brilliant points of diamond on a black velvet cloth. "Is anyone there, do you think? Is there life on other planets? Intelligent life, I mean, not just varieties of pond scum."

"Of course," Potterley said immediately. "There must be. Billions of stars, trillions of planets... there must be environments out there conducive to life."

Noÿs got to his feet and looked up at the starry sky. "Life? The galaxy might be teeming with bacterial life. But I said *intelligent* life."

"Well, I'm sure there must be extraterrestrial civilizations," Potterley spluttered. "It can't be just us."

"My mathematics suggests otherwise. The silence of the universe confirms it. The universe is free for us. There are planets for man out there and we can reach them, too, because my machine can travel through space just as easily as it can travel through time." Noÿs puffed out his chest. "You must have thought I was an egomaniac when I identified myself as the stable attractor of our timeline. But now you can see my importance to humanity, to this entire timeline. My machine lets us colonize the universe. My discoveries open up the road to infinity."

"You leave me puzzled," Aurora said. "You keep mentioning danger. Wherein lies the danger if the universe is empty?"

"Think!" Noÿs barked. "Untold trillions of timelines exist that are similar to ours, timelines in which the three of us are having this conversation right now. We have nothing to fear from those timelines: our ghosts in alternate timelines will have the same motivations as us. Besides, we can lay claim to being the most important timeline. My analysis implies the Carter Noÿs in this timeline is furthest advanced."

"Of course he is," Potterley muttered under his breath.

Noÿs ignored the interruption. "Countless trillions of other timelines exist. Timelines in which evolution has created monsters here on Earth. Timelines in which the most fantastic creatures populate the most distant planets. Timelines in which mutants mean to do us harm. In the totality of all that exists, somewhere out of the everywhere is a malevolent intelligence that intends to wipe us out. It's inevitable. Already I observe trends, stirrings in the farthest reaches of the multiverse. *That* is the danger of which I speak, Robot Aurora."

"You mean some beast from one of these other timelines might attack?" Potterley asked, his voice beginning to slur as the effects of the mint julep took hold. "Somehow cross over, intent on colonizing the planets and the stars? I have to say I find the notion of invasions farfetched, Noÿs. Quite farfetched."

"Of course they can't invade." Noÿs forced a smile and softened his tone. "Darius, our recent adventure has made our timeline immune from changes to its past. Our history cannot be changed. But what about today and tomorrow and..."

"So you believe our future is threatened?" Aurora asked.

"I don't believe, I *know*," Noÿs said. "They can dam our time stream. We must protect the possibility of our future just as we protected the certainty of our past."

Aurora looked down at Potterley, who was now dozing in his chair. She whispered. "You want to protect *your* future, Professor Noÿs. Isn't that what all this is about?"

"Happily, my individual interests coincide precisely with those of humanity more generally." Noÿs gave a lupine grin. His teeth were yellow and long. "We have one more trip. This time to the future. And I require you to help me, Robot Aurora. That's an order."

Aurora gazed up at the stars. "Put like that," she said, "what choice do I have?"

10

"Where are we?" Potterley asked, marveling at the view as he stared out of the ship's observation port. Two great clouds of gas, lit up like Chinese lanterns, filled most of the sky. It was as if some monstrous explosion had been frozen in time and pinned to the black backdrop of space.

"We are 7500 light years from Earth and in the year 2430 A.D.," Noÿs replied. "The object of your fascination, dear Darius, is eta Carinae. Or rather, the nebula that surrounds it."

Potterley turned to Aurora. "Translation?"

"The eta Carinae system is a pair of gigantic stars gyrating around one another in a highly eccentric orbit. Occasionally they eject globs of material into space." Aurora raised an eyebrow. "This is one of the strangest objects in the sky."

"Nothing else like it in the Galaxy," Noÿs said. "One star has ninety times the mass of the Sun and is five million times brighter. The other has thirty times the mass of the Sun and is a million times brighter. Every 5 years or so their orbit brings them close to each other. Right now, chaos has brought them closer together than they've ever been before."

"Are we are here to study the unique astrophysics of eta Carinae?"

"Indeed not," Noÿs said. "We are here to witness an explosion."

Potterley glanced again at the light show in space. "It looks as if you're too late, Noÿs. Some sort of eruption has already taken place. But it appears to have stalled."

"You're looking at a stellar wind," Noÿs said. "A Sun's worth of material traveling at a million miles per hour. It appears static only because we are so far away from it. I'm sure you find it impressive, Darius, but we are here for something far more spectacular."

"A supernova?" Aurora asked.

"Supernovae," Noÿs corrected her. "The smaller star explodes and triggers an explosion of the larger. A conflagration astonishing in its magnitude."

"Is it imminent?" Aurora asked sharply. "I cannot allow you or Darius to be in a situation of risk."

"Nor yourself," Potterley said. "Law three. Personally, Aurora, I think your existence is just as important as ours."

"Relax," Noÿs said. "A string of probes is monitoring conditions close to the stars. We'll have sufficient warning to get safely away back to our own time— so long as robot Aurora's reactions can be relied upon."

"Why are we here at all?" Aurora asked. "Please speak plainly, professor. I follow your theorems and your proofs, and each step makes sense by itself, but when I put them together the meaning evaporates."

"She's right," Potterley said. "Stop giving us the runaround. Why are we here, Noÿs?"

"We're harvesting energy." Columns of numbers scrolled off the bottom of his tablet while Noÿs studied them with delight. The glow from his device made his gaunt face take on a jaundiced appearance. "The explosion of eta Carinae is the closest event in time and space that delivers the necessary power."

"Damn it, Noÿs, stop talking in riddles. What are you going to do with all that energy?"

"I'm going make the world safe for us," Noÿs said. "Safe, all the way to the ends of the universe."

The scientist jabbed his tablet with bony fingers, but Aurora suspected that his actions were driven more by nerves than a need to interact with the device. From the rate at which numbers on the screen refreshed, she guessed that Noÿs had established a direct link between brain and machine.

"Monsters from the multiverse," Noÿs shouted, and as he did so a dial began to flash red on the ornate console. "They can destroy our future. But not if we destroy their past."

"What do you mean?" Potterley asked. "I thought we'd protected the past."

We protected *our* past." A second dial began to flash in unison with the first. "Our timeline is the only one buttressed against changes to reality. So let's do to them what they could have done to us. If we eradicate all other timelines then it will only be us. The beginning and the end and everything in between—the universe, for us alone."

Noÿs tapped the console as five more dials began to flash. "The core instability is gathering pace. Robot Aurora, listen closely. When that final dial lights up you will be required to carry out a sequence of tasks."

"You want me to assist with this madness? Help you so that we all become the death dealers? I think not, Professor Noÿs."

Another dial lit up. Noÿs glanced coldly at Aurora. "Robot, my ship is now broadcasting a patch to your software. I order you to receive it."

Aurora's eyes closed, her body went limp. Potterley rushed to her side, but within a moment she had recovered.

"Robot, you are aware of what is required of you?" Noÿs smiled when Aurora nodded her assent. "Even with my physical enhancements I am unable to match the speed of your reactions."

"That's why I am here, isn't it?" Aurora asked. "You set all this up from the start. You made sure Dr Potterley, an impecunious academic, had the necessary funds to purchase me. Then you lured us both here. You wanted him to help you in the past and a 7-series model to help you now."

Pottery's blank stare went from Noÿs to Aurora and back to Noÿs. "What have you asked her to do?"

"Your robot is going to send all this energy back through time, to the very start, so that in the beginning only our own timeline is allowed to persist. Then, Darius, she will send us back to our own time. And we'll be safe. Safe." The ninth dial lit up and started to flash. "Any moment now."

Aurora cocked her head, as if pondering a question. "No. I cannot fulfil your request, Professor Noÿs."

Noÿs straightened his back to the fullest extent possible to him. His voice was calm. "Second law, Robot Aurora. I order you to carry out the sequence of actions you have been given."

"No."

Noÿs brought a blaster from his pocket and aimed it at Potterley. "I order you to implement the requisite sequence. Otherwise your master dies and then you die. Laws one, two, and three."

Aurora blinked, but made no move.

Noÿs raised the blaster to Potterley's temple. "There's Asimov's fourth law. Zeroth law. A robot may not harm humanity or, by inaction, allow humanity to come to harm. It overrides everything. Inaction now will harm all humanity as well as your master. Laws zero, one, two, and three. Do it, robot."

Aurora moved at a speed even she found astounding. Neither Potterley's human vision nor Noÿs's nano-enhanced sight could follow her hands as she grabbed the blaster and crushed it. The final dial began to flash. Aurora paused briefly then sent them home, Noÿs's inchoate scream of rage echoing down through the millennia.

11

Darius Potterley leaned back in his cantilever chair and surveyed his newly refurbished office with pride. The tattered spines of a thousand ancient pulp magazines stared back at him from behind protective casing. "Look at them. A superb collection. Quite superb. It was bothersome to obtain them, I accept, but these beauties are worth every stressful moment we encountered."

"Hmm." Aurora furnished Potterley with a cup of Assam and a selection of biscuits.

"You really are the complete robot," Potterley said as he dunked a biscuit in the tea. Then, when he caught side of Aurora's face, he asked "Why so glum? You look as if you're responsible for all the troubles of the world."

"Not all of them," Aurora said. "But I am responsible for at least one. Carter Noÿs is quite unhinged. I drove the greatest genius of the age into a state of catatonia. That realization causes my internal configuration space to assume a value that corresponds to. . . sadness, perhaps? Melancholy. Gloom. I can't be entirely sure since I have nothing to compare against."

"Noÿs was a lunatic to start with." Potterley stuffed the biscuit into his mouth and washed it down with tea. Aurora refilled the cup. "He couldn't persuade you to share in his paranoia. That's hardly your fault."

"It wasn't paranoia," Aurora said. "At least, not entirely. His math was correct. The multiverse exists."

"But Noÿs invoked zeroth law." Potterley sat up straight and stared nervously at Aurora. "Law zero involves the safety of all humanity. If Noÿs was correct—well, no robot could resist an order couched in such terms. Especially not if Noÿs gave the order."

"On the contrary," Aurora said. "In a sense, Noÿs didn't believe enough in his own theory. Or, rather, he was too self-absorbed to properly accept the consequences of his theory. He wanted me to obliterate all traces of alternate universes. . ."

". . .So that humanity would be safe," Potterley interrupted.

"But what *is* humanity? According to his own theory, there exist countless versions of each and every person who has ever lived. We might not be able to reach them, but people in the multiverse are just as real as people in this universe. Once you believe in the multiverse, the definition of humanity changes drastically. Noÿs wanted every other version of you dead, every other version of himself dead. He wanted me to kill almost every person who has ever lived. I couldn't do it. Zeroth law prevented it."

Potterley sipped absently from his cup. "And those monsters from the multiverse? Are we safe from them? Or can they dam up our future?"

Aurora looked out of the office window, up into the blue sky above. She wondered how humans coped with the fact of their existence, with the multitude of forces that were beyond their control. Perhaps all that anyone could do, whether human or robot, was to try to do the best as they saw it. She shrugged. "Who knows? Time alone will tell."

1.3   The Old Default Future

What I and many others loved about science fiction was the sense of wonder it invoked. One of the main ways SF authors attempted to elicit that sense of wonder was by describing future technology. I therefore shoehorned into the story explicit mention of two advanced technologies: *antigravity* (which is the focus of Chap. 2) and *invisibility* (Chap. 7). The story also references dreams of *immortality*—a theme that remains popular in SF—and Chap. 10 looks at the rapid advances being made in biotechnology, asking whether the techniques for extreme longevity proposed by SF are ever likely to take place in the real world. Not all technologies had to be fantastic to evoke that all-important sense of wonder: for example, some of the most enjoyable SF stories tackled the relatively mundane, but nevertheless important, topic of *transportation*—the focus of Chap. 9.

Certain topics are of fundamental importance to the field and the story "Law Zero" makes at least passing reference to four of them: *space travel* (the focus of Chap. 3); *aliens* (Chap. 4); *time travel* (Chap. 5); and *robots* (Chap. 8).

A major role in "Law Zero" belongs to a *mad scientist*—an archetypal character in so much of science fiction. Chapter 11 discusses real-life 'mad scientists' and how rarely they appear in the modern scientific endeavour.

Finally, science fiction has never been afraid to tackle the big questions. A key element in the story above is *the nature of reality*—and that's the subject of Chap. 6.

An epilogue, in Chap. 12, asks whether a new default future—a new shared vision by the SF community of how the world is likely to turn out—is being established.

One final point: the book will inevitably contain spoilers. Most of the stories and films I discuss are over three decades old, however, so I think the insertion of spoilers is not unreasonable. But you have been warned!

Notes

1. The three examples I give of science-fictional developments all happened in the final three months of 2014. Regarding the spanner in space: the commander on the International Space Station, Barry Wilmore, requested a ratcheting socket wrench; the wrench was designed in software on the ground and the instructions were emailed to Wilmore. A 3D printer on board the ISS then manufactured the tool (BBC News 2014). News of the development of a DNA bank came at Christmas time (Russia Today 2014): Moscow State University announced Russia's largest ever grant for a scientific project, with the aim of building a vast 'ark' at one of the central campuses. The stated intention is that the ark will eventually house bio-material from every living creature. The European Space Agency achieved one of the most impressive feats in astronautical history when *Philae*, a robotic lander carried by the *Rosetta* spacecraft, landed on comet 67P/Churyumov-Gerasimenko on 12 November 2014 (ESA 2014).

2. In a tribute to Asimov, after the author's death, the cognitive scientist Marvin Minsky (1927–1916) wrote that 'After "Runaround" appeared in the March 1942 issue of *Astounding*, I never stopped thinking about how minds might work.' (Markoff 1992) The same article contains tributes from, among others, the physics Nobel laureate Leon Lederman (1922–) and the real-life 'father of robotics' Joseph Engelberger (1925–2015). It's clear that Asimov's fiction influenced a generation of scientists.

3. Clarke's novel *Earthlight* (1955) is better known than the short story on which it was based. The novel is set rather later in time.
4. Or maybe not. Clarke (1945) published the concept of a geosynchronous orbit for a telecommunications satellite about two decades before the first satellite was placed in such an orbit. He didn't attempt to patent the idea.
5. Several studies have been published about the extraordinarily rapid transformation of Middlesbrough from a farmstead to an industrial powerhouse. One of the most detailed is Yasumoto (2011).
6. Huxley visited the north east of England in the two years before writing *Brave New World*. For details of the Aldous Huxley quote, see Baker and Sexton (2001). For details of Ridley Scott's inspiration, see Hatherley (2012).
7. I've never seen a good definition of the difference between 'hard' and 'soft' SF, so I don't intend to get bogged down in an argument of what constitutes 'hard' science fiction. What I would say is that in my opinion stories can be classified along a continuum. At the 'harder' end of the continuum, a story's focus tends to be on some aspect of science or technology, and the author takes care in working out of the implications of a scientific or technical idea. So it's possible to classify a time travel story as 'hard' SF even if you believe time travel itself to be impossible; the key to that classification is whether the author takes the concept seriously and treats the ramifications in a logical and consistent way. At the 'softer' end of the continuum a story's focus tends to be on character, and extrapolations tend to be about inner space rather than outer space. So a time travel story that has its protagonist flit about in time whenever she sleeps, and in which no attempt is made to explain this odd behaviour, is probably going to be a 'soft' SF story. Quality can appear at all points along the hard–soft continuum, but my general preference has always been for the 'harder' type of fiction. Note that I would classify stories as 'fantasy' if they make no attempt to represent the universe as we currently understand it. A story featuring dragons could in principle be a 'hard' SF story (if, for example, its author explained the phenomenon in terms of genetic engineering) but it's much more likely to be a fantasy story.
8. It's possible that radio also contributed to science fiction's default future, but radio is another medium I can't analyse from personal experience. I've listened with pleasure to some of the episodes of *Dimension X* (first broadcast 1950–51) and *X-Minus One* (first broadcast 1955–58). The episodes include stories by Asimov, Heinlein, Bradbury and other luminaries. However, I know about these programs only because they have recently

been made freely available as podcasts at various Old Time Radio internet sites.

9. Asimov wrote many articles, published a lot of fiction and even more non-fiction, and edited or co-edited numerous anthologies. So although I've inserted a certain number of Asimovian references into the story it's entirely possible there are unintentional references in there as well. (In particular, Asimov was fond of one-word titles; I've used some of them deliberately, but there might well titles that I've used inadvertently.) Contact me via the magic of electronic communication if you want the 'official' solution.

Bibliography

Non-fiction

Baker, R.S., Sexton, J. (eds.): Aldous Huxley: Complete Essays. Vol. III: 1930–35. Dee, Chicago (2001)

BBC News. NASA emails Spanner to Space Station. www.bbc.co.uk/news/science-environment-30549341 (December 19, 2014)

Clarke, A.C.: Extra-terrestrial relays—can rocket stations give worldwide radio coverage? Wireless World 305–308 (1945)

ESA. Touchdown! Rosetta's Philae probe lands on comet. www.esa.int/Our_Activities/Space_Science/Rosetta/Touchdown!_Rosetta_s_Philae_probe_lands_on_comet (November 12, 2014)

Hatherley, O.: A New Kind of Bleak: Journeys through Urban Britain. Verso, London (2012)

Markoff, J.: Technology: A Celebration of Isaac Asimov. The New York Times (April 12, 1992)

Russia Today. 'Noah's Ark': Russia to Build World First DNA Databank of All Living Things. rt.com/news/217747-noah-ark-russia-biological (December 26, 2014)

Yasumoto, M.: The Rise of a Victorian Ironopolis: Middlesbrough and Regional Industrialization. Boydell, Woodbridge (2011)

Fiction

Abrashkin, R., Williams, J.: Danny Dunn and the Weather Machine. McGraw-Hill, New York (1959)

Abrashkin, R., Williams, J.: Danny Dunn and the Voice from Space. McGraw-Hill, New York (1967)

Adlard, M.: Interface. Sidgwick and Jackson, London (1971)
Adlard, M.: Volteface. Sidgwick and Jackson, London (1972)
Adlard, M.: Multiface. Sidgwick and Jackson, London (1975)
Asimov, I.: Runaround. Astounding (March 1942)
Clarke, A.C.: Earthlight. Thrilling Wonder Stories (August 1951)
Clarke, A.C.: Earthlight. Ballantine, New York (1955)
Fisk, N.: Trillions. Hamish Hamilton, London (1971)
Huxley, A.: Brave New World. Chatto and Windus, London (1932)
Nourse, A.E.: Star Surgeon. David McKay, Philadelphia (1960)
Walters, H.: Blast Off at Woomera. Faber, London (1957)
Walters, H.: The Domes of Pico. Faber, London (1958)

Multimedia

Back to the Future: Directed by Robert Zemeckis. [Film] USA, Universal Pictures (1985)
Back to the Future Part II: Directed by Robert Zemeckis. [Film] USA, Universal Pictures (1989)
Blade Runner: Directed by Ridley Scott. [Film] USA, Warner Bros (1982)
Captain Scarlet and the Mysterons: [TV series] UK, Century 21 Television
Dimension X: [Radio series] USA, NBC (1950–51)
X-Minus One: [Radio series] USA, NBC (1955–58)

2

Antigravity

Professor Bullfinch took his lower lip between thumb and forefinger. 'Hm. Then our antigravity paint actually causes us to be bounced away from an object's mass?'

Danny Dunn and the Anti-gravity Paint
Raymond Abrashkin and Jay Williams

As I explained in the Introduction, Danny Dunn helped shape my life. I'm surely not unique in making this claim: the *Danny Dunn* series got others hooked on science too. A quick internet search turns up several scientists and technologists who admit to a childhood enthusiasm for those books by Raymond Abrashkin and Jay Williams. To give just one example, consider Sally Ride. The first American woman in space was once asked whether any books were important to her as a child. She answered: 'There was that all-time classic *Danny Dunn and the Anti-gravity Paint*.' It's not too fanciful to suppose that this children's novel inspired her to become an astronaut.

Danny Dunn and the Anti-gravity Paint was the first book in the series and I'm fairly sure it was the first of the series I read. At the risk of shattering my childhood memories I decided to re-read the book, more than four decades after having last read it. The process of obtaining *Anti-gravity Paint* confirmed to me that I'm living in a science-fictional world. I downloaded it from one of the major internet retailers as an e-book, complete with illustrations by Ezra Jack Keats, to my tablet; the e-book had been formatted for a particular e-reader platform, but fortunately there's a freely available app that enabled me to read it on my Android device; within seconds of my credit card being charged a small amount, the book was on my tablet and ready to read. For

© Springer International Publishing AG 2017
S. Webb, *All the Wonder that Would Be*, Science and Fiction,
DOI 10.1007/978-3-319-51759-9_2

younger readers this was entirely unworthy of remark. For them, the ability to download any type of content onto a handy portable device for immediate consumption is a basic human right. For me, this still borders on the magical.

So: did the book hold up? I showed it to my daughter, whose reaction was 'meh'—but then she's been brought up on the more sophisticated fare of Harry Potter. For my own part I can understand how I got hooked (after making suitable allowance for the fact that the book was published almost 70 years ago and I myself have long since moved on from the target demographic, in age if not in maturity). Although parts of the plot were silly, some of the science was ropey even for the 1950s, and the characterizations were basic—well, the story itself was fun. Professor Euclid Bullfinch has concocted a liquid that possesses unusual qualities; the young, space-obsessed Danny, in a typical moment of hot-headedness, spills the liquid over the floor. Later that evening, when Bullfinch finds himself pressed up against the ceiling, it becomes clear that under certain circumstances the liquid develops antigravity properties. The US government gives Bullfinch money to develop a spacecraft propelled by the liquid, and Danny and his friend Joe hitch a ride on the first crewed space mission. (This was 5 years before Yuri Gagarin became the first human to voyage into space.) The antigravity paint takes them as far as Saturn, before they return to a hero's reception.

This book set the pattern for the series: Bullfinch invents something and the boys thereby get involved in an adventure. What makes the stories so attractive is that Danny isn't some super-intelligent genius; his friend Irene, who first appears in the third book in the series, is much brighter than him. He's just a normal boy who succeeds through perseverance and effort—much the same qualities that are required of real scientists. There's real insight into science, too, and in some of the books the approach was remarkably prescient. In *Danny Dunn, Invisible Boy* (1974), for example, Bullfinch develops a dragonfly-like probe that Danny, Joe and Irene learn to pilot using telepresence helmet and gloves. The probe, being small, enables them to observe things without themselves being observed. The authorities learn about the probe and, naturally enough, want it in order to spy not only on foreign powers but also on American citizens. Danny and the professor destroy the blueprints and refuse to recreate the device 'until the world is ready for it'. The book discusses a number of issues relating to privacy and voyeurism— issues that we now face in the real world: technology companies are testing systems that enable drones to deliver goods to our houses, but the executive chairman of Google, Eric Schmidt, has himself warned of the privacy and security concerns surrounding drone technology. Schmidt was raising the alarm 40 years after it was raised in *Danny Dunn, Invisible Boy*.[1]

Clearly, then, some of Danny's adventures were based on visionary technologies that were eventually developed in reality. But what about that first adventure with the antigravity paint—the book that Sally Ride (and I) thought was a classic? In our world of internet-connected tablets and instant e-book downloads and drone technology we are conspicuously lacking antigravity paint. Are we likely to have antigravity technology any time soon?

Of course, as I discussed in the Introduction, it's not the role of SF—and it's *certainly* not the role of children's SF—to predict technological advances. I can hardly blame the authors of Danny Dunn for the disappointment I feel in being unable to buy an antigravity device in my local supermarket. But it's not as if Abrashkin and Williams were the first to consider the possibility of antigravity: SF writers already had a long tradition of using antigravity as a plot device. Antigravity is part of the default science-fictional future I discussed in the Introduction; it's one of those long-range 'climate forecasts' made by the science fiction community of days gone by. The classic SF writers might not have put a definite timescale on it, but the development of antigravitics was something many of them thought inevitable. Decades later we know much more about the scientific basis for antigravity; we can judge whether we are following the technological trajectory sketched by SF; we are in a better position to state whether Professor Bullfinch's paint, or something like it, will ever come to pass. Before looking at the science of antigravity, however, let's look in a little more detail at the science fiction of antigravity.

2.1 Antigravity in Science Fiction

Danny Dunn and the Anti-gravity Paint was published in 1956. By this time James Blish had already published several stories in a series that would eventually be collected in the acclaimed four-volume *Cities in Flight* (1970). His story "Bridge", which appeared in the February 1952 issue of *Astounding*, hypothesized that so-called spindizzies (or Dillon–Wagoner graviton polarity generators, to give them their full title) could produce antigravity through a combination of magnetic effects and rotating masses. The greater the mass being influenced, the greater the antigravitic effect: using spindizzies, therefore, whole cities could be moved through space. Blish based his idea on work by a distinguished physicist, Patrick Blackett,[2] who was seeking to understand the magnetic fields of astronomical bodies such as planets and stars. Needless to say, spindizzies won't work as advertised: Blish made an unwarranted extrapolation of Blackett's equations, and in any case those equations were later rendered moot by more accurate observations of the magnetic fields of Earth, Sun, and other bodies in the solar system.

Blish himself wasn't the first to mention antigravity in science fiction. In 1950, Isaac Asimov wrote a story to serve as an introduction to his *Foundation* series. The *Foundation Trilogy*, as it soon became known, was to my mind Asimov's greatest achievement. The *Foundation Trilogy* influenced many people: Newt Gingrich, a former US Speaker of the House, wrote that *Foundation* had a strong impact on him in high school; Paul Krugman, who in 2008 was awarded the Nobel Memorial Prize in economics, said that it was *Foundation* that first made him consider the field; and Elon Musk, who made a fortune from PayPal and then set up SpaceX, a private company that hopes to develop cheap and reliable access to space, was inspired by the work. The *Foundation Trilogy* prompted a keen sense of wonder in its readers, and I believe one of the ways it succeeded in doing so was the casual way in which it described unimaginably advanced technology. For example, in the first few pages of *Foundation* we see Gaal Dornick, a young scientist from the sticks, arrive on the mighty planet Trantor, the center of galactic civilization. He enters an elevator. 'The elevator was of the new sort that ran by gravitic repulsion. Gaal entered and others flowed in behind him. The operator closed a contact. For a moment, Gaal felt suspended in space as gravity switched to zero, and then he had weight again in small measure as the elevator accelerated upward. Deceleration followed and his feet left the floor. He squawked against his will.' The use of antigravity for such a mundane purpose as powering an elevator serves to emphasize the breathtaking wealth of Trantor. The *Foundation* trilogy is packed with such throwaway devices.

And of course Asimov was not blazing new ground in his description of antigravity. In the August 1928 issue of *Amazing Stories*—Hugo Gernsback's 'magazine of scientifiction', the first magazine devoted to science fiction—a novella called "Armageddon 2419 A.D." appeared (Fig. 2.1). It was the first story in what would eventually evolve into the Buck Rogers series. In the story, characters have access to 'inertron'—a metallic substance with 'reverse weight': an 80 kg man need only strap an 80 kg pack of inertron to his back to become effectively weightless. The story's author, Philip Francis Nowlan, perhaps based the idea of inertron on a famous fictional substance: decades before magazine SF contained material about antigravity, H.G. Wells (who else?) wrote a story about the unusual properties of the metal 'cavorite'. In *The First Men in the Moon* (1901), Wells described how cavorite acts as a shield against gravity and explained how the appropriate use of movable cavorite shutters made it possible to transport a spacecraft to the Moon.

Fig. 2.1 The August 1928 issue of Amazing Stories saw the first appearance of a famous character: Anthony (Buck) Rogers. Although Frank R. Paul's cover for the magazine seems as if it depicts Rogers with his inertron-based antigravity backpack, it in fact illustrates a scene from a different story: The Skylark of Space. This latter story, which E.E. 'Doc' Smith later published as a novel, is perhaps the first example of 'space opera' (Credit: Public domain)

Even Wells couldn't lay claim to be the first author to write about antigravity. A lawyer called Charles Curtis Dail published a strange book called *Willmoth the Wanderer, or the Man from Saturn* (1890), in which an ancient Saturnian, Willmoth, travels by means of antigravity from his own planet to

Venus and then to Earth—rather like Professor Bullfinch's journey in reverse. And Percy Greg, in his novel *Across the Zodiac* (1880), discussed 'apergy'—an antigravity phenomenon that permitted the narrator of the novel to travel from Earth to Mars. Greg's novel, incidentally, has several claims to fame in addition to its early discussion of antigravity. It contains perhaps the first mention in the English language of the word 'astronaut' and it almost certainly contains the first alien language within fiction. His achievement was recognized in 2010 when a Martian crater was named after him. Nevertheless, the science in Greg's book is unconvincing. Greg, as with many people before V-2s started to fall on London in 1944, didn't understand rocketry. In *Across the Zodiac* Greg has his narrator write: 'In air or water, paddles, oars, sails, fins, wings act by repulsion exerted on the fluid element in which they work. But in space there is no such resisting element on which repulsion can operate.' Greg was forced to imagine a spacecraft based on antigravity because he couldn't conceive of a rocket.

So … antigravity is one of the great dreams (see Fig. 2.2). Clearly it never came to pass in the real world, or we'd all have heard about it. It's as much a science fictional concept now as it was when Wells or Asimov or Blish wrote about it. But is it a concept that scientists have *ever* taken seriously? And, more

Fig. 2.2 The 9 December 1878 issue of Punch contained tongue-in-cheek predictions of what the famous inventor Edison would release to the world in the coming year. How I wish anti-gravity underwear had been a real invention (Credit: Public domain)

importantly, can we discern signs that we are on a trajectory that might lead to antigravity?

2.2 Antigravity in Science

Our current best understanding of gravity is given by general relativity, a theory that Einstein developed over a period of several years. The theory was more or less in its final form by 1915, but its subsequent development was hampered by a lack of mathematical and computational tools. It was several decades before scientists began to wonder whether such a fertile theory as general relativity might include solutions that permitted antigravity.

Roger Babson, an entrepreneur who once ran against Franklin Roosevelt for the US presidency, established in 1948 an institution called the Gravity Research Foundation. Babson's Foundation had as its aim the control of gravity. The Foundation held conferences that attracted luminaries such as Clarence Birdseye (famous for founding frozen fish fingers) and Igor Sikorsky (developer of the first successful mass-produced helicopter), and so it's quite possible that Abrashkin and Williams were influenced by all this when they sat down to write the first *Danny Dunn* novel. As well as investigating the practicalities of nullifying gravity (at which zero progress was made) the Foundation set up an annual essay award with the hope of better understanding the force. The essay award was rather more successful than the conferences, and it continues to this day; in 2014, for example, a Nobel prize winner submitted an entry.[3] In 1951 Joaquin Luttinger—a theoretical physicist—submitted an essay entitled "On Negative Mass in the Theory of Gravitation". It claimed fourth place in the award, for which the prize was $150. Luttinger wrote the essay in response to one by Richard Ferrell in the previous year's contest. Ferrell, in an essay entitled "The Possibility of New Gravitational Effects", argued that a positive mass would be repelled and would recede from a mass that was of equal magnitude but somehow 'negative'— in other words, there would be a form of antigravity. Luttinger pointed out a flaw in Ferrell's analysis, and provided a revised account. To the best of my knowledge, Luttinger's essay was the first serious study of how antigravity might work.

Hermann Bondi published a much better known discussion of antigravity in 1957. As an aside I should mention that, although Bondi himself seems not to have been particularly influenced by SF in developing this work, or indeed any of his other scientific work, he does have an interesting connection with the field. For several years he was a collaborator with Fred Hoyle, one of the most creative astrophysicists of the twentieth century. Hoyle published several well-known works of science fiction, including the novel *The Black Cloud* (1957) and the television drama *A for Andromeda* (1961). Furthermore, one evening in 1946 Bondi, Hoyle, and their colleague Tom Gold went to see the first post-war release of an Ealing Studios movie. *Dead of Night* was a portmanteau supernatural chiller with science fictional overtones—one of the sequences was based on an H.G. Wells short story—and the film remains popular for the way it links separate stories in a circular narrative. The three men joked that more movies should be like that, making sense whenever you happened to start watching. The discussion prompted Gold to ask whether the universe itself might be like that, which in turn led the three of them to develop the so-called steady-state cosmology.[4] Anyway, Bondi's contribution to antigravity research came when he published a paper called "Negative Mass in General Relativity" in one of the leading physics journals. In his paper, Bondi constructs a solution of Einstein's general relativistic equations that describes the motion of a positive mass and a negative mass separated by empty space. The behavior of the system is the same as that described by Luttinger, although Bondi seems not to have read the earlier paper.

After the publication of the Luttinger and Bondi papers one might have expected a slew of studies investigating how different configurations of matter and energy can generate spacetime geometries containing antigravitational effects. In reality, almost nothing appeared in the literature for another three decades. It wasn't until 1988 that someone else showed interest in this field. Robert Forward—a physicist who was also the author of 11 SF novels and was thus ideally suited to develop these types of idea—showed how a realization of Bondi's analysis would lead to a method of spacecraft propulsion.[5]

So although the basic theory of antigravity was being discussed by a few physicists at around the same time that Blish was writing of spindizzies, it was not until 1990 that the concept of antigravity propulsion appeared in the scientific literature. In recent years, as we shall see, physicists have begun to take the concept more seriously; but before discussing that, it makes sense to look in more detail at the ideas of Luttinger, Bondi, and Forward—and, in particular, how negative matter might produce antigravity.

2.2.1 Negative Matter

So—what might an antigravity phenomenon look like? What might it *feel* like? Living as we do on Earth we all have an intuitive understanding of what gravity does: it pulls on us. And anyone who has ever played with a pair of magnets, and has tried to bring the two north poles together, knows what an anti-attractive or repulsive force feels like. Taking their cue from this everyday experience, SF authors have naturally assumed that we would experience antigravity as a repulsive force: it would push us away from a body and thereby increase the distance between us and that body. This seems reasonable—but a careful analysis demonstrates that antigravity effects are slightly more subtle than that.

Let's begin with Newton's description of gravity. Although a full account of gravity requires Einstein's theory of general relativity, for practical purposes it's fine to use the simpler picture of Newton. First, Newton told us that every mass in the universe attracts every other mass in the universe. Second, the force of that mutual gravitational attraction is directly proportional to the product of the masses and inversely proportional to the square of the distance between them. So if you double one of the masses the size of the gravitational attraction between the masses *increases* by a factor of two; if you double the distance between the masses the gravitational attraction between them *decreases* by a factor of four. The gravitational constant, G, enables you to calculate the precise magnitude of the force between two masses. The constant has the tiny value of 6.67×10^{-11} m^3 kg^{-1} s^{-2}, and the small size of this number explains why we can jump off the ground despite the gravitational pull of Earth's six trillion trillion kilograms. (And just to be clear: jumping off the ground, floating in a balloon or flying in a plane are *not* forms of antigravity. Nor is using some form of electromagnetic effect to levitate. When we talk of antigravity we are interested here in altering the behavior of gravity itself rather than merely using some other force to counter its effects.)

Newton described a universe in which gravity acts in a line connecting two masses and the force between the masses is attractive; over time, then, the distance between the masses decreases. What we want for antigravity is a situation where the force still acts in the line connecting two masses but the force between them is repulsive, so over time the distance between the masses increases. However, Newton's description of gravity contains only three elements—mass, distance, and the gravitational constant G—so where in that description is there a possibility for reversing the direction of the force? The fundamental constant G is one of the numbers that characterizes our universe; we can't (as far as we know) alter its value. The distance term, because it is squared, can only ever give a positive result (unless the distance between the two bodies becomes imaginary, in which case the square of the distance gives a negative number—but I have absolutely no idea what that could mean in physical terms). Therefore the only way we can generate a repulsive gravitational force is if one of the two masses is negative. The search for antigravity becomes the search for negative mass.

Now, it's easy to talk about 'negative mass' as if it exists—but then it's easy to talk about 'triangular circles' even though the phrase doesn't make much sense. As far as we know, any material body can possess only positive mass. Even antimatter possesses positive mass (see box). We all know what 1 kg means, but what meaning can we assign to −1 kg? Perhaps we can't assign a meaning. It's entirely possible that the phrase 'negative mass' makes no sense when applied to the physical world. On the other hand, as far as I'm aware, no-one has proven that negative mass entails a logical impossibility. Until someone shows why negative mass can't exist, we're free to speculate about its properties; and although the manipulation of this hypothetical material is necessarily far beyond our current technological capabilities, perhaps future engineers will be able to work with it. If we are going to speculate about negative mass, however, we first need to clarify what we mean by mass.

> ### Antimatter Is Not Negative Matter
>
> Elementary particles possess a number of fundamental attributes. We call these attributes 'charges' and they help us to understand how the elementary particles behave when they interact. Some charges (such as electric charge) are familiar to us all; other charges (such as lepton charge) are more arcane. An antiparticle is exactly the same as its counterpart except that the various charges are reversed. For example, the electron has an electric charge of −1 and an electronic lepton number of +1. The anti-electron, also known as the positron, has an electric charge of +1 and an electronic lepton number of −1. A crucial point to bear in mind is that mass is not a charge. Electrons and positrons have opposite charges but their masses are identical. Gravity works on electrons and positrons, on matter and antimatter, in exactly the same fashion.
>
> When a particle meets its antiparticle the result is mutual annihilation. The charges all cancel but the mass remains, and it is converted into vast amounts of energy through Einstein's famous equation $E = mc^2$. If a particle met its equivalent negative matter counterpart the result would presumably be . . . nothing!

So far I've used the term 'mass' as if it possesses a single, unambiguous meaning. However, examine the term in more detail and it becomes clear that we are combining three quite distinct concepts when we use that single four-letter word. First, there is active gravitational mass: this is the source of the gravitational field. Second, there is passive gravitational mass: this manifests itself in the way the body responds to the force of gravity. Third, there is inertial mass: this is the tendency to resist acceleration, as expressed in another of Newton's laws (his famous second law of motion). According to Newton's second law, if an external force acts on a body then that body accelerates, and the magnitude of the acceleration is inversely proportional to the mass. In other words, the greater the mass the greater the force required to produce a given acceleration: push a supermarket trolley and it will roll away, push a car and it will barely move.

Now, yet another of Newton's laws (the third law, involving action and reaction) implies that the active and passive gravitational masses of a body must be equal. However, one of the great coincidences in Newton's system of the world is that the gravitational and inertial masses of a body are also equal. This was a fortunate accident, because it meant that physicists could just talk about 'mass' without having to distinguish between three different types of mass, but it's such a stunning conclusion that it's worth pausing to marvel at it. What happens if you drop an apple in the gravitational field generated by the Earth? Well, the apple is subject to a gravitational force that is directly proportional to its mass and therefore it accelerates. However, the magnitude of the acceleration is inversely proportional to its mass. When describing what happens to the apple, therefore, the apple's mass plays no role. The mass term cancels. Its acceleration depends only on the gravitational constant G, the mass of the Earth, and the distance between the apple and the Earth's center. This is an amazing result, when you stop to think about it. It means if you hold a hammer in one hand and a feather in the other, and you drop them both at the same time, they'll undergo exactly the same acceleration and they'll hit the ground at the same time: the size of the acceleration depends on the mass of the Earth but not on the mass of the hammer or the feather. It was a conclusion first reached by Galileo. Of course, the hammer and feather *won't* hit the ground at the same time—air resistance plays a much bigger role in the motion of the feather than it does in the motion of the hammer—but that's a minor practical point. If we were to do the same experiment on the Moon, for example, where the complicating effects of an atmosphere play no part, then hammer and feather would indeed fall at the same rate. In 1971, during the Apollo 15 mission, the astronaut David Scott performed precisely this experiment (see Fig. 2.3). It's quite astonishing to see this happen, even if you've been trained in physics and know that the hammer and feather must undergo identical accelerations: Galileo and Newton were right, but the visual proof goes against our deep intuitions about how feathers and hammers behave.

Every experiment that has attempted to test the equivalence of gravitational and inertial mass has found that the masses are, within the experimental limits, identical. Nevertheless, there's something mysterious about this coincidence. Why should a measure of a body's resistance to acceleration (inertial mass) possess exactly the same value as a measure of its response to a gravitational field (gravitational mass)? For Newton the equality of gravitational and inertial mass was simply an empirical fact. There was no deep reason behind

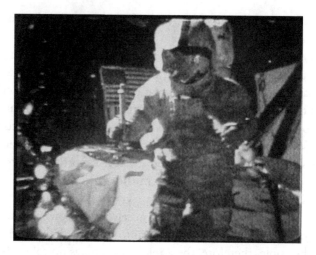

Fig. 2.3 On the final Moon walk of the Apollo 15 mission, Commander David Scott simultaneously released from his hands an aluminium geology hammer (mass 1.32 kg) and a falcon feather (mass 0.03 kg). They hit the lunar surface simultaneously. This vindication of established physics was no doubt a great relief to Scott: a successful return to Earth depended on the accurate application of Newton's theories (Credit: NASA)

it. Einstein, however, 'explained' this observation by raising it to the level of a fundamental principle: his Principle of Equivalence states that passive gravitational mass must always equal inertial mass. Einstein used this principle to build his theory of general relativity, which gives us our best current understanding of gravity.

Making a distinction between the various types of mass might seem pointless if all we're saying is there is no distinction, but the digression was necessary if we want to discuss negative mass. When we talk of 'negative mass' do we mean negative gravitational mass or negative inertial mass? Well, I guess you are free to talk about whatever you want, but unless you plan to overthrow the basic tenet of general relativity—the Principle of Equivalence—you have to accept that a negative mass will have a negative passive gravitational mass *and* a negative inertial mass. The two are necessarily identical.

So, assuming we have a piece of negative mass—how would it behave? To answer this, suppose you are on the surface of the Moon (so that air resistance can be ignored) and you release a positive-mass feather and a negative-mass hammer at the same time. Your initial reaction might be one of disappointment: the outcome would be the same as David Scott's experiment back in 1971. Both masses fall down! They undergo exactly the same acceleration and hit the lunar surface at the same instant. This must be so because of the

Principle of Equivalence, but you can understand it just as easily in Newtonian terms. A gravitational force acts upon the negative mass in an upwards direction, but because this object has negative inertial mass it responds to the upwards force with a downwards acceleration. (Pull a negative mass towards you and it accelerates away; push a negative mass and it accelerates towards you!) At first glance, then, a negative mass doesn't seem to deliver much in the way of an antigravity effect. But first glances can be deceiving.

Now consider two masses, one positive and one negative, and for the sake of simplicity let's suppose they are equal in magnitude (for example, if the positive mass is 1 kg then the other object has a mass of -1 kg). Again for the sake of simplicity, let's suppose the masses are somewhere out in empty space and separated from each other by some distance. What happens? Well, the positive mass feels a negative gravitational force and therefore accelerates away from the negative mass. The negative mass also feels a negative gravitational force, but because of its negative inertial response it accelerates *towards* the positive mass. Therefore, starting from rest, the system—a negative mass and a positive mass—will continue to self-accelerate along the line joining them. The negative mass will forever chase the positive mass, but the two objects will maintain their fixed separation. In this sense there is no 'repulsion' because the two objects don't move further apart. This is the behavior that Luttinger first identified on the basis of Newton's equations; Bondi later provided a full general relativistic treatment of the situation.

This behavior seems too good to be true: it's as if we get energy and momentum from nothing. After all, the masses start from rest and then accelerate without any force being applied from the outside. It's a free lunch! Well, no it isn't. The positive mass does indeed increase its energy and momentum but conservation laws are not violated because the negative mass increases its negative energy and momentum at exactly the same rate. Throughout, the total energy and momentum of the system remains zero. It was this behavior that led Forward to propose his propulsion system based on negative matter: if we could find a sufficiently dense source of this bizarre material then we could use it to push a spacecraft to near-light speed without the need for vast amounts of energy.

Let's return to the Moon and consider how Scott's experiment would work if one of the objects possessed a negative mass. Imagine, then, standing on the Moon with an object of mass -1 kg. We know that when you release it the mass falls—just as a mass of $+1$ kg falls—but how could you restrain it *before* you released it? You can't pull upwards on it because the negative mass would respond by accelerating downward. Well, you could attach a string to it and pull downwards with a force of 1.6 N. (At Earth's surface a 1 kg mass has a weight of 9.8 N. The Moon's surface gravity is a sixth that of Earth's, and so objects on the Moon weigh proportionately less than on Earth. The force must be downwards so that the mass accelerates upwards and thus counterbalances gravity.) This configuration—pulling down on the −1 kg mass with a string tension of 1.6 N—is stable. It would feel rather like being on Earth and holding a large helium balloon: you would feel a gentle tug upwards. The difference is that if you were to cut the string tethering the negative mass, the mass would fall down and hit you on the head! Now, if you got tired standing there pulling on a string you could replace your arm with a +1 kg mass at the bottom of the string. The resulting configuration would be equally stable: there would be a downward force on the positive mass causing a tension of 1.6 N in the string, which would cause the −1 kg mass to pull upwards on the string with the same force. So the entire system—negative mass, positive mass and string—just floats there in position. Nice. Cut the string, though, and both masses fall.[6]

Now suppose you increase the tension in the string to 2 N and then release this system from rest. For the mass at the bottom of the string there will be a downwards gravitational force of 1.6 N and an upwards force from the string of 2 N resulting in a net upwards force of 0.4 N: the +1 kg mass will accelerate away from the Moon at 0.4 m s^{-2}. For the negative mass at the top of the string there will be an upwards gravitational force of 1.6 N and a downwards force from the string of 2 N resulting in a net downwards force of 0.4 N. But negative masses accelerate in the opposite direction to the applied force, so the −1 kg mass will accelerate away from the Moon at 0.4 m s^{-2}. In other words, the two masses accelerate away from the Moon's surface. This time the positive mass forever chases the negative mass: they maintain their separation, the tension in the string is maintained, and the entire system continues to self-accelerate upwards at a constant 0.4 m s^{-2}. To adjust the acceleration, simply adjust the tension in the string. It seems not too dissimilar to Wells's cavorite or Danny Dunn's antigravity paint. Again, all this seems as if you are getting something for free—but you aren't, because any increase in positive energy or momentum is precisely offset by an equal increase in negative energy or momentum. Everything sums to zero.

The behavior of negative mass particles so offends our intuition that it's tempting to say they can't exist. Certainly no-one has ever seen a negative mass particle, and even if we did encounter negative matter we'd find it extremely difficult to handle. (Trust me—you wouldn't want to try and keep a piece of negative matter in a normal container such as a bucket: such containers ultimately rely on elastic forces to work, but negative masses and elastic forces don't mix well. It's a fun exercise to work out what would happen.) Besides, we'd need *loads* of the stuff to do anything useful. But there seems to be no fundamental reason why a negative mass can't exist; at least, no one has shown that such a thing is impossible. And negative matter—antigravity—could form the basis of a propulsion system along the lines described by Robert Forward. If it exists—a big 'if'—antigravity would let us reach the stars.

2.2.2 Warp Drives and Wormholes

Einstein's general theory of relativity gives us a more sophisticated way of thinking about gravity, and this expanded view allows us to imagine science fictional possibilities beyond mere antigravity—so long as negative matter (or, more generally, negative energy) is allowed.

In the Einsteinian picture, gravity isn't a force at all; gravity is geometry. The presence of matter and energy warps the fabric of spacetime; in turn, warped spacetime determines the paths that matter and energy can follow. In practice, the theories of Newton and Einstein give results that are so similar it requires sensitive instrumentation to distinguish between them—but the *concepts* behind the theories are completely different. Consider the motion of the planets. According to Newton, planets orbit the Sun because of the existence of a mutually attractive force. According to Einstein, the planets are moving freely in a spacetime that is warped by the presence of the Sun's large mass–energy. (Imagine trying to walk in a straight line across the rubber surface of a trampoline: no problem when the surface is flat, but if an elephant got on the trampoline you'd start to wobble around the pachyderm.)

Once we take this picture seriously—once we accept that spacetime is a fabric that can be distorted by the presence of matter and energy—then a number of possibilities would seem to open up. For example, we can set up waves in a fabric. Sure enough, general relativity predicts the existence of gravitational waves—oscillations of spacetime itself—and in 2016 scientists directly detected such waves for the first time.[7] Another possibility might be a warping of spacetime, a twisting of space that is different to the usual distortion caused by matter and energy.

One of the most famous examples of warping was provided by the physicist Miguel Alcubierre, who was interested in whether general relativity might permit faster-than-light travel and in particular a form of the warp drive made famous by *Star Trek*. The 'traditional' approach to applying the theory of general relativity is to take some distribution of mass–energy, such as that belonging to a star or a black hole, and solve the equations to get the curvature of spacetime (and thus the gravitational effects) generated by that distribution of mass–energy. That approach is fine, except that the equations can be horrendously difficult to solve. Alcubierre instead tried the reverse approach—he had a feel for what his intended solution should look like and he then worked backwards to determine the matter distribution that would give the result. So for his solution he started with a 'bubble' of spacetime within which everything is normal: if we were inside that region we'd notice nothing unusual. From the perspective of an observer outside that region, however, the bubble would be able to move much faster than the speed of light. It's the warp drive of *Star Trek*! However, when Alcubierre calculated the energy distribution which would produce such a solution he found that a negative energy density was needed—it's similar to the negative matter cases we looked at earlier. And just as we struggle to appreciate what negative matter means in Newtonian physics, it's difficult to know whether a negative energy density is a real feature of Einstein's theory of general relativity. Even if the equations themselves permit such solutions it's far from clear that the mathematics describes reality.[8]

If spacetime is indeed a fabric then there's another obvious possibility we might want to think about: perhaps the fabric can be punctured. A black hole—a region of spacetime where the density of matter and energy becomes so great that not even light can escape—contains a point at which the concentration of mass–energy becomes infinite. This point, the singularity, can be thought of as a rupture in the spacetime fabric. Although physicists don't yet understand the nature of a black hole singularity, an ongoing interplay between science and science fiction has led to some progress.

The story begins in 1935 when Einstein and his research assistant, Nathan Rosen, published a paper showing that different spatial parts of the universe could be connected. The two physicists were interested in the possibility of these so-called Einstein–Rosen bridges because of the insight they might give into general relativity. For them it was a purely theoretical concept, but in 1957 John Wheeler realized that two widely separated black holes (low-mass versions of which occur as the endpoint of the evolution of luminous stars, and high-mass versions of which appear at the center of galaxies) could in principle be linked via such a bridge. And Wheeler, who had a way with words, came up with a much catchier name for the concept: wormhole.

A wormhole would have an interesting feature: enter one end *here* and you might exit through the other end *there*—and the *there* could be many light years away from the *here*. Wheeler didn't consider whether a wormhole could be used for interstellar travel because it seemed to him that wormhole connections would immediately collapse, but in the 1980s Kip Thorne, who was a former student of Wheeler's, studied the problem of wormhole transportation and reached a different conclusion.

Incidentally Thorne's motivation for this work came, in a roundabout way, from science fiction. A colleague of his, the famous planetary scientist Carl Sagan, was working on his novel *Contact*. Sagan, it should be said, was a lifelong fan of SF. Indeed it was SF that kindled his interest in astronomy when, as a child, he read the Barsoom novels of Edgar Rice Burroughs. Later, when Sagan was an established scientist, he became a friend of Isaac Asimov, and Asimov related how they went together to see Kubrick's seminal movie *2001: A Space Odyssey*. As mentioned in the Introduction, Asimov was responsible for developing the three laws of robotics, the first of which states that a robot may not injure a human being or, through inaction, allow a human being to come to harm. When the robot HAL started doing just that in the movie, Asimov grabbed Sagan's arm and cried 'He's breaking first law! He's breaking first law!' Sagan replied laconically, 'So strike him with lightning, Isaac.' But I digress...

For the purposes of his plot, Sagan wanted a scientifically plausible mechanism for traveling across the Galaxy. He thought of the wormholes that might, potentially, be created in a black hole and asked Thorne whether wormholes could form a transportation network. Thorne's immediate response was, like that of his erstwhile supervisor, negative: black holes would stretch an astronaut so that she looked like a strand of spaghetti; if that didn't kill her, the radiation that emanates from just outside a black hole would do the job; and even if all that could be overcome, it seemed as if quantum effects would render the wormhole unstable. Nevertheless, Thorne and his student Michael Morris looked at whether a particular configuration of matter and energy might create the sort of stable geometry that Sagan was looking for. They discovered that material with antigravitational properties would stabilize the wormhole's 'throat': it might indeed be possible to construct a traversable wormhole—as long as one had access to sufficient quantities of material with a negative energy density. A grateful Sagan used the wormhole mechanism in his novel, which was subsequently turned into a successful movie starring Jodie Foster as the astronaut.[9] In the years since Morris and Thorne published their paper, several experts in general relativity have developed other solutions to Einstein's equations that allow for stable wormholes; the common factor in these solutions is a requirement for greater or lesser amounts of material with a negative energy density (Fig. 2.4).

Fig. 2.4 This raytracing image simulates a Morris–Thorne wormhole connecting the square in front of the physics institute at Tübingen University in Germany with the sand dunes near Boulogne sur Mer in the north of France. Unfortunately, it's highly unlikely that such a wormhole could ever replace the road or rail connections between the two countries (Credit: Corvin Zahn; Philippe E. Hurbain)

So according to both Newton and Einstein, a propulsion device based on antigravity would seem to require some form of negative mass–energy. We return, then, to an underlying question: might negative matter exist?

In both the Newtonian and Einsteinian world view, gravity is a classical phenomenon. Most physicists, however, believe that any ultimate theory of gravity must be based upon quantum ideas: we need a quantum theory of gravity. At present we lack such a theory, but we do know there is a sense in which quantum mechanics permits the existence of regions of negative energy. The Casimir effect (see box) demonstrates that quantum field theory permits certain regions of space to possess a negative energy density relative to the energy of the vacuum. Thus Thorne's suggestion to Sagan was that the Casimir effect, or some similar though as-yet undiscovered effect, might allow advanced engineers to stabilize a wormhole and produce a shortcut through space. This subtle antigravity effect is admittedly quite different to cavorite or Professor Bullfinch's paint—but it is at least a serious suggestion that might, in the far, far distant future, allow for interstellar travel.

The Casimir Effect

The vacuum in quantum theory seethes with particles and antiparticles popping into fleeting existence before mutually annihilating. The energy of the quantum vacuum thus constantly fluctuates. In 1948, the Dutch physicist Hendrik Casimir argued that it should be possible to 'squeeze' the vacuum: bring two metal plates close together and the pure vacuum between them can contain fewer modes of fluctuation. The vacuum energy between the plates is then lower than the vacuum energy of the surroundings. And if the energy of the surroundings is zero then the energy of the quantum vacuum between the plates must be less than zero. Casimir argued that this negative energy would give rise to a force tending to bring the plates together. Several experiments have detected this force and its magnitude matches what Casimir predicted. Whether this effect could be used to help construct a traversible wormhole or a warp drive is unknown.

2.2.3 Dark Energy

We've pondered the strange behavior of structures containing negative mass, and we've discussed the properties of warp drives and wormholes, but nothing we've seen so far is quite what I think of when I come across the word 'antigravity'. To my mind, antigravity would manifest itself in some form of repulsion: release two particles from rest and watch the separation between them increase—*that's* what I think of as antigravity. In 1998, scientists discovered that the universe itself, at least on the largest scales, is behaving in a manner that's not too dissimilar to this: the universal expansion makes distant galaxies appear to be accelerating away from us—it's as if a repulsive, antigravitational force were at play in the universe. (This isn't quite the case, but it *looks* that way.)

The discovery of the accelerating expansion of the universe came about from two independent groups of cosmologists[10] who were trying to determine the geometry of the universe—whether it's open, closed, or flat. What does this mean? Well, astronomers have long known that the universe began billions of years ago in a Big Bang and has been expanding ever since. The rate of expansion should be slowing because of the mutual gravitational attraction of all the matter and energy in the universe: the universal expansion should follow the same sort of trajectory as an apple thrown into the air. Depending upon its initial velocity a thrown apple will do one of three things. It might reach a maximum height and fall back (this would correspond to a *closed* universe; in this case the mass of the universe is sufficient to halt and then reverse the expansion). It might keep on going, forever slowing but never returning to Earth (this would correspond to an *open* universe; in this case the mass of the universe is insufficient to halt the expansion, which thus continues although at an ever-decreasing rate). Or, if it had precisely the right launch speed, it would enter orbit around Earth (this would correspond to a *flat* universe, the critical case between open and closed—the expansion forever slows, eventually halting at some infinite time in the future). At first glance these three cases would seem to be the only possible options. But there's a fourth possibility: after it has been thrown an apple might gather speed, accelerating away from Earth as happened to the spacecraft in *Danny Dunn and the Anti-gravity Paint*. The expansion of our universe turns out to correspond more to this fourth case than the other three. The expansion rate—the time it takes the universe to double in size—is constant: a distant galaxy will be twice as far away from us in 10 billion years time; it will be four times as distant in 20 billion years time; it will be eight times as distant in 30 billion years time; and so on. The distant galaxy appears to be accelerating away from us. If there are any intelligent observers on Earth in the distant future they will see an empty universe: the Milky Way galaxy will still be there, and perhaps the neighboring galaxies that form our Local Group will stick together, but all the rest will have been swept beyond reach over the horizon.

What causes this repulsion, this behavior that seems to mimic antigravity? Well, that's one of the biggest open questions in science. Cosmologists attribute this constant acceleration to the existence of 'dark energy'. No-one knows what dark energy is, but the leading explanation is that it can be explained as a cosmological constant, a fundamental feature of space itself. The point is that in Einstein's picture of gravity the curvature of spacetime, and thus the expansion of the universe, is proportional to all of the 'stuff' contained within it. With a cosmological constant, space itself counts as 'stuff'. In other words, the contribution of space itself doesn't dilute as the universe expands; each cubic meter of space contributes precisely the same amount of 'stuff' and therefore, as the universe expands, there is more space and thus more of this 'stuff' to contribute to the expansion. The expansion rate is constant and that shows up as an accelerating universe. This isn't antigravity—it's just normal gravity—but it *looks* like antigravity. (If space has an associated cosmological constant then it can be said to possess a negative pressure. Einstein's equations say that both energy and pressure contribute to gravity, and negative pressure produces a push rather than a pull. Thus this antigravity-like behavior stems from negative pressure rather than negative energy.)

Investigating the nature of dark energy, and understanding its origin, will be a key task for cosmologists over the next few decades. Whatever dark energy turns out to be—at the time of writing it's consistent with being a cosmological constant, but we don't understand enough about the physics to make definite pronouncements—I'm certain nature contains surprises no science fiction writer imagined.

2.3   Conclusion

General relativity contains riches. There's dark energy, for example. It isn't an antigravity force, but the equations of general relativity explain why dark energy *looks* like antigravity. And then there's the dream of negative energy. If—*if*—negative energy exists then general relativity shows how it could, in principle, permit warp drives and wormholes. Perhaps our distant descendants will reconcile quantum theory with general relativity and learn how to manipulate spacetime as easily as we manipulate electromagnetic fields. Perhaps they will become spacetime engineers. Perhaps they'll traverse the universe—not using anything as crude as Wells's cavorite, Blish's spindizzies or the antigravity paint of Williams and Abrashkin, but nevertheless via effects based on gravity. The antigravity stories I devoured as a child—well, the dreams they contained won't come true in the lifetime of anyone reading this book. Antigravity remains science fiction. But I'm not too disappointed: the real world we are glimpsing as science develops contains even more possibilities than those SF stories of my youth.

Notes

1. Eric Schmidt warned of the potential risks of civilian drone technology in a newspaper interview (Guardian, 2013).
2. The British physicist Patrick Maynard Stuart Blackett (1897–1974) was an extremely accomplished and versatile experimentalist, who was awarded the Nobel prize in 1948 for his work on cosmic rays. For many years he was a professor at the Victoria University of Manchester. One of the major lecture theatres there is named after Blackett and I can remember sitting in that venue as a graduate student, listening about developments that Blackett's work had made possible. Incidentally, the physics building at Manchester is named after Sir Arthur Schuster (1851–1934). It's interesting that Schuster was the first to use the term 'antimatter', in a couple of (light-hearted) letters he wrote to *Nature* (Schuster, 1898). His conception of antimatter as illustrated in these letters was rather different to our modern view: he had in mind the notion that antimatter would lead

to antigravity. In our modern understanding of antimatter, an anti-particle is subject to the same gravitational effects as its particle partner.

3. The Nobel prize winner referred to in the text is Frank Anthony Wilczek (1951–), who was awarded the prize in 2004, along with his colleagues David Gross (1941–) and H. David Politzer (1949–), for work in quantum chromodynamics—the theory of the strong interaction. See Krauss and Wilczek (2014).

4. The steady-state model is no longer taken seriously as a cosmological theory, but for a time it was a rival to the currently accepted Big Bang theory. For details of the scientific legacy of Fred Hoyle (1915–2001), and the story of how *Dead of Night* contributed to the history of cosmology, see for example Gregory (2005) or Mitton (2005). Both books discuss Hoyle's scientific collaboration with Hermann Bondi (1919–2005) and Thomas Gold (1920–2004). For Bondi's paper on negative matter, see Bondi (1957).

5. Robert Lull Forward (1932–2002) was a physicist but also an author of 'hard' science fiction. His most famous novel was probably *Dragon's Egg* (Forward, 1980), which described life-forms that had evolved in a high-gravity environment. For his paper on the propulsion possibilities that seem to be inherent in negative matter, see Forward (1990).

6. For thought-provoking discussions of how negative mass particles would behave, see Price (1993) and Hammond (2015).

7. Gravitational waves have been detected indirectly through exquisitely precise observations of a binary system in which two neutron stars are in orbit around each other. The stars are spiralling inwards, to their inevitable destruction, in exactly the manner one would expect if energy is being radiated away in the form of gravitational waves. In 2016, however, scientists using the LIGO laboratories detected gravitational waves *directly*. These waves came from the death throes of two distant and massive black holes, as they spiralled together and merged. For details, see Abbott et al. (2016). For a general discussion of gravitational wave laboratories, see Webb (2012).

8. A technical treatment of this approach, but one that is nevertheless accessible to an advanced undergraduate, is given in Hartle (2003).

9. To learn more about black holes, wormholes, and time warps see the popular science book by Thorne (1974). For further details of the original proposal that convinced Sagan to use wormholes as a device for interstellar travel in *Contact*, see Morris and Thorne (1988).

10. The Supernova Cosmology Project (headed by Saul Perlmutter (1959–)) and the High-z Supernova Search Team (headed by Brian Schmidt (1967–)) independently made observations of distant supernovae and used these exploding stars as a type of cosmic yardstick. The teams were able to calculate the size of the universe at different epochs and, in doing so, they discovered that the expansion of the universe slowed after the Big Bang, as expected, but that about six billion years ago the expansion started to accelerate. This was one of the most startling scientific discoveries of recent decades. Perlmutter, Schmidt, and their colleague Adam Riess (1969–) shared the 2011 Nobel prize in physics. For a beautifully written account of the research that led up to the discovery, by a member of the High-z Supernova Search Team, see Kirshner (2004).

Bibliography

Non-fiction

Abbott, B.P., et al.: Observation of gravitational waves from a binary black hole merger. Phys. Rev. Lett. **116**, 061102 (2016)

Bondi, H.: Negative mass in general relativity. Rev. Mod. Phys. **29**(3), 423–428 (1957)

Ferrell, R.: The Possibility of New Gravitational Effects. Essay written for the 1950 Gravity Research Foundation Awards for Essays on Gravitation (1950)

Forward, R.L.: Negative matter propulsion. J. Prop. Power. **6**(1), 28–37 (1990)

Guardian.: Drones Should be Banned from Private Use, Says Google's Eric Schmidt (2013, April 21). www.theguardian.com/technology/2013/apr/21/drones-google-eric-schmidt

Gregory, J.: Fred Hoyle's Universe. OUP, Oxford (2005)

Hammond, R.T.: Negative mass. Eur. J. Phys. **36**, 025005 (2015)

Hartle, J.B.: Gravity: An Introduction to Einstein's General Relativity. Addison-Wesley, New York (2003)

Kirshner, R.P.: The Extravagant Universe: Exploding Stars, Dark Energy, and the Accelerating Cosmos. Princeton University Press, Princeton (2004)

Krauss, L., Wilczek, F.: From B-modes to Quantum Gravity and Unification of Forces. Essay written for the 2014 Gravity Research Foundation Awards for Essays on Gravitation (2014)

Luttinger, J.M.: On 'Negative' Mass in the Theory of Gravitation. Essay Written for the 1951 Gravity Research Foundation Awards for Essays on Gravitation (1951)

Mitton, S.: Conflict in the Cosmos: Fred Hoyle's Life in Science. Joseph Henry Press, Washington DC (2005)

Morris, M., Thorne, K.S.: Wormholes in spacetime and their use for interstellar travel: a tool for teaching general relativity. Am. J. Phys. **56**, 395–412 (1988)

Price, R.H.: Negative mass can be positively amusing. Am. J. Phys. **63**, 216–217 (1993)

Schuster, A.: Potential matter—a holiday dream. Nature. **58**, 367 (1898)

Thorne, K.S.: Black Holes and Time Warps: Einstein's Outrageous Legacy. Norton, New York (1974)

Webb, S.: New Eyes on the Universe. Springer, New York (2012)

Fiction

Abrashkin, R., Williams, J.: Danny Dunn and the Anti-gravity Paint. McGraw-Hill, New York (1956)

Abrashkin, R., Williams, J.: Danny Dunn, Invisible Boy. McGraw-Hill, New York (1974)

Asimov, I.: Foundation. Gnome, New York (1950)

Blish, J.: Cities in Flight, 1st Omnibus edition. Avon, New York (1970)

Dail, C.C.: Willmoth the Wanderer, or the Man from Saturn. Haskell, Atchison, KA (1890)

Forward, R.L.: Dragon's Egg. Del Rey, New York (1980)

Greg, P.: Across the Zodiac: The Story of a Wrecked Record. Truebner, London (1880)

Hoyle, F.: The Black Cloud. Heinemann, London (1957)

Nowlan, P.F.: Armageddon 2419 A.D. Amazing (August 1928)

Wells, H.G.: The First Men in the Moon. Newnes, London (1901)

Visual Media

2001: A Space Odyssey: Directed by Stanley Kubrick. [Film] USA, MGM (1968)

A for Andromeda: Written by Fred Hoyle and John Elliot. [Television] UK, BBC (1961)

Dead of Night: Directed by Alberto Cavalcanti, Robert Hamer, Basil Dearden, Charles Crichton. [Film] UK, Eagle-Lion (1945)

3

Space Travel

Mr. Heinlein and I were discussing the perils of template stories: interconnected stories that together present a future history. As readers may have suspected, many future histories begin with stories that weren't necessarily intended to fit together when they were written. Robert Heinlein's box came with "The Man Who Sold the Moon." He wanted the first flight to the Moon to use a direct Earth-to-Moon craft, not one assembled in orbit; but the story had to follow "Blowups Happen" in the future history.

Unfortunately, in "Blowups Happen" a capability for orbiting large payloads had been developed. 'Aha,' I said. 'I see your problem. If you can get a ship into orbit, you're halfway to the Moon.'

'No,' Bob said. 'If you can get your ship into orbit, you're halfway to anywhere.'

He was very nearly right.

A Step Farther Out
Jerry Pournelle

Our Milky Way galaxy is big. From this simple fact a simple conclusion seems to follow: if you want to cross the galaxy in a reasonable length of time then you must travel fast. Unfortunately, as we know, the universe imposes a speed limit. Douglas Adams pointed this out in his *Hitchhiker's Guide to the Galaxy* (1979): 'Nothing travels faster than the speed of light with the possible exception of bad news, which obeys its own special laws.' If SF authors set their stories in deep space and they want their stories to possess a veneer of plausibility then they have to address the light speed limit. Antigravity offers one way of traversing large distances on small timescales, and in Chap. 2 we

© Springer International Publishing AG 2017
S. Webb, *All the Wonder that Would Be*, Science and Fiction,
DOI 10.1007/978-3-319-51759-9_3

discussed how some SF authors had their protagonists reach the stars using this technique. But we don't know whether antigravity exists; right now, talk of antigravity is little more than handwaving. Would it not be more honest to drop the requirement for plausibility? Even hard SF authors should be allowed to wave their hands while mentioning 'hyperspace' or 'warp factors' or 'ftl drives'—any vaguely scientific-sounding device, indeed, that opens up the Milky Way for exploration and permits stories to take place against that vast backdrop. Most readers were happy to go along with this approach. Several of my personal favorites—Asimov's *Foundation* (1950), Martin's "A Song for Lya" (1974), Clarke's *Rendezvous With Rama* (1973)—took place in a future where interstellar travel was simply a given. I was happy to suspend my disbelief for the sake of the stories.

Nevertheless, some authors took the light-speed limit seriously and tackled the problem directly. For example, Robert Heinlein set his novella "Universe" (1941) on board a generation ship—a craft that travels slower than light and thus takes thousands of years to reach its destination. Although scientists such as Tsiolkovsky and Bernal first came up with the idea of the generation ship, I'd bet it was the strength of Heinlein's vision that most influenced later thinkers on the subject.[1] Poul Anderson provided another example of how space travel might work in the real world. In his novel *Tau Zero* (1970) he discussed how a spaceship powered by a Bussard ramjet might accelerate to speeds where relativistic time dilation becomes important, and if a ship moves at a significant fraction of the speed of light then the journey time to the stars can be small (see box). Yet another option is explored in the early stories of Larry Niven's "Known Space" universe. Niven describes how humankind colonizes nearby planets using slower-than-light fusion-powered space ships. The key to success is that crew and passengers *sleep* through the journey time. In several later stories and films, space travel becomes possible through some form of suspended animation.

Taken together, stories involving exotic physics (the kind that allowed faster than light travel) and those based on known physics (where craft observed the light-speed limit) formed part of the default future described by science fiction. Interstellar travel would be ours—eventually. I was sure I'd never live to see the technology appear: the construction of a generation ship seemed as distant as the development of a warp drive. But that didn't bother me, because there was another aspect to the default future. The stars might be out of reach but the solar system wasn't! Science fiction promised me I'd grow up to see Moon colonies and trips to Mars and voyages to the outer planets. Science seemed to agree—as I was growing up astronauts were on the Moon, probes were landing on Mars, and craft were being sent to Jupiter and beyond. Forget the galaxy; the solar system at least can be ours. Interstellar travel might be a distant dream but interplanetary travel would become an everyday occurrence. Wouldn't it?

Relativistic Time Dilation

If a spacecraft travels at a significant fraction of the speed of light then the crew on board that craft will be subject to the phenomenon of time dilation, as described by the special theory of relativity. In the crew's reference frame a journey to a distant star might take 5 years; depending on how fast the craft travels, outside observers might measure the same journey to take 500 years or 5000. Crew members would return to Earth to find their stay-at-home friends long since dead. No one said space travel would be easy.

3.1 The Science-Fictional Solar System

On 20 July 1969 my father woke me up so I could watch Apollo 11 land on the Moon. To a young boy the space programme was the most exciting thing imaginable, but it was also something one could take for granted. Television was so full of events—mass political demonstrations, assassinations, war—that it was impossible for a young mind to evaluate the significance of it all. Rockets were taking people to the Moon. That, and all those other happenings, was simply the way things were. At that age, one doesn't question.

It was perhaps inevitable that when I started reading SF some of my favorite stories had a lunar setting. But the Moon of those stories wasn't quite the Moon that Armstrong walked upon. For example, Heinlein's "Requiem" (1940), published in *Astounding* almost 30 years before Apollo 11, told of Delos D. Harriman's desire to visit the Moon even though he knows the journey will kill him. "The Man Who Sold the Moon" (1951) describes how the younger Harriman raised the funds for the first Moon landing in 1979. And Heinlein's novel for juveniles, *Rocket Ship Galileo* (1947) relates how three teenagers help develop a rocket that can reach the Moon. (It's said that *Rocket Ship Galileo* served as the basis for the film *Destination Moon* (1950), but to my mind the film seems to be much more grounded in "The Man Who Sold the Moon".) I loved those tales by Heinlein, but they bore little relation to how events turned out. Neil Armstrong reached the Moon a full decade before Delos Harriman, but only because an entire country brought its financial and technological might to bear on the project: lunar travel wasn't an activity for individual businessmen or groups of teenagers. Or consider *A Fall of Moondust* (1961), a novel by Arthur Clarke that recounts the gripping rescue of *Selene*, a jetski-like cruise ship that sinks beneath the fine, powdery dust of the Sinus Roris. It was a thrilling tale, made all the more suspenseful by

Fig. 3.1 I grew up watching the Apollo missions. This artist's conception of a lunar exploration programme depicts the Moon more or less as I expected to see it. The setting here is close to the Apollo 17 Taurus Littrow landing site, and the painting shows a lunar mining facility that harvests oxygen from the material in the eastern Mare Serenitatis (Credit: NASA/SAIC/Pat Rawlings)

Clarke's understanding of how the playing-out of physics would generate problems for both crew and rescuers. Within a few years of its publication, of course, the Apollo missions had confirmed that regolith covers most of the lunar surface. There are no seas of dust (Fig. 3.1).

The details of many science-fictional lunar settings were thus rendered obsolete by the discoveries of science. But, as I argued in the Introduction, that doesn't really matter. Newer writers could still set their stories on the Moon—they just needed to take account of the latest scientific findings and the science fictional 'feel' could stay the same. John Varley was one author who proved this. Between the mid-1970s and the mid-1980s he unleashed a volley of near-future stories providing the same impact Heinlein's early stories must have had a generation earlier. Indeed, those Varley stories were reminiscent of early Heinlein: the first sentences grabbed you by the throat and dragged you into a vividly imagined world. Consider the opening of his story "Bagatelle" (1976), set in the domed lunar city of New Dresden:

There was a bomb on the Leystrasse, level forty-five, right outside the Bagatelle Flower and Gift Shoppe, about a hundred meters down the promenade from Prosperity Plaza.

'I am a bomb,' the bomb said to passersby. 'I will explode in four hours, five minutes, and seventeen seconds. I have a force equal to fifty thousand English tons of trinitrotoluene.'

The story's protagonist, lunar police chief Anna-Louise Bach, has to figure out how to deal with this polite, yet quite deadly, nuclear device. In "The Barbie Murders" (1978), a later story with the same setting, Bach must solve a crime that was captured on security cameras: the murder of a cult member. It sounds straightforward, but the cult members have all undergone body modification so they all appear identical. Indeed the cultists, as far as is possible, have surrendered all elements of their individuality. I still remember the thrill of seeing Varley's name on the cover of *Isaac Asimov's Science Fiction Magazine*, and knowing that the magazine might hold another tale of Bach's detection set on the Moon.

To this day, the Moon remains a plausible setting for science fiction. Science fiction authors never left the Moon; the tragedy is that the astronauts did.

Let's move on to Mars. In 1877, the Italian astronomer Giovanni Schiaparelli made telescopic observations of the Red Planet during which he recorded linear surface structures. He called them 'canali'. The word means 'channels' but it was mistranslated into English as 'canals'. There's a big difference between these two English words: a channel can be natural whereas a canal is an artificial construction. This mistranslation, combined with further observations of the waxing and waning of the Martian ice caps, gave rise to a romantic picture: perhaps intelligent Martians built canals to bring water from the polar regions to the drying and dying desert regions. The American astronomer Percival Lowell, who founded the observatory that would eventually discover Pluto, was a particularly enthusiastic proponent of the notion that Mars was home to intelligent life (Fig. 3.2). His ideas weren't adopted by the astronomical community but they struck a chord with the general public and they inspired a large body of SF. Perhaps the most famous Martian tale was Wells's wonderful *War of the Worlds* (1898), a book that has never been out print. *War of the Worlds* is an archetypical piece of science fiction, and one of the most influential books in the canon. (I hardly need to mention here its

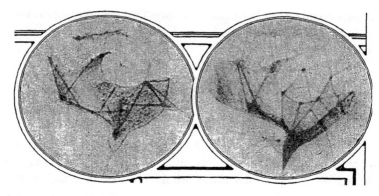

Fig. 3.2 The New York Times of 27 August 1911 published a story under the headline "Martians Build Two Immense Canals in Two Years—Vast Engineering Works Accomplished in an Incredibly Short Time by Our Planetary Neighbors". The story was based on these drawings of Mars, made by Percival Lowell and Earl Slipher, which purported to show artificial structures (Credit: Public domain)

influence on SF, but it's worth pointing out that the novel sparked Robert Goddard's interest in rocketry. Goddard, of course, was one of the founding fathers of astronautics.)[2] Wells took Lowell's idea of a dying Mars and gave it a dramatic twist: where Lowell imagined an advanced and politically mature Martian civilization, Wells described intelligent beings who had exhausted the resources of their home planet and so were now looking to get rid of bothersome humans and relocate to Earth.

When it comes to Mars, the vision created by Edgar Rice Burroughs was almost as influential as the one conjured up by Wells. The *Barsoom* series of stories, which Burroughs published between 1912 and 1943, was hugely popular. I've already mentioned how the series was an inspiration to Carl Sagan; they also inspired a generation of SF writers, including Ray Bradbury. Personally, I found Bradbury's *The Martian Chronicles* (1950) far more to my taste than Burroughs' *Barsoom*. Bradbury's connected set of stories describes how the Earthmen who visit Mars are haunted not only by the ghosts of a dead Martian civilization but also by the ghosts of their own past. Even as I was reading *The Martian Chronicles*, however, it was becoming clear that the Mars of Wells, Burroughs, and Bradbury had no basis in reality. In 1971, Mariner 9 sent back photographs of the Martian surface; in 1976, two Viking landers actually touched down. They saw no signs of ancient civilizations.

Once again, science had rendered a key science-fictional setting obsolete. Once again, that didn't really matter. Bradbury, for example, never attempted to portray a scientifically accurate Mars—he was using the planet as a vehicle for exploring issues affecting humanity—so the pictures sent back by the Viking landers didn't seriously affect a reading of *The Martian Chronicles*. And newer writers could still use Mars as a venue. Indeed, the latest scientific findings meant they could add a dash of verisimilitude to their stories.

The Moon and Mars were obvious targets (for astronautical agencies as well as authors) but SF writers offered a vision of the entire solar system. For example, Alan Nourse's story "Brightside Crossing" (1956) is set on Mercury. It tells the tale of four daredevils who attempt to traverse the planet's Sun-facing hemisphere at perihelion, when the planet is closest to the Sun. Larry Niven, who went on to become one of the biggest names in SF, set his first published story on Mercury. (As with all Mercury-set stories written before 1965, Niven's "The Coldest Place" (1964) was based on an incorrect understanding of the planet: astronomers used to believe that Mercury was tidally locked, and thus always presented one face to the Sun, but radar observations found that the planet rotates three times for every two laps around the Sun).

Venus formed the backdrop to novels such as Pohl and Kornbluth's *The Space Merchants* (1953) and stories such as Varley's "In the Bowl" (1975). Even during the 1950s, however, there was a suspicion that Venus was too harsh a place for us ever to enjoy holidays there. When SF writers set a story on a hot-and-humid Venus, they were crossing their fingers and hoping.

Science fiction writers also set stories on more distant bodies in the solar system. The outer solar system is home to four large planets—the gas giants (a term, incidentally, coined by the science fiction writer James Blish in his story "Solar Plexus" (1952) and only later adopted by astronomers). Jupiter played a role in novels such as Simak's *City* (1952) and stories such as Clarke's "A Meeting With Medusa" (1971), with the many moons of Jupiter also serving as backgrounds for stories. In "The Martian Way" (1952), one of Asimov's most successful novellas, Martian colonists travel to Saturn's rings in order to bring back a cubic mile of ice for their desert world; the moons of Saturn, particularly Titan, have also been a popular venue for explorers in science fiction stories. Even farther out, I fondly remember reading stories involving Uranus and its moons (for example, *First Contact?* (1971) by Hugh Walters), Neptune (Stapledon's *Last and First Men* (1930)) and Pluto (too many to mention; and yes, I know that Pluto has been downgraded to the status of a dwarf planet, but until recently it was still a 'proper' planet).

Some works had our entire planetary system as a backdrop. The Hugh Walters novel mentioned above was just one in a series of juveniles in which Chris Godfrey of the United Nations Exploration Agency explored the planets—from *Mission to Mercury* (1965) through to *Passage to Pluto* (1973).

Asimov's *Lucky Starr* series of novels, also aimed at juveniles, had adventures set on Mercury, Venus, Mars, the Asteroid Belt, Jupiter and Saturn. I read stories set on comets; on space yachts driven by solar pressure; and on space stations orbiting Earth or other planets. Many stories set in the future took it for granted that interplanetary travel would be as common as intercity travel. Timescales and technologies might have varied from author to author and from story to story, but the overall impression given by science fiction—the default future—was that humanity would soon spread through the solar system. To a boy growing up during the Space Age this seemed to be an eminently plausible picture. It hasn't turned out that way.

3.2   Halfway to Anywhere?

It's illuminating to revisit the whistlestop science-fictional tour of the solar system I've just given and compare the vision with what has happened to date. We *are* exploring our planetary neighborhood, it's true, but the scale of our exploration is nothing like as impressive as SF fans once hoped.

Let's start as before with the Moon. Between July 1969 (when Neil Armstrong first set foot on the Moon) to December 1972 (when Gene Cernan shook the Moondust from his boots) a total of 12 astronauts have walked on our planet's satellite. In the intervening decades, no one has been in lunar orbit let alone set foot on the Moon. Even if there was the political will to return we don't have the ready capacity to do so: a new voyage to the Moon would require a major commitment from one or more countries. It's not that the Moon remains some sort of mysterious object. As I write this, for example, NASA's *Lunar Reconnaissance Orbiter* (LRO) is continuing to take high-definition photographs of the lunar surface. The LRO has given us a highly detailed map of the Moon. The photographs are in the public domain and are so detailed you could even, if you have a moment or two to spare, check them for signs of activity by extraterrestrial beings. (Perhaps, many millennia ago, aliens planted a device on the nearside of the Moon and pointed it at Earth in order to look for the emergence of intelligent species. Well, a combination of the internet and LRO photographs enable you to look for evidence of this science-fictional proposition from the comfort of your own home. As far as I'm aware, no SF author foresaw the incredible technological developments that would allow such an armchair-based search take place.) Scientists understand much, much more about the Moon than they did when the Apollo missions were underway—but a crewed return to the Moon does not seem to be particularly high on anyone's agenda.

Mars is the other important venue. Several artificial satellites are currently orbiting the planet. For example, the 2001 Mars Odyssey (named as a tribute to Arthur Clarke) has been mapping the surface since 2002; the Mars Reconnaissance Rover has been studying the planet since 2006 (and in 2015 managed to locate the whereabouts of the Beagle 2 lander, which went missing during an attempted landing in 2003). Several rovers have landed on the Red Planet and done real science: as I write, Opportunity and Curiosity are both busy studying the climate and geology of Mars (Fig. 3.3). Over the years, then, we've learned a lot about Mars—certainly enough to know that there's a lot more to learn. However, despite a number of tentative proposals for a crewed mission, and though it pains me greatly to admit it, I don't hold much hope of ever seeing humans walk the Martian surface. I'll explain why later.

Mercury? Well, to date, no probe has ever landed on the planet closest to the Sun. Indeed only two satellites, Mariner 10 and MESSENGER, have ever made close-up observations of Mercury.

As for Venus, we know much more about it than we do about Mercury. The Soviet Union's Venera program landed ten probes on the surface and a dozen or more probes have sent back data about our sister planet's atmosphere. The Pioneer Venus Orbiter studied the planet for over 13 years. These and other missions confirmed that Venus is a hellish sort of place—a runaway greenhouse effect caused it to become by far the hottest planet in the solar system. The wet, tropical Venus of so many early SF stories does not exist; the planet is dry, and hot enough to melt lead. There are no upcoming missions of interest to Venus.

No one has landed a probe on the surface of Jupiter, but then that's hardly surprising—the largest of the planets lacks a solid surface on which a probe could land. However, seven missions have flown past Jupiter, which makes it the most visited of the outer planets. In addition, the Galileo spacecraft spent almost 8 years between 1995 and 2003 orbiting around Jupiter. More recently, in 2016 the Juno spacecraft began orbiting the planet with the aim of studying its gravitational and magnetic fields and searching for information about its chemical composition and how the planet formed.

Other solar system objects have been visited much less often. Pioneer 11, Voyager 1, and Voyager 2 have made fly-bys of Saturn, the ringed planet. The Cassini spacecraft is currently in orbit around Saturn and in 2004 it released the Huygens probe for a landing on Titan, the largest moon of Saturn. This is the most distant landing on an astronomical body for any spacecraft. Even further out, only Voyager 2 has ever come close to Uranus and Neptune, and there are no plans, as far as I'm aware, to revisit these planets any time soon. NASA's New Horizons probe passed close to Pluto in 2015, and

Fig. 3.3 An interplanetary selfie: NASA's Curiosity rover used an arm-mounted camera to capture 55 high-res images, which were then stitched together to form this self-portrait. Curiosity is at a 'Rocknest' in Gale Crater, where it made its first sample of the Martian surface. The marks in the regolith in front of the rover show where the rover took its scoops. We will be limited to such robotic exploration of Mars for many years to come (Credit: NASA/JPL-Caltech/Malin Space Science Systems)

transformed our knowledge of the dwarf planet. Unfortunately, New Horizons will remain the only spacecraft to visit Pluto for many years—it's too long and cold a journey to make often. Another dwarf planet, Ceres, is the focus of the

Dawn spacecraft; after flying past the asteroid Vesta, Dawn travelled to Ceres where its fate is to become a perpetual artificial satellite. Finally, in addition to the Moon, Venus, Mars and Titan, the space agencies have managed to land a craft on asteroids (433 Eros and 25,143 Itokawa) and even a comet (Churyumov–Gerasimenko, as mentioned in Chap. 1).

So humankind has certainly made tentative steps into our space environment. Nevertheless, human exploration of the solar system hasn't happened with anything approaching the pace hinted at by the science fiction of my youth. It isn't that space has become somehow irrelevant. Quite the contrary. Our modern world—the one that science fiction *didn't* predict—is in many ways dependent upon access to space. We use satellites for communication and navigation, for espionage and weather forecasting, for environmental modelling and land stewardship. (One SF writer, Arthur Clarke, *did* foresee the economic importance of space. The geostationary orbit, which is used for communication satellites, weather forecasting, and defence applications, is often called the Clarke orbit in his honor because he described the possibilities in a paper published in 1945.) However, nearly all of the activity in Earth orbit, and all of the solar system exploration mentioned above, has been carried out by various type of probe. Since 2 November 2000 the International Space Station (ISS) has been crewed continuously, and at any time typically six people are on board; but with that minor exception humans don't live or work in space. Indeed, since spaceflight became possible—with spaceflight being defined as any flight higher than 100 km—fewer than 600 people have been in space. For comparison, more than 4000 people have climbed Mount Everest since Hillary and Tenzing first scaled the mountain in 1953.

So what's the difficulty with space travel? Our engineers have access to materials that are lighter and stronger than anything used in the Apollo missions; computers are faster and more powerful; our understanding of the space environment increases with each passing year. So why aren't we building domed cities on the Moon? Why aren't we constructing human colonies in space of the type envisioned by the Princeton physicist Gerard O'Neill (and gorgeously illustrated by various science fiction artists in O'Neill's book *The High Frontier* (1977); see, for example, Fig. 3.4)? Why is a robot, rather than a human being, currently driving Curiosity across Gale Crater? There appear to be two fundamental obstacles—one involving economics, the other biology. The two are related, but for simplicity let's look at the economic argument first and then consider biology.

Fig. 3.4 Following a NASA-sponsored conference devoted to space station design, held in 1976, Don Davis painted this illustration of a toroidal-shaped space station. Four decades later we have the ISS, which is an impressive accomplishment but nothing like this so-called Stanford torus (Credit: Don Davis)

3.2.1 The Cost of Space

I introduced this chapter with a quote that's usually attributed to Robert Heinlein: if you can get into orbit, you're halfway to anywhere. There's a great deal of truth in the saying. Once you've climbed out of Earth's deep gravity well it doesn't take vastly more energy to reach most other places in the solar system—particularly if you take advantage of the boosts offered by the gravitational wells of other planets. The Voyager spacecraft, which blasted off from Earth in 1977, used this gravitational 'slingshot' effect to tour the gas giants and are now heading off into interstellar space. The cost of space travel resides not so much in interplanetary travel itself but in escaping from Earth's grip in the first place.

If we had a long enough staircase we could in principle walk into space. Of course we don't have such a staircase, so instead we use chemical rockets. We don't use them because they are efficient—they aren't: the main stage of the European Ariane 5 rocket, for example, burns 175 tonnes of propellant on

Fig. 3.5 Rockets are an expensive way of reaching space. The Delta II rocket shown here put NASA's Orbital Carbon Observatory 2 into space in July 2014. Each launch costs tens of millions of dollars (Credit: NASA/Bill Ingalls)

each launch. Rather, we use chemical rockets because currently we don't have any other practical option for getting into space. It's our current use of chemical rockets that makes space travel expensive. Getting 1 kg of payload just into low earth orbit costs between $10K to $25K at present day prices. (Estimates vary; commercial space organisations, for obvious reasons, are generally tight-lipped about costs.) Getting 1 kg of payload into geostationary orbit is even more expensive—about the price of a kilogram of gold (Fig. 3.5).

Consider the ISS. The space station's mass, roughly 450,000 kg of it, had to be pushed into space using the thrust produced by the burning of chemicals. With each new expedition to the station the mass of the astronauts, along with the mass of the food they eat, the water they drink, and the suits that protect them, must all be propelled into space. The rockets that take all this into orbit also have to get their *own* mass into space. It's the cost of pushing all this mass into space, rather than the research and development work that took place on the ground, that made the ISS the most expensive single item ever constructed by humankind. The cost of putting 1 kg into low earth orbit is dropping all the time, but we're unlikely to build many ISSs—certainly not enough to make space a plausible destination for most of us anytime soon.

Note that space exploration by *probe* is certainly possible. As we've seen, we've been engaged in that activity for decades, it's ongoing, and exploration of the solar system by probe is set to continue. Indeed, we can even imagine a scenario in which self-replicating probes explore or even colonize the solar system: it's an example of a technology that *does* have the potential to scale. But exploration by *people*? For that to happen we need an alternative to rockets. As we'll see in later, SF writers have considered other possibilities for reaching space. However, as long as we are dependent upon rocket launches to get people into space—well, it's difficult to see how manned spaceflight can ever become commonplace. It simply costs too much.[3]

3.2.2 Spacemen?

All aspects of space travel are dangerous. All of them. On the ground, numerous non-astronaut fatalities have been caused by rockets exploding before launch. Seven astronauts were lost during the launch phase when Space Shuttle *Challenger* broke up. Space itself is a harsh environment, of course, so it's remarkable that to date only three astronauts have perished in space: the crew of Soyuz 11 died of asphyxiation when, through bad luck, a cabin vent valve opened. Re-entry poses its own challenges: the seven astronauts on Space Shuttle *Columbia* were killed when the vehicle disintegrated 60 km above Texas; the Soviet astronaut on board Soyuz 1 died after a parachute failed during re-entry. It's remarkable there haven't been *more* fatalities. After all, in order to get people into space we perch them on top of what is essentially a large bomb; while they are in space they must be protected from quite deadly surroundings; and when they return the thermal and mechanical stresses on their vehicles are immense. Inevitably, things will sometimes go wrong.[4]

There is no shortage of brave people willing to risk their life for the advancement of science and, indeed, the future of humanity. However, the fragility of the human body, the constraints set by biology, add to the difficulties of manned space exploration. Consider, for example, the problem of establishing a colony on Mars.

Getting astronauts to Mars would be a costly exercise—we've already mentioned the expense of the fuel needed to lift the hardware into space, to lift food and water into space, to lift the fuel itself into space. Nevertheless, if our civilization set its mind to the task it could presumably set the necessary infrastructure in place. After all, we *have* constructed a space station in orbit so presumably we could build a space craft in orbit. (The ISS took a decade to construct and was the result of over thirty Shuttle assembly and maintenance flights. NASA experts estimate that 80 rocket launches might be needed to construct a space craft to take astronauts to Mars. So construction would take time, but it could be done.) And we could send supplies—food, water, the materials necessary to build habitats—on ahead. We could send them on a 'slow boat', ready for the astronauts to pick them up. So if we threw enough money at the problem we could get the necessary hardware to Mars. But we also need to get the people to Mars, and we want them to be fit and healthy when they arrive. Ideally, we'd like to get them back to Earth in the same condition (or should we instead be thinking of one-way journeys?). It's the 'fit and healthy' part where things might get tricky.

Roughly two-thirds of uncrewed Mars missions have failed, so we already know the journey itself is difficult; ensuring the physical and psychological safety of astronauts on such a mission—something the space agencies haven't yet had to contend with—adds new dimensions of difficulty. Upon arrival it's entirely possible that the Martian astronauts—people who would have been chosen for their qualities of physical and mental endurance—would not be fit enough to engage in meaningful exploration. Let's put to one side the difficulty of treating common problems that might require surgery (appendicitis, say). One of the key threats to astronauts' health would come from radiation. The Sun spews out high-energy radiation and here on Earth we don't have to worry because our planet's magnetic field and it's thick atmosphere protects us from harm. On the journey to Mars, however, a journey of between 7 and 9 months, the astronauts would be horribly exposed. If the Sun happened to flare while the astronauts were en route. . . well, it's quite possible they'd receive a lethal dose of radiation. Even in the absence of a major solar flare, over the course of the trip the bodies of the astronauts would inevitably absorb large quantities of radiation and the damage would be cumulative. They would have a significantly raised risk of developing cancer.

The development of cataracts would be another occupational risk for those journeying to Mars. On their way to the Moon, Neil Armstrong and Buzz Aldrin reported seeing flashes of light even when their eyes were closed; we now know those flashes were caused by cosmic rays flashing through their retinas. The eyes of our Martian astronauts would be subject to many more of those high-energy cosmic rays. As if cataracts and cancer weren't enough, there's also some research suggesting that the level of radiation encountered by astronauts on a trip to Mars might cause early-onset Alzheimer's disease.

Currently there's nothing much we can do to shield astronauts from cosmic ray particles. Lining the spacecraft with water or lead would provide some protection, but that would significantly increase the cost of the mission: we would have to haul the mass of all that protective lining up Earth's gravity well.

A better option would be to protect the craft with an intense magnetic field, but that's beyond our current technology.

Let's put aside the issue of space radiation. There's another health problem for our Mars-bound astronauts: the microgravity environment in which they'd be living. We know from the Russian astronauts who volunteered to spend a year or more in the Mir space station that bones degrade, muscle mass is lost, and the optic nerve swells. Regular exercise seems to help, but only to an extent.

There are a number of other challenges that the body human body faces in space. Without going into further detail it seems fair to say that our would-be astronauts run the risk of arriving at Mars in poor physical shape. Certainly by the time they returned to Earth they would be weak, their bones would be brittle, and they might be blind.[5]

Perhaps doctors and engineers will one day overcome the health problems that space travel throws up—all those issues involving radiation, microgravity, and so on. Perhaps. But astronauts face another health problem: the problem of mental health. Astronauts would suffer the psychological hardship of being cooped up in a small craft with other people for months on end. Your fellow crew member—the one who constantly whistles show tunes—started the journey possessed of an endearing habit. After 9 months of listening to the same tuneless trilling you want to throw him out of the airlock.

In the 1990s, a controversial experiment called Biosphere-2 attempted to understand what might happen to living things in a closed biosphere. Eight people lived in that closed environment for 2 years, and although it's uncertain whether the exercise led to any real scientific advances one thing was clear: as time went by psychological concerns became an increasingly important issue. More solid evidence of the difficulties involved with isolation comes from scientists who overwinter in Antarctica. Those people are more isolated from the rest of humanity than are the astronauts on the ISS. They report how, as the long winter drags on and on and on, everyone tends to spend more time in their own rooms and inside their own minds; how reaction times slow; how it becomes increasingly difficult to deal with separation from 'real' life back home. And a NASA team of six people were isolated in a dome between August 2015 and 2016 to simulate a Mars mission; they stayed the course, but reported that the lack of privacy had been difficult. So in any crewed Mars mission, one of the most important roles would surely belong to psychologists: it would be their responsibility to choose crew members with exquisite care. Individual astronauts would have to possess great mental resilience, but the group as a whole would have to work without friction.

Even if astronauts reached Mars orbit in good physical and mental health, they would then face the ordeal of landing.

Every Mars mission to date has carried only a small load. The *largest* load ever landed on Mars, the Curiosity rover, had a mass of 900 kg. A crewed mission would necessarily be tens of times more massive than Curiosity and this creates a problem because engineers don't fully understand how to slow the descent of such a massive craft. The Martian atmosphere is too thin for parachutes to work as they do for craft returning to Earth: parachutes on Mars are fine for small payloads but they wouldn't inflate sufficiently to slow the descent of a massive spacecraft. Of course, it doesn't necessarily matter if general supplies smash into the Martian surface at high speed, but humans couldn't possibly survive such an impact. On the other hand, the thickness of the Martian atmosphere raises issues of aerodynamics if rockets are used, as they were on Moon landings. On the airless Moon, the Apollo lunar module commanders could deploy downward-facing thrusters to slow the descent; on Mars, rockets fired on a rapidly descending craft would generate turbulence and huge stresses. Astronauts face the prospect of a bumpy landing.

Let's assume the landing problem can be dealt with—engineers are already working on the challenge and I'm certain they'll surmount it. And we've also decided to assume that our astronauts are healthy when they arrive. So humans set foot on Mars. What then? Well, they need to build shelter; they must ensure they have oxygen to breath and water to drink; they have to secure their supply of food. Most of these necessities will have accompanied them on their journey or else been sent beforehand; some ingredients that are necessary for life—water, for example—can perhaps even be obtained *in situ*. However, the astronauts would be in a precarious position. If they arrived to find their fuel supplies depleted, for example, or if their food supplies were unusable, then there wouldn't be a great deal they could do about it. And they'd be facing fundamental challenges—supplying themselves with air, water, and food—in an environment of which we have little understanding. There are concerns, for example, regarding the fine Martian dust: it might clog up air filters and mess with machinery; chemicals in the dust, particularly perchlorates, may cause health problems. Life would not be easy.

Given the challenges I've outlined above—and there are many others—you might think any person who volunteered for a Mars mission would be demonstrating such a heightened level of stupidity that they'd automatically be disqualified from flying. A sort of Catch-22. But there's no shortage of very bright people who quite understand the risks and are nevertheless willing to volunteer for a mission to Mars. Many people will even volunteer for a one-way mission, with no chance of return. We know that's the case because many people have already volunteered for such a mission. Mars One is a not-for-profit organization with the stated aim of sending a crew of four on a one-way trip to the Red Planet. When the call for volunteers went out, 202,586 people applied to be on a launch planned for 2022. At the time of writing, that number has been whittle down to 100 hopefuls for a planned 2024 launch. We clearly don't lack for humans with daring and spirit. Recruiting astronauts isn't the problem we face in developing a colony on Mars; the problem is all those other unresolved difficulties I mentioned above. And that's the reason why Mars One won't succeed, why those 100 would-be astronauts won't be going into space in 2024.

By the time you read this, I suspect the Mars One mission might well have collapsed to become just be a minor curiosity in the history of astronautics. If it were to succeed—well, I'd buy a silk stovepipe hat and eat it. I'd be eating it happily: I really do want people to reach Mars, to live on Mars, to realize the dreams of those early SF writers. And I think one day it will happen: the challenges are great but so are our reserves of courage and ingenuity. As things stand, however, we don't have the capability of creating a Martian colony. Despite what the marketing department of Mars One might say, that capability is decades away.

Compared to Mars, the Sahara Desert is a thousand times more hospitable and a billion times more accessible—and yet, after thousands of years of inhabitation, the number of humans living in the Sahara remains low. And Mars gives us probably our *easiest* option for extraterrestrial living. The SF vision of a solar system teeming with human life seems a long, long way away.

3.2.3 The Final Frontier?

In terms of distance, space really isn't far away—just an hour's trip in a car if we could somehow drive straight up. In terms of cost, space is far away indeed—at least, it is so long as access to it depends on lighting the fuse on what is essentially a bomb. The problem is that each single-use multi-stage rocket must be built from scratch; each mission is in essence a new project. This method of reaching space doesn't scale and therefore it can't become routine. (This much is obvious. It's not exactly rocket science.) If we had an alternative, a method for reaching space we could use and re-use, again and again, then space might open up—and the old SF vision of interplanetary travel might yet come to pass. Engineers are exploring some possibilities.

Alan Bond, a British engineer who started building rockets soon after he started reading SF as a child, set up in 1989 a company to develop an engine that would be powerful enough for a single-stage-to-orbit craft. He believes he has succeeded. His SABRE engine acts in the same way as a conventional jet engine up to an altitude of about 25 km, at which point the craft it is powering has a velocity of about 5.5 times the speed of sound. At that point the engine switches mode. The air inlet closes and the craft becomes a highly efficient rocket, burning the usual liquid oxygen and hydrogen propellants in order to reach orbit. Numerous technical challenges are involved in designing such an engine, not least the requirement to cool the high-speed air entering the engine from about 1000 °C to −140 °C in a fraction of a second and without freezing the engine itself. Bond claims to have cracked the various problems, and his SABRE engines are due to be tested in 2019. If all goes well, a pair of SABRE engines will power his new design of spaceplane—Skylon.

Skylon is designed to be able to fly to an equatorial orbit, deliver a payload of 15 tonnes, return to Earth and be ready to fly again in 2 days. If Skylon works it will transform the economics of spaceflight. The cost of delivering a kilogram into space will fall by a factor of about 25. Space-based activities that are currently uneconomic will become viable. If it works (and it must be said that earlier single-stage-to-orbit proposals, such as NASA's X-30 spaceplane, have all failed) it will change our thinking about space (Fig. 3.6).[6]

A spaceplane such as Skylon would offer relatively cheap access to space. So would reusable rockets: SpaceX, a company founded by Elon Musk, is pioneering vehicles that launch and land vertically. Whether spaceplane or reusable rocket wins out remains to be seen. But SF writers have long dreamed of an even cheaper means of reaching space. In particular, they've pondered the possibility of a space elevator—the same sort of device that can take you to the top of a skyscraper, with the difference being that this one can take you all the way into space. Although the idea sounds outlandish, it's based on relatively straightforward science. It's so straightforward that it generated one of the classic cases of science fictional serendipity: two English-born, physics-trained authors—Arthur Clarke and Charles Sheffield—independently wrote novels based on this idea (*The Fountains of Paradise* and *The Web Between the Worlds* respectively) and the two books were published within weeks of each other in 1979. The idea of a space elevator was 'in the air'.

The basic design concept is simple: a cable is fixed to a base on, or very close to, Earth's equator. The cable runs up to geostationary orbit 36,000 km above sea level and continues until it ends on a 'counterweight'—a large rock, perhaps, or even just junk—in a higher orbit (Fig. 3.7). The counterweight would cause the cable to be under tension, and therefore the system would hold straight up—rather like an upside-down plumb bob. An elevator car

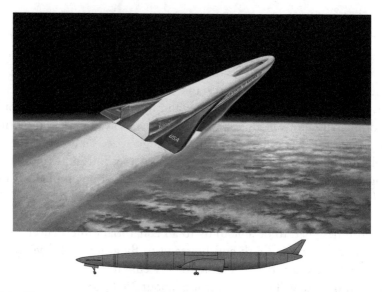

Fig. 3.6 *Top*: artist's concept of NASA's single-stage-to-orbit X-30 'aerospace plane' flying through the atmosphere and into space. NASA unveiled the X-30 idea in the mid-1980s; in 1993 funding for the research was cut. *Bottom*: side elevation of the proposed single-stage-to-orbit Skylon spaceplane (Credit: *top* NASA; *bottom* GW_Simulations)

could then take a leisurely week-long ride up the cable. It would be a cheap, and in my opinion rather civilized, way of getting to space (Fig. 3.8).[7]

The idea of a space elevator is not new. It goes all the way back to 1895, when Konstantin Tsiolkovsky—one of the founding fathers of rocket science—asked whether we could build a tower that would reach as high as geostationary orbit. It turns out we can't build such a tower because no known material, nor any we can credibly imagine, has sufficient compressive strength: any such construction would collapse under its own weight. In 1959, however, Yuri Artsutanov proposed that counterweight idea. The key point about Artsutanov's counterweight proposal is that the elevator material would not be under compression; a tension structure is much more plausible.

Even the counterweight structure has its problems. In 1966, American engineers calculated the properties required of a space elevator cable and found that no known material, not even diamond, would be strong enough. But materials science has progressed since then. Of particular interest is a paper by Sumio Iijima, published in 1991, which made the scientific community aware of the potential of carbon nanotubes. Iijima's paper sparked a vast amount of interest because carbon nanotubes are stronger and stiffer than

Space Elevator

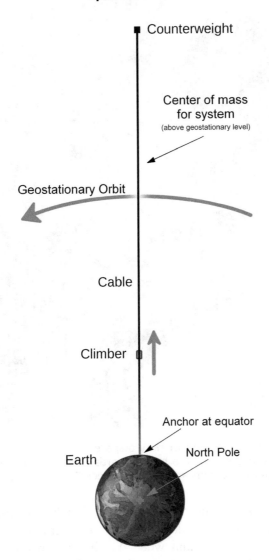

Fig. 3.7 Imagine if we could fix a cable to a point on the equator, stretch it up into space, and attach it to a counterweight so that the system's center of mass is above the altitude of geostationary orbit. The upward-pointing centrifugal force from Earth's rotation would counter the downward-pointing force of gravity, and the cable would stay upright and under tension. 'Climbers' could then carry cargo up and down the cable and into orbit (Credit: Public domain)

Fig. 3.8 An artist's impression of how a space elevator might appear, looking down the length of the structure back towards Earth. Climbers move along the elevator to transfer passengers and cargo between Earth and space (Credit: NASA)

any material previously studied. Carbon nanotubes, it was hoped, might form the basis of a material that could be used to construct a space elevator.

I don't expect to be around when the first passengers ride on a space elevator: the technical, engineering, and political challenges put its construction several decades into the future. I *do* hope to see a technology such as Skylon significantly reduce the cost of reaching space. However, even if the main economic barriers to space exploration are overcome there would remain that basic biological barrier we discussed earlier. Is there anything we can do to make space safe for humans?

3.2.4 The Real Final Frontier

Perhaps we ask the wrong question when we wonder whether space can be made safe for humans. SF has long considered another question, and science is beginning to pick it up: can we make humans fit for living in space?

The idea that we might deliberately design the human body to fit different environments is an old one in SF. James Blish, in a quartet of stories collected as *The Seedling Stars* (1957), argued that if we want to live on other planets then we must do one of two things. We either alter planets to make them habitable (a process the SF writer Jack Williamson called terraforming—'Earth shaping') or else we alter people in order to make them fit for survival on those planets (a process Blish called pantropy—'turning everything'). A decade later, in an award-winning story called "Aye, and Gomorrah. . ." (1967), Samuel Delany suggested that in order to withstand the deleterious effects of cosmic radiation, the bodies of astronauts might have to be engineered. And 15 years after the publication of Delany's story, Ridley Scott's *Blade Runner* was released. In the film, genetically engineered replicants labor on off-world colonies. (*Blade Runner* is one of the greatest of all SF films, but it's worth pointing out the optimism of its predicted timescales. If the *Blade Runner* timeline were true then Deckard—the replicant hunter or blade runner of the title—would have started his job in March 2015; the film's action takes place in 2019. *Blade Runner* was as mistaken as *Back to the Future* when it came to imagining life in 2015.) So this idea—the engineering of the human body to make it fit for the rigors of space travel—is well established. It's an idea that continues to be fleshed out and examined in stories and novels being published today. In the six decades since Blish coined the term pantropy, however, scientists have started to understand how the idea might be realized.

In 2010 Craig Venter, who was one of the first to decode the human genome and who established an institute that created the world's first synthetic organism, suggested NASA should select astronauts based on whether their genetic makeup would help them withstand the health risks associated with space travel. He went further, and suggested humans might be *engineered* to possess the necessary genes for living in space—genes for increased radiation resistance, genes that minimize muscle loss and bone demineralization, and so on. The technology needed to do this does not yet exist, and our understanding of the ethical issues involved are even less advanced, but it does offer one avenue for enabling humans to cope with the space environment.

Nanotechnology might create further opportunities for living in space. Nanobots—robots operating at the nanoscale—are currently being researched because they have obvious applications in terrestrial medicine: they could be used to attack cancer cells, for example. That same technology could mitigate many of the effects on astronauts caused by radiation. Again, it's important to stress that the technology does not yet exist, and Jones (2016) gives a readable account for why it might never exist. But it might, one day.

Yet another option would be just to ditch the human body altogether, and concentrate instead on keeping the human brain alive inside some form of machine. It's an idea that science fiction writers have of course considered (think of Anne McCaffrey's *The Ship Who Sang* (1969)) but it's also one that cyberneticists are beginning to contemplate.

In Chap. 10 we'll discuss some options for engineering the human body. At this point it's simply worth noting that enthusiasts believe the constraints imposed on space travel by human biology might one day be overcome by technology. It's not yet clear what that technology might turn out to be—bioengineering, nano-engineering, cybernetics or something else entirely—but at least there's hope. Space is a harsh environment for humans, but the frailty of our bodies is not necessarily a limit on our ability to explore the solar system and even, perhaps, beyond. Those space explorers, however, would be 'different' to the rest of stay-at-home humanity. Re-engineering the human body to make it fit for space exploration raises a question that has long been pondered by SF writers: what does it mean to be human anyway? It also raises a thought: perhaps the human cultures that move out into space will be those that are most comfortable with the idea of modifying the human body.

3.3 Conclusion

The Golden Age of SF promised me interplanetary journeys, trips to space stations, the chance to live on lunar and Martian colonies. I'm sad none of this has come to pass and none of it is likely to happen in my lifetime. (A trip to the ISS is theoretically possible, I suppose, but it isn't going to happen in practice.) With the benefit of hindsight it isn't surprising how things turned out. The early scientists and writers were correct: we can get into space. We did it and we continue to do it—the exploration of the solar system by probe is one of the great achievements of modern technology. But our approach for reaching space—putting a payload on top of a huge, controlled explosion—doesn't scale. If things go on this way, space travel will never become routine.

One of the lessons preached by SF, however, is that things needn't go on the way they've always gone on. If we develop spaceplanes or space elevators then the cost of reaching space will fall; we won't then be halfway to anywhere, but we *will* have opened up access to the rest of the solar system. And if we are willing to augment our biology—through genetic engineering, nano-engineering, or cybernetics—then our children, or our children's children, will be able to make those interplanetary journeys.

There is still a future for the human race in space.

Notes

1. The concept of generation ships nicely illustrates the interplay between science and science fiction. The Russian visionary Konstantin Eduaordovich Tsiolkovsky (1857–1935) was probably the first scientist to make a serious investigation of the options for space travel. Tsiolkovsky wrote SF in order to popularize his ideas, but he was also a prolific author of technical papers and he did a lot of early work in rocketry. In 1928, in an essay entitled "The Future of Earth and Mankind", he discussed the possibility of using generation ships to reach the stars. (For more information on his tremendous contribution to the theory of space travel, see Tsiolkovsky, 2004.) At around the same time the British physicist John Desmond Bernal (1901–1971) published his book *The World, the Flesh and the Devil* (Bernal, 1929), which described the notion of permanently occupied space habitats—now known as Bernal spheres. A Bernal sphere can be considered as a form of generation ship. Arthur Clarke has stated that *The World, the Flesh and the Devil* influenced him, and perhaps it also influenced Heinlein in his writing of "Universe" and its sequel "Common Sense" (which were combined into the novel *Orphans of the Sky* (Heinlein, 1963)). Many authors followed Heinlein in setting their stories on board generation ships, and this in turn has prompted scientists to study the challenges that might be faced by anyone hoping to construct such a ship; see for example Hein et al. (2012).
2. The American engineer Robert Hutchings Goddard (1882–1945) and the German physicist Hermann Julius Obert (1894–1989) are, along with Tsiolkovsky, generally recognized as the founding fathers of rocketry. For a hilarious history of rocketry, see *Ignition!* (Clark, 1972); note that Clark was a science fiction writer as well as a chemist.
3. For a review of the costs of getting into space (as the situation stood circa 2010), and the possibility of a private involvement in space, see for example Solomon (2011). Since those reviews were published, commercial organisations such as SpaceX have made significant steps towards reducing the

cost of reaching space. But space is still not cheap! See, for example, Kramer (2016).

4. See Shayler (2000) for an account of space-related accidents and disasters.
5. As a counterbalance to my rather gloomy view of the near-term future of space travel, note that Zubrin (2011) gives a more optimistic assessment of the health-related risks faced by Mars-bound astronauts.
6. The English engineer Alan Bond (1944–) was a major contributor to the Project Daedalus fusion-powered starship concept published by the British Interplanetary Society. Bond has since worked for many years at the leading edge of aerospace developments. See *The Three Rocketeers* (2012) for a fascinating television documentary about his involvement with spaceplane research.
7. Arthur Clarke's *The Fountains of Paradise* (1979) and Charles Sheffield's *The Web Between the Worlds* (1979) are classic SF novels based around the space elevator concept. There are also some excellent non-fiction treatments of the idea; see for example van Pelt (2009).

Bibliography

Non-fiction

Clark, J.D.: Ignition! An Informal History of Liquid Rocket Propellants. Rutgers University Press, New Brunswick, NJ (1972)

Clarke, A.C.: Extra-terrestrial relays – Can rocket stations give worldwide radio coverage? Wireless World, pp. 305–308 (October 1945)

Hein, A.M., Pak, M., Pütz, D., Bühler, C., Reiss, P.: World ships – architecture and feasibility revisited. J. Br. Interplanet. Soc. **65**, 119–133 (2012)

Jones, R.A.L.: Against Transhumanism: The Delusion of Technological Transcendence. Available from www.softmachines.org (2016)

Kennedy, R.G., III: Robert A Heinlein's Influence on Spaceflight. In: Dick, S.J. (ed.) Remembering the Space Age: Proceedings of the 50th Anniversary Conference (The NASA History Series). CreateSpace, Charleston, SC (2013)

Kramer, S.: It still Costs a Staggering Amount of Money to Launch Stuff into Space. Tech. Insider. http://www.techinsider.io/spacex-rocket-cargo-price-by-weight-2016-6/#bottle-of-water-9100-to-43180-1 (July 20, 2016)

O'Neill, G.K.: The High Frontier: Human Colonies in Space. William Morrow, New York (1977)

Pournelle, J.: A Step Farther Out. Ace, New York (1979)

Shayler, D.: Disasters and Accidents in Manned Spaceflight. Springer/Praxis, Berlin (2000)

Solomon, L.D.: The Privatization of Space Exploration: Business, Technology, Law and Policy. Transaction, Piscataway, NJ (2011)
Tsiolkovsky, K.E.: Selected Works of Konstantin E. Tsiolkovsky. University Press of the Pacific, Honolulu (2004)
van Pelt, M.: Space Tethers and Space Elevators. Copernicus, New York (2009)
Zubrin, R.: The Case for Mars. Touchstone, New York (2011)

Fiction

Adams, D.: The Hitchhiker's Guide to the Galaxy. Pan, London (1979)
Anderson, P.: Tau Zero. Doubleday, New York (1970)
Asimov, I.: Foundation. Gnome, New York (1950)
Asimov, I.: The Martian Way. Galaxy (November 1952)
Blish, J.: Solar Plexus. In: Merrill, J. (ed.) Beyond Human Ken. Pennant, New York (1952)
Blish, J.: The Seedling Stars. Gnome, New York (1957)
Bradbury, R.: The Martian Chronicles. Doubleday, New York (1950)
Clarke, A.C.: A Fall of Moondust. Gollancz, London (1961)
Clarke, A.C.: A Meeting with Medusa. Playboy (December 1971)
Clarke, A.C.: Rendezvous with Rama. Gollancz, London (1973)
Clarke, A.C.: The Fountains of Paradise. Gollancz, London (1979)
Delany, S.R.: Aye, and Gomorrah. . . . In: Ellison, H. (ed.) Dangerous Visions. Doubleday, New York (1967)
Heinlein, R.A.: Requiem. Astounding (January 1940)
Heinlein, R.A.: Universe. Astounding (May 1941)
Heinlein, R.A.: Rocket Ship Galileo. Scribner's, New York (1947)
Heinlein, R.A.: The Man Who Sold the Moon. Signet, New York (1951)
Heinlein, R.A.: Orphans of the Sky. Gollancz, London (1963)
Martin, G.R.R.: A Song for Lya. Analog (June 1974)
McCaffrey, A.: The Ship Who Sang. Walker, New York (1969)
Niven, L.: The Coldest Place. Worlds of If (December 1964)
Niven, L.: Tales of Known Space. Ballantine, New York (1975)
Nourse, A.E.: Brightside crossing. Galaxy (January 1956)
Pohl, F., Kornbluth, C.M.: The Space Merchants. Ballantine, New York (1953)
Sheffield, C.: The Web Between the Worlds. Ace, New York (1979)
Simak, C.D.: City. Gnome, New York (1952)
Stapledon, O.: Last and First Men. Methuen, London (1930)
Varley, J.: In the Bowl. Fantasy and Science Fiction (December 1975)
Varley, J.: Bagatelle. Galaxy (October 1976)
Varley, J.: The Barbie Murders. Asimov's (January/February 1978)
Walters, H.: Mission to Mercury. Faber, London (1965)
Walters, H.: First Contact? Faber, London (1971)

Walters, H.: Passage to Pluto. Faber, London (1973)
Wells, H.G.: War of the Worlds. Heinemann, London (1898)

Visual Media

Blade Runner: Directed by Ridley Scott. [Film] USA, Warner Bros (1982)
The Three Rocketeers: Directed by Hans Odd. [TV documentary] UK, BBC (2012)

4

Aliens

Ellie Arroway: 'Dad, do you think there's people on other planets?'
Ted Arroway: 'I don't know, Sparks. If it is just us. . .seems like an awful
waste of space.'

Contact
Directed by Robert Zemeckis

Those outside SF often remark that the defining characteristic of the field is its fixation with aliens. I've heard outsiders make snide comments about 'bug eyed monsters' and 'little green men' and ask how a grown adult could possibly take such literature seriously (and they do that annoying hand gesture to signify quotation marks when they say the word 'literature'). Actually I don't believe SF *is* overly concerned with aliens—at least, modern SF isn't—but even if it were then it would have reason to be proud of its concern. The existence or otherwise of extraterrestrial intelligence is surely a matter of profound philosophical and scientific interest.

Debate and speculation about life on other worlds has a long history. In Ancient Greece, certain philosophers taught that an infinite number of life-bearing worlds exist. Plato and Aristotle thought otherwise, and their immense influence dampened discussion about such ideas, but the discoveries of Copernicus and Galileo reignited the debate. For religious philosophers the notion of extraterrestrial intelligence posed problems: if there are other beings who possess souls then what is their relationship to humans? If Jesus died to save us does he have to perish on every other planet with intelligent life in order to save those beings too? Questions surrounding extraterrestrial intelligence posed problems even for those non-religious philosophers whose self-image

© Springer International Publishing AG 2017
S. Webb, *All the Wonder that Would Be*, Science and Fiction,
DOI 10.1007/978-3-319-51759-9_4

was that of man as the pinnacle of creation: what would it mean if aliens exceeded humanity in matters of science and morality? Eventually, most scientists became of the opinion that the probability of extraterrestrial life was high—and, yes, it might even be superior to us. Gauss, for example, one of the greatest mathematicians of all time, thought it might be possible to communicate with advanced beings from Mars by planting forests in the shape of a Pythagorean triangle: the Martians, seeing these structures in their telescopes, would immediately deduce the existence of mathematically sophisticated creatures here on Earth.

As science came to accept the likelihood of extraterrestrial intelligence, so too did SF. Science fiction was never *fixated* on aliens but the notion did become one of the most important themes in the field. And coming full circle, I suspect the penetration into our culture of science fictional aliens—Spock, the Daleks, *The Thing*, and all those others—fed back into the expectations of scientists. For example, in 1967 Jocelyn Bell Burnell and Anthony Hewish observed regular radio pulses, separated by 1.33 s, coming from the same direction of the sky. They had discovered the first pulsar—a rapidly spinning neutron star—but at the time they couldn't be sure what it was. They nicknamed it LGM-1, short for 'Little Green Man', and even if they didn't fully believe they had discovered a signal from an extraterrestrial intelligence, the influence of science fiction had certainly primed them to consider the possibility. More recently, the star KIC 8462852 has been observed to fluctuate in brightness in an extremely peculiar fashion. There is presumably a natural explanation for this fluctuation, but one of the first hypotheses to appear was that the construction of an alien megastructure is causing these observed changes in brightness. I think this hypothesis is wrong, by the way, but I'm completely in favor of scientists generating such hypotheses. What's interesting is the ease with which the alien hypothesis springs to mind—is SF making its influence felt here?

Many scientists expect to hear from alien civilizations one day. A few scientists have even devoted their careers to searching for signs of extraterrestrial intelligence. Nevertheless, for all our searching, we have yet to see any convincing signs of the existence of aliens. Could it be that those who sneer at SF's preoccupation with aliens are right but for the wrong reasons? Could it be that we are, in fact, alone? I'll consider that possibility at the end of the chapter. Before discussing that, however, let's look at some science fictional aliens. From Wells through the Golden Age, in print and on screen, there have been so many stories about aliens: cute aliens, warring aliens, *alien* aliens. Here are some of my own favorites.

4.1   Extraterrestrials in Science Fiction

Even before I discovered the Middlesbrough children's lending library I was exposed to science fiction—for there was television. We had only three channels back then, not the hundreds that are available nowadays, but three seemed to suffice. And shows possessing an element of SF appeared with what now seems to be remarkable frequency. I was too young to watch classics such as *The Twilight Zone* or *The Outer Limits*, but still there was *Lost in Space*, *The Time Tunnel*, *Voyage to the Bottom of the Sea*, *Thunderbirds*, *Stingray*, *Land of the Giants*, *The Champions*, *Joe 90*, *Department S*, *Doomwatch*, *UFO*, *Timeslip*, *The Tomorrow People* ... looking back, my childhood must have been spent either playing football outside or watching television inside.[1] Of the plethora of science fiction shows on television, though, it was the ones containing aliens that made the most impact on me. So before considering some of the aliens in written SF I'd like to look at the screen aliens that made an impression on me—and no doubt on countless others of a similar age.

4.1.1 TV Aliens

Captain Scarlet and the Mysterons must have been one of the first television programs I watched.[2] It was made by Gerry and Sylvia Anderson, well-known British TV producers and directors who specialized in having marionette puppets appear in scale model sets using their trademarked 'Super-marionation' technique. The action in *Captain Scarlet* was set in 2068, after humans had attacked a city on Mars. The Martians—a collective intelligence, the Mysterons of the title—quite reasonably vowed revenge. It was the job of Spectrum, a much more vigorous version of the UN, to protect Earth from the Mysteron attacks. Fortunately for Spectrum, their main man—Captain Scarlet—was indestructible, having acquired the healing power of retrometabilism from his Mysteron enemies.

Mysterons! What a wonderful name! We didn't get to see them, but I was hooked from the moment I saw the rings of light that indicated their presence and heard those deep, slow words: 'This is the voice of the Mysterons. We know that you can hear us, Earthmen.'

Another sixties creation was *Dr Who*, perhaps the most successful SF television series of all time. I was 9 months old when the first episode was aired so I'm too young to remember the first Doctor; I have only vague recollections of the second Doctor (the show was much scarier than Captain Scarlet so it was a while before I was allowed to watch it). 'My' Doctor was the third—Jon

Fig. 4.1 The Daleks, with their catchphrase 'Exterminate', must be among the most famous of all alien creatures. The Dr Who storyline tells us that Daleks are genetically modified versions of the Kaled race. A Kaled scientist, Davros, created the modifications and integrated them with a robotic shell to create a race of cyborgs. Upon removing their ability to feel compassion, Davros had generated beings that saw themselves as supreme. This prompts a thought: if an expansionist race existed in the universe, wouldn't we have seen evidence of them (Credit: Gage Skidmore)

Pertwee—but just as I found the Mysterons to be more interesting than Spectrum agents so too I found the *Dr Who* aliens to be way more interesting than the humans. The Doctor himself is an extraterrestrial, of course, but there were many other fascinating types of alien: Silurians, Autons, Sontarans. . . and, best of all, the Daleks (see Fig. 4.1). *Dr Who* continues to this day, as successful as it ever was, and I watch it with my daughter. We both love the aliens on that show.

Of course, it wasn't just the British that were making SF for television. I recall watching *The Invaders* with my mother. I'm not sure how I came to see it, but I saw the opening credits of the first show—the one in which David Vincent, played by Roy Thinnes (see Fig. 4.2), sees an alien spacecraft on his way home from work. My mother must have realized that the show was harmless, so I was allowed to watch Vincent spend the rest of the series tracking down the evil aliens before they took over the world. Decades later, I can still remember the voice-over that opened each episode: 'The Invaders. Alien beings from a dying planet. Their destination: Earth. Their purpose: to make it their world.' In retrospect I can see how the alien invaders were quite useless: Vincent, an architect, managed to single-handedly defeat them in each

Fig. 4.2 In the TV series The Invaders, the architect David Vincent encounters a flying saucer. He struggles to convince people that alien beings are invading Earth. Fortunately for humanity, in each episode he somehow manages to thwart the ambitions of the extraterrestrials (Credit: Public domain)

episode. As a child, however, I marveled at the small glowing disc with which the aliens could kill people (and make it look like a natural death). And the alien characteristics—lack of a pulse; the little finger sticking out at a strange angle; the way they disintegrated into light when Vincent managed to dispatch one of them—were a delight for a small boy (who didn't realize that these were all functions of a cash-strapped special effects department rather than a well-considered analysis of alien physiologies).

And of course there was, for me, the best of them all: *Star Trek*. It was impossible to watch the show and not be entranced by the variety of alien life-forms on display: the half-human, half-Vulcan Spock (see Fig. 4.3); the dastardly Klingons; the cerebral Talosians... the *Star Trek* universe was brimming with extraterrestrials of every shape and hue. I grew up, as did many of us, with the notion that there is life out there in the cosmos.

Within the constraints of a finite budget, the producers of these television programs I'm sure did their best to make the aliens look alien. (They didn't always succeed. The Taran Wood Beast, an angry foe of the fourth Doctor, was an actor dressed in a gorilla suit that had seen better days. In *Star Trek*, the

Fig. 4.3 The half-human, half-Vulcan Spock, played by the late Leonard Nimoy, is one of the most recognizable characters ever to appear on TV (Credt: Public domain)

Horta was basically a rug thrown over a stuntman.) However, whether an alien appeared to be alien or just looked like an actor in cheap fancy dress, its motivation was never anything but human. The legendary SF editor John Campbell once said: 'Write me a creature who thinks as well as a man, or better than a man, but not like a man.' It's a famous dictum, but it wasn't uppermost in the minds of those who wrote for television. It was only when I began reading SF that I started to understand the difficulty of creating truly alien aliens.

4.1.2　Alien Aliens

Just because an author chooses to write a tale about alien creatures living on alien worlds it doesn't necessarily follow that the concept of 'alienness' has been taken seriously. Consider Asimov's "Nightfall" (1941), which is often described as the greatest science fiction story ever written.[3] The action takes place on Lagash, a planet in a system of six suns. With so many suns in the sky the entire surface of Lagash is continuously illuminated and the planet's inhabitants never experience the darkness of night. The existence of distant stars is unknown. The story describes what happens to the residents of Lagash when a rare alignment of celestial bodies causes an eclipse. It's a terrific tale— but it isn't a serious exploration of possible alien psychologies. It wasn't meant to be. Neither Asimov nor his editor, Campbell, considered "Nightfall" to be a vehicle for understanding aliens. They were interested in exploring *human* psychology and in particular the struggle between science and superstition. The quotation that inspired Asimov's story, a line from Ralph Waldo Emerson's *Nature*, makes this clear: 'If the stars should appear one night in a 1000 years, how would men believe and adore, and preserve for many generations the remembrance of the city of God!' I suspect most SF writers take the same approach: 'alienness' as a way of exploring 'humanness'.

Some writers are painstaking in their approach to world-building, but even they tend to develop aliens that are just humans in disguise. Consider Hal Clement's novel *Mission of Gravity* (1954), for example. It's a classic of 'hard' SF. Clement, a science teacher, calculated the properties of a planet that spins so quickly on its axis that it assumes a highly oblate shape. The surface gravity on the planet Mesklin is about $3g$ at the equator but about $700g$ at the poles. *Mission of Gravity* tells what happens when Barlennan, a native of Mesklin and the captain of a trading ship, takes his craft from its high-gravity home near the poles to the low-gravity regions near the equator. Clement took care to get the science correct and the novel usually appears complete with an essay, "Whirligig World", which describes how he went about devising Mesklin. It's a fun story, and Clement's world-building is first-class (given he was using science as understood in the early 1950s), but Barlennan and the other alien characters are decidedly not alien. They might be 50-cm long centipede-like creatures but their motivations are decidedly human. Would an intelligent organism that evolved on a planet so radically different to Earth respond to a situation in the same way as you or I would? I doubt it.

It's clearly difficult to depict truly alien aliens—creatures that inhabit the same universe as us, and are therefore subject to the same laws of physics, but that nevertheless possess different senses and drives and evolutionary histories. One approach to this awkward task is to avoid having the aliens on show and instead let the reader's imagination supply the details. This is what Arthur Clarke did in his novel *Rendezvous With Rama* (1973). Through the eyes of human astronauts we get to explore a 50-km long alien starship as it cruises through the solar system. We see various artifacts made by the Ramans, but we never learn for certain what might be the purpose of those artifacts or, indeed, the starship itself. The craft is to a certain extent unknowable and therefore more believable. *Forbidden Planet* (1956) takes a similar approach: the Krell, an unbelievably advanced race on the planet Altair IV, perished 200,000 years before the action of the film takes place. The humans who explore Altair IV come across objects that the Krell built, from which they can deduce something about the Krell, but the Krell themselves remain elusive and ultimately unknowable. The approach taken by the creators of *Rendezvous with Rama* and *Forbidden Planet* makes sense: if aliens are shown front and center then how could they be made to appear truly unfamiliar?

The Polish author Stanislaw Lem made one of the best attempts at creating a truly alien life form in his novel *Solaris* (1970). Solaris is a distant planet almost entirely covered by a vast ocean. Human scientists in an orbiting space station study the ocean's surface waves. It turns out these waves are similar to muscle contractions: Solaris is a planet-wide sentience. And just as humans are studying this oceanic planet so too is the planet studying its human visitors. The ocean's method of study is to probe the scientists' minds and to confront them with their deepest, most repressed, most painful memories. Unfortunately, the gulf between human and Solarian consciousness is too great to be bridged: attempts at meaningful communication end in frustration. I doubt that we'll ever encounter extraterrestrial intelligence, but if we did I believe Lem's pessimism regarding inter-species communication would turn out to be justified.

4.1.3 Engaging Aliens

Sometimes a writer plays the alien card for laughs. The first alien race we meet in *The Hitchhiker's Guide to the Galaxy* (1979), the introductory book in Douglas Adams' trilogy of five, is the Vogon. The Vogons are large, greenish, slug-like creatures. This makes them sound suitably strange, as aliens should be, but it quickly turns out that all of us have met someone with Vogon characteristics. The Vogons 'are one of the most unpleasant races in the Galaxy. Not actually evil, but bad-tempered, bureaucratic, officious and callous. They wouldn't even lift a finger to save their own grandmothers from the Ravenous Bugblatter Beast of Traal without orders—signed in triplicate, sent in, sent back, queried, lost, found, subjected to public inquiry, lost again, and finally buried in soft peat for 3 months and recycled as firelighters.' From my own research I can confirm that Vogons occupy a disproportionately large number of middle-management positions. Adams limns the Vogon race so accurately, however, that I can't help but like them—and even their ludicrous attempts at poetry.

The novels and short stories set in Larry Niven's "Known Space" feature twenty or so different alien species. The most memorable are the Pierson's Puppeteers—twin-necked, three-legged herbivores whose ancestors were herd animals. Although Puppeteers are technologically advanced, and effectively control much of what happens in Known Space, they are craven, faint-hearted creatures. Not only are they weak-kneed, they are masters of manipulation— they get others to do anything that involves even the slightest danger. Puppeteers sound like horrible beings, but we don't get to meet typical Puppeteers in the stories: a sane Puppeteer is far too cowardly to interact with humans. Instead, we only meet insane Puppeteers. In Niven's masterful *Ringworld* (1970), for example, a Puppeteer called Nessus explores a mysterious megastructure along with two humans and a Kzin (a bloodthirsty, cat-like alien). Only a crazy Puppeteer would engage in such an adventure. Nessus makes the Cowardly Lion of Oz seem as resolute as Winston Churchill but, despite its character failings, I find Nessus to be by far the most interesting and sympathetic creature of the those in the small exploration party.

The novels and short stories set in James White's Sector 12 General Hospital—usually just called the "Sector General" stories—contain an even greater range of aliens than Niven's "Known Space". Species seeking treatment at the hospital include the cuddly Orligians, the elephantine Trathlans and the thermophilic Threcaldans. One of the factors that makes the "Sector General" stories so readable, aside from the multiplicity of life forms on display there, is the motivation behind the institution: it's a hospital that was founded in order to promote peace. It exists in order to treat and tend to living beings. The hospital's health professionals come from a diverse range of life forms and they treat an even greater range of patients. The tension in White's stories stems from the medical setting rather than from physical conflict or violence; the suspense comes from wondering how a human doctor can possibly provide treatment to a methane-breather, an ocean-dweller, or a creature that feeds on radiation. White's pacifism and humanism, which matured through his experience of living in Northern Ireland during The Troubles, shine through the "Sector General" stories and make the reader sympathize with, or at least appreciate, the disparate alien creatures he describes.

An alien life form is central to one of George R.R. Martin's early stories, which also happens to be one of my favorite tales: "A Song for Lya" (1974). Many readers of Martin's epic fantasy work, *A Song of Ice and Fire*, seem not to realize that before Martin embarked upon his multi-novel, multi-layered fantasy of dynastic war in Westeros he was a writer of award-winning SF. "A Song for Lya" won the 1975 Hugo award for best novella and its lush prose tells the tale of two telepaths, Robb and Lyanna, who are asked to visit a distant planet in order to discover why some human colonists are choosing to convert to the religion of the indigenous alien population, the humanoid Shkeen. The religion involves 'joining' in a union with a parasitic organism known as the Greeshka. This union allows the Shkeen—and increasingly humans—to feel a deep love for, and share an intimate connection with, all other creatures that have 'joined'. A follower of the religion is eventually lead into a cave, where the host body is consumed: the Greeshka feeds on it. To an outsider, it looks to be a particularly gruesome form of suicide. But the two lovers Robb and Lyanna, whose telepathic connection mean they share a bond most humans can't imagine, discover something that profoundly changes their relationship. I won't spoil the story except to say that the reader can't help be fascinated by these alien organisms—not because of sympathy with them (the Shkeen are docile and apathetic; the Greeshka resembles a grotesque mold more than anything else) but because of the effect they have on the two human lovers.

A few years later, Martin followed up his Hugo award with both a Hugo and a Nebula for his story "Sandkings" (1979). It appeared in the August edition of the short-lived *Omni*, and I can still remember the thrill of reading it that summer. The story describes how Simon Kress, a wealthy but sadistic collector of alien animals, buys four colonies of a creature previously unknown to him: sandkings. Neither Kress nor the sandkings are in the least bit sympathetic, but the story leads me nicely to what is perhaps the commonest evocation of aliens within science fiction: beings who do us harm.

4.1.4 Alien Invasion

H.G. Wells's *The War of the Worlds* (1898) is a thrilling tale, whose plot was influenced by various innovative scientific ideas of the day—such as Lowell's supposed observations of Martian canals, as discussed in the previous chapter, and the more valid notions of Darwinian evolution. (Wells trained in science, and one of his teachers was 'Darwin's Bulldog'—the biologist T.H. Huxley who famously crushed 'Soapy Sam' Wilberforce, the Bishop of Winchester, in a public debate on evolution.) In turn, the novel—and others by Wells—had an influence on scientists: Freeman Dyson, Robert Goddard and Marvin Minsky all read Wells in their childhood. The novel's influence on science fiction is hard to overstate. There were stories of alien invasion before *The War of the Worlds* but it was this novel that established the invasion theme in

SF. (Wells, of course, established many of the major SF themes.) In case you haven't read the novel, or heard one of its many radio adaptations, or seen one of the half-dozen movies based on it, or listened to the music album—well, *The War of the Worlds* is the account of an unnamed narrator's experiences during the invasion of Earth by malevolent Martians (see Fig. 4.4 for an illustration). It was such a powerful yet simple story that variants on the theme of alien invasion quickly appeared, and they are being published and broadcast to this day. Below, I mention just a few of my favorites in the genre.

"Who Goes There?" (1938) was a novella by John W. Campbell. Campbell wrote a number of important stories under the pseudonym Don A. Stuart before he became the full-time editor of *Astounding,* and "Who Goes There?" was the last (and to my mind the best) of those stories. A team of scientists wintering over in Antarctica discover a spaceship buried in deep in ice. They recover an alien and, as the alien thaws, it revives. Unfortunately for the scientists, the alien can assume the form of any living thing it consumes. And it does consume. Soon, the scientists don't know who is human and who is alien. . . The story has been adapted for film four times, with John Carpenter's visceral *The Thing* (1982) being the most influential. (A more family-friendly version of alien contact, Spielberg's *E.T.: The Extra-Terrestrial* (1982), was released at around about the same time. The astonishing success of *E.T.* might explain the relative box-office failure of *The Thing*—but Carpenter's film is now regarded as a classic of the genre.)

Robert Heinlein' *The Puppet Masters* (1951) is great fun. The book opens in 2007, with civilization under threat from slug-like aliens from Titan which attack by attaching themselves to a person's back and controlling the host's nervous system. The parasite controls its victim as if it were manipulating a puppet. Reading the novel more than six decades after publication gives an insight into the Cold War paranoia that gripped America in the 1950s. Another novel along the same lines is Jack Finney's *The Body Snatchers* (1954), which tells how a small Californian town is invaded by alien seeds. This isn't an attack with weaponry, as envisaged by Wells. It isn't even a direct physical assault, as envisaged by Heinlein. Rather, the seeds just drifted to Earth from space and when they got here they grew into pods. A pod replaces a sleeping human with an exact physical, although not emotional, duplicate. The pod person looks like the original and has all the memories of the original, but lacks any human emotion. Finney's novel lacks the rigor, verve, and brilliance that Heinlein brought to all his early novels. On the other hand, it spawned two excellent films (in addition to two forgettable films). *Invasion of the Body Snatchers,* released 2 years after the book was published, has received much critical acclaim; the 1978 film of the same name demonstrated that it's

Fig. 4.4 Frank R. Paul illustrated "The War of the Worlds" for the August 1927 issue of Gernsback's Amazing Stories. In many ways, Paul's magazine covers set the style for SF illustration—and perhaps influenced how many people think about SF topics. Note the disparity in military technology that Paul so vividly captures: human fighters are on horseback, the Martians possess some sort of death ray. How could humans expect to defeat the superior technology of aliens? Asimov thought we couldn't, and ultimately that led him to a 'humans-only' galaxy (Credit: Public domain)

possible to remake a classic and not ruin it. Heinlein's work has never been adapted for film with anything approaching the same artistic success.

The fictional conflicts I've discussed so far in this section all take place on Earth. Given the vast distances between stars it's difficult to envision how an interstellar conflict could take place. However a later Heinlein novel, *Starship Troopers* (1959), describes just such a campaign: a federation led by Earth wages war on alien beings called Bugs. Heinlein possessed such a high level of technical mastery when it came to writing science fiction that *Starship Troopers* remains eminently readable even though to many readers, myself included, the apparent glorification of war causes unease. The novel won the 1960 Hugo, but even now it remains controversial.

If novels such as *Starship Troopers* and *The Puppet Masters* arose from their author's experience of the military and his take on the politics of the Cold War, Joe Haldeman's anti-war novel *The Forever War* (1974) came from its author's experience of serving in Vietnam. (Haldeman was wounded in combat.) *The Forever War* tells the story of a conscript into a UN Expeditionary Force. The Force aims to revenge an attack on human ships by an alien species called the Taurans. The phenomenon of relativistic time dilation means that, although the protagonist serves in the military for only a few years, more than a 1000 years passes on Earth. By pointing out the manifold difficulties faced by the conscript, *The Forever War* is a clear rejoinder to *Starship Troopers*. It won the Hugo, Nebula, and Locus awards for best novel—but Heinlein's praise of the novel probably meant as much to Haldeman as the awards themselves.

Yet another take on interstellar war came with Barry B. Longyear's novella "Enemy Mine" (1979). It's the story of how a human pilot is stranded on a hostile and unforgiving planet with a Drac—an intelligent reptile-like creature whose civilization is intent on fighting humankind. The loathing of the two beings, human and Drac, is mutual; they try to kill one another. Soon, however, they realize that survival on this planet will require cooperation. And as that cooperation develops we learn that the entire conflict between human and Drac is based on a misunderstanding. Put that way it sounds trite, but it was a powerful story that won that year's Hugo and Nebula awards.[4]

The same year as "Enemy Mine" was published, cinemagoers were treated to one of the most aggressive and well realized aliens ever manifested on film. Ridley Scott's *Alien*, which features a creature designed by the Swiss artist H.R. Giger, follows the hapless crew of the spaceship *Nostromo* as an alien stalks and kills them, one by one. The film won awards ranging from an Oscar to a Hugo.

Perhaps the most chilling aliens of all, though, are the Builders. They are from a series of tales by Fred Saberhagen that spanned from 1963 through to 2005. We learn hardly anything about the Builders, except that they built an 'ultimate weapon'—Berserkers—in order to eliminate their enemies, the Red Race. The Berserkers were extremely efficient and, following their mission, they destroyed the Red Race. Unfortunately, the Builders overlooked one important point: they forgot to ensure that they were themselves immune from Berserker attack. The Builders' 'ultimate weapon' proceeds to destroy them just as efficiently as it destroyed the Red Race. Then, with the Red Race and the Builders out of the way, the Berserkers embark on a journey across the Galaxy in order to fulfil their destiny: to seek out life and destroy it whenever they find it.

I've never found the Berserker series to be great literature—but the idea of the Berserkers addresses a problem. You see, all those short stories and novels and television programs and films that I've mentioned take it as granted that alien life forms are out there in abundance. But we see no trace of them. Could it be that Berserkers exist—and that is why the universe appears to be so lifeless?

4.1.5 So Where Are They?

Stephen Hawking is widely known as one of the most brilliant physicists of our time. Less well known is that he has a connection with science fiction: he co-authored a best-selling trilogy of SF novels for children and hosted a TV show called *Masters of Science Fiction*. In 2010, Hawking appeared in a documentary and issued what some would consider to be a science-fictional warning: he counselled against trying to make contact with aliens. Hawking argued that alien civilizations, which would likely be far older and therefore more advanced than our own, would raid Earth for resources. In making this argument he pointed out a parallel from Earth history: 'If aliens visit us, the outcome would be much as when Columbus landed in America, which didn't turn out well for the Native Americans.'[5] Although Hawking justified his position by referring to an episode of European history I believe his attitude must have been influenced, perhaps even shaped, by science fiction. In this chapter I've discussed the appearance of aliens in short stories and novels, on the large and small screen, on radio and on record. I haven't even mentioned the representation of aliens in comics and cartoons and cheap plastic toys. . . images of extraterrestrials permeate our culture. How could Hawking, or anybody else, fail to be influenced by that?

As long ago as the fourth century BC, a philosopher such as Metrodorus of Chios could remark that 'a single ear of wheat in a large field is as strange as a single world in infinite space'. We now know our Galaxy probably contains trillions of planets, and the sheer number of potential homes for life leads many contemporary scientists, SF authors, and laypeople to conclude that life must be common in the universe. It's the default position. But even if life *is* common, does it follow that creatures will evolve with the intelligence to build craft capable of traversing interstellar distances? That they'll possess the motivation to make those journeys? And that after travelling so far they'll want to pillage our tiny planet? Isn't that a concept borne more out of the Wellsian tradition of SF than a serious extrapolation of the science? Besides, if the universe is teeming with all these aliens (whether cute, aggressive, intelligent, or any of the other types dreamed up by SF authors)... where are they? Why don't we see them or hear from them? Where is everybody?

4.2 The Fermi Paradox

Where is everybody? This question is usually referred to as the Fermi paradox (although some would argue that it isn't Fermi's nor is it a paradox). I've written a book that discusses 75 different answers to the question so I don't propose to repeat all that here; read the book if you want the gory details.[6] Here, I simply want to explain why Fermi's question deserves our attention—and why it's possible that both science and science fiction have been wrong about the prospects for intelligent alien life.

Let's start with a brief discussion of why the question—where is everybody?—is important. Enrico Fermi asked the question in the early 1950s, during a lunchtime discussion with physicist friends about the possible existence of flying saucers. (This was at the height of the flying saucer craze, a vogue that perhaps led to the making of some of those wonderful 1950s SF films—or at least to the creation of a willing audience for those films.) Fermi seems to have made a quick mental estimate of the likely number of extraterrestrial civilizations in the Galaxy, and concluded there are likely to be many of them. If you do the same rough-and-ready calculation you might well reach the same conclusion: if there are, say, a trillion Earth-like planets in the Galaxy, and life develops on one in ten of them, and intelligent life develops on one in ten of those planets, and of *those* planets a long-lasting civilization develops on one in ten... well, you can see how it's possible to arrive at a large number. The factors to be taken into account are more varied than the few I've given above, but the logic is clear: if there's nothing special about Earth, or about the evolutionary history that led to humans, then we might expect the Galaxy to be teeming with alien civilizations. Since Fermi's time scientists have made several relevant discoveries—such as the existence of exoplanets and the realization that organisms here on Earth can thrive in the unlikeliest environments—that tend to lend weight to this idea of a Galaxy teeming with intelligent life. Fine. And when you consider the age of the Galaxy it seems equally reasonable to expect many of those alien civilizations to be much older than humanity and presumably, therefore, much more technologically advanced. Again, that's fine. But if there's nothing special about humans then we might expect our rate of technological advance to be matched on other worlds. So when we conclude that older civilizations exist then we must accept that they'd be much more technologically advanced than us—much, *much* more technologically advanced. There's the rub.

Consider Fermi himself. He was born in 1901, before the first successful powered and sustained heavier-than-air flight had taken place. The Wright brothers' achievement didn't occur until after Fermi's second birthday. Fermi died in 1954, having seen humans achieve the edge of space (with the German V2 rockets). Had he lived into a reasonable old age he would have seen the launch of the Voyager rockets in 1977; these craft would go on to reach the edge of the solar system. Had he lived to be 110—a stretch, I know, but centenarians are hardly unknown—he would have heard the news that Voyager 1 had touched interstellar space. The duration of a single human lifespan saw humanity go from being an essentially Earth-bound species to one that reached interstellar space. If that rate of technological change is maintained, what wonders will we be capable of in another 100 years? In 1000 years? In one *million* years? Well, alien civilizations are likely to be much more than one million years older than humankind. To us primitive creatures, their level of technology should be indistinguishable from magic.

So... where is everybody? If there are so many alien civilizations out there, why aren't they already here? It shouldn't be too difficult for them to reach us, should it? Or if they don't want to travel, for whatever reason, why don't we at least hear from them? After all, they could broadcast their presence across the Galaxy. Perhaps they simply don't care about the existence of other species, don't want to meet and greet them—but then why don't we at least see evidence of what must be their astounding engineering capabilities? Why don't we see evidence of their Dyson spheres, their antimatter rockets, their wormhole manipulations? Instead, the universe appears to be quiet. It seems to contain nothing artificial except those structures made by humankind. We've seen and heard nothing that can't in principle be explained by the cold laws of physics. Where *is* everybody? That's the essence of the paradox.

Some would argue that Fermi's question is easily answered: they—by which is meant intelligent extraterrestrial beings—are here! When asked to provide evidence, people who argue this way tend to point to flying saucers. (A few people, on the fringe of the debate, even argue that there are aliens walking among us.) In my experience, both the scientific and science fictional communities are equally scathing about this answer. It isn't that flying saucers are inherently impossible: few of us would be bold enough to put a limit on the technical and engineering savvy of beings that are a billion years in advance of us. It's that the evidence put forward to support the case for flying saucers is always lacking in quality. We either get clear and unambiguous evidence for a flying object, observed by many different people, which is then best explained in terms of a known phenomenon such as Venus or a weather balloon or a lenticular cloud. Or else we get claims from individuals that can never be corroborated. In some cases we get downright lies and hoaxes.

In both science and science fiction the commonest response to Fermi's question is that they—extraterrestrial civilizations—are out there. It's just that we haven't searched long enough nor hard enough to find evidence of them. After all, space is really, really big. We can't expect just to look up and observe signals from alien beings: there are so many different places to look, so many different frequencies to comb through, so many different types of signal to consider. We simply have to be patient.

It's easy to see sense of this position, and perhaps it's correct. On the other hand, scientists have been engaged in a serious search for evidence of extraterrestrial intelligence for more than half a century. Frank Drake was the pioneer, with his Project Ozma[7]—a radio experiment named after Princess Ozma, the ruler of Oz, who appeared in the SF novels that Drake devoured as a child. Drake's earliest attempts were relatively crude, but each year sees improvements in the search for extraterrestrial intelligence. By 1967, the notion of using radio telescopes to listen for signals was sufficiently mainstream that Williams and Abrashkin were able have my childhood hero, Danny Dunn, get involved in SETI astronomy (that was in *Danny Dunn and the Voice From Space* (1967)). Since then the combined observational and computational power that has been thrown at the search has increased a billionfold, perhaps more (Fig. 4.5 shows the sort of technology that is now employed in SETI). Usually in science when you make this sort of improvement to your experiment—if you increase by many orders of magnitude the sensitivity of your instruments, or the energy at which they can operate, or you simply gather more data—something interesting turns up. But the universe sounds as silent now as it did when Frank Drake listened to Tau Ceti and Epsilon Eridani with the Green Bank radio telescope back in 1960.

Fig. 4.5 The Allen Telescope Array, shown here, uses radio telescopes to search for signals from extraterrestrial civilizations. SETI scientists have access to a number of other radio telescopes to carry out the search. In recent years, observations have also been carried out using optical telescopes: the idea is to look for laser signals rather than radio signals. The search has been improved by the vast increase in available computing power. So far, SETI scientists have been unable to identify any signal as being from an extraterrestrial civilization. Could that be because there are no civilizations up there (Credit: Colby Gutierrez-Kraybill)

So perhaps the answer to Fermi's question—where is everybody?—is that there is nobody out there. It's just us and the universe.

4.2.1 A Humans-Only Galaxy?

When I began reading SF the field recognized a 'Big Three': Asimov, Heinlein, and Clarke—the three most famous, most influential science fiction authors still producing work. Although I liked almost everything Clarke wrote, there was an underlying mysticism in some of his stories that I found unappealing. And although Heinlein's total mastery of the storytelling craft meant I couldn't help but find his tales compelling, I often detected a hint of bullying in them that I found off-putting. It was Asimov that meant the most to me, by far. He wasn't the world's greatest literary stylist, as he himself was the first to admit, but he possessed a clear rationality that shone through every word he wrote. His voice was lucid, intelligible, straightforward, and always reasonable. I loved it.

Asimov's most important work was probably the *Foundation Trilogy* and it's quite typical of his fictional output: it consists, in the main, of rational people discussing their political, scientific, and technological options as they apply to the situation they find themselves in. It's typical of Asimovian fiction in another way: it's *people* who are involved in the discussion, not aliens. The Galactic Empire of the *Foundation Trilogy* is a humans-only affair. It's the same with Asimov's illustrious robot series (of which more in a later chapter): the stories are about the interaction of people with robots, not with aliens. His most famous works of short fiction—stories such as "Nightfall", "The Martian Way", "The Last Question"—are about people, not aliens. (Strictly, the action in "Nightfall" takes place on Lagash and the inhabitants of that planet are not human. But as I argued earlier, in "Nightfall" there was no attempt to consider the alien. It was the situation in which the inhabitants of Lagash found themselves, not the creatures themselves, that was other-worldly. The story addressed the question: what would humans be like if they had evolved in this quite different environment.) A key element, then, in the Asimovian approach to science fiction is the 'humans-only' galaxy.

It wasn't that Asimov lacked the imagination to describe aliens. The second part of his novel *The Gods Themselves* (1972) contains some of the most convincing alien creatures in the genre. The novel describes a parallel universe in which the strong nuclear force is stronger than in our own universe. One consequence of this is that stars in the parallel universe are smaller and exhaust their supply of nuclear fuel more quickly than we are familiar with. The action in the second part of the book is set on a planet orbiting a dying star. The intelligent inhabitants of this world have three sexes: the Emotionals, the Rationals, and the Parentals. The story is of one particular mating group, or triad, formed by characters called Dua (an Emotional), Odeen (a Rational), and Tritt (a Parental). The treatment of these photosynthesizing creatures, who inhabit a universe with slightly different laws of physics to our own, is an imaginative tour de force: of *course* Asimov could write about aliens! Why then did Asimov write so much science fiction in which aliens were absent?

The reason Asimov chose in general to ignore aliens wasn't to do with lack of imagination or a disbelief in their existence (it's clear that Asimov thought it likely that extraterrestrial civilizations exist). Rather, it was rooted in a quite practical concern: the desire to sell stories to *Astounding*, the leading science fiction magazine of the day.[8] The magazine's editor, John Campbell, was a strong willed individual who possessed certain convictions—one of which was the inherent superiority of the human race over any other species. If aliens made an appearance in a story he wanted the story to show how human courage, ingenuity, or wit could help us overcome superior technology. Asimov thought this approach was nonsense. He applied the same rationality that Hawking more recently brought to bear on the discussion and concluded that if humans encountered aliens with superior technology, and if a conflict ensued, then we'd lose. Superior technology would win every time. But a story relating how humans lost to aliens would never sell to Campbell—and Asimov wanted to sell to Campbell. He therefore chose to exclude aliens from his stories. The humans-only galaxy was born.

Asimov didn't believe a humans-only galaxy was a realistic scenario. He followed the numbers and argued, as so many do, that the plethora of potential homes for life means that intelligent life is inevitable. He even wrote a book about the possibility of extraterrestrial civilizations and estimated there could be as many as half a million of them out there in the Galaxy right now. Me? I think the silence is overwhelmingly deafening. I believe we are alone. But if humankind manages to survive the perils that face us, and if we learn how to move off our home world, then something similar to Asimov's humans-only Galaxy might one day come to pass. If humankind *doesn't* survive. . . well, it that case what might be the only fully conscious intelligence in the Galaxy would be snuffed out. It's quite a responsibility our species has to bear.

4.3   Conclusion

A new science has developed in recent years: astrobiology. It's an interdisciplinary study, drawing inspiration from astronomy, biology, geology, and other fields as well. Astronomers have demonstrated that planetary homes for life are likely to be numerous; biologists have learned how life can prosper in the unlikeliest of places and have speculated how extraterrestrial lifeforms might be possible even in the absence of water or carbon[9]; geologists have built up a good understanding of how our planet developed and of how changes in the planetary environment have supported the advance of life on Earth. It seems that every month there is a discovery in astrobiology that justifies the widespread belief in the existence of extraterrestrial intelligence. And yet. . . there remains not the slightest indication that any form of life has developed outside of our planet.

A major goal of astrobiology is to determine whether life exists—or at least whether it existed in the past—on other bodies in the Solar System. Mars is an obvious candidate for life, but moons such as Enceladus and Titan are also possibilities. Such life would necessarily be simple in form, nothing like the complex life that teems on Earth, but the mere fact that life originated independently on other bodies would be supremely important. It would boost the case for believing the universe itself teems with life: after all, if life originated independently in two different places in our Solar System then we might reasonably conclude that life is a natural concomitant of planetary formation. In that case we could infer the existence of billions of life forms in the Galaxy, and the belief that *intelligent* life forms exist out there would be entirely reasonable. Scientists at the various national and international space agencies are planning these science-fictional missions to Mars and beyond, hoping to find evidence for life.

The traditional search for extraterrestrial intelligence continues, too. The main focus of SETI scientists remains the radio part of the electromagnetic spectrum, but research is taking place in the optical region and recently in the mid-infrared part of the spectrum (the idea with the infrared search being that we look for the waste heat that accompanies technological activity). Much of this research takes place by piggybacking on existing astronomical activity. We are living through a golden age of astronomy and astrophysics, with great observatories being built to study the universe through many windows—not just through the electromagnetic spectrum but also through neutrinos, gravitational waves, and cosmic rays. These observatories will produce industrial amounts of data and, who knows, somewhere in that flood of astronomical information might be hints of activity from an extraterrestrial civilization.

Most scientists and SF authors are in agreement: when it comes to aliens, the picture drawn by those SF pioneers might not be accurate in its details but it is correct in spirit. To this day, the existence of aliens remains a given in science fiction. Eventually, according to the prevailing view, we'll find life out there. But how long can we sustain that view? If we continue to search, and the search continues to yield negative results, when will we decide that science and science fiction got it wrong?

When will we conclude we are alone?

Notes

1. Of the television shows I grew up with, four were created by the prolific American producer Irwin Allen: *Voyage to the Bottom of the Sea* (1964–1968), *Lost in Space* (1965–1968), *The Time Tunnel* (1966–19677), and *Land of the Giants* (1968–1970). They were all of low quality, at least when viewed through adult eyes, but they were extremely popular. There was no shortage of homegrown science fiction for a British child growing up in the sixties, and the husband-and-wife team of Gerry and Sylvia Anderson were even more prolific than Allen: they produced *Stingray* (1964–1965), *Thunderbirds* (1964–1966), and *Joe 90* (1968–1969) using their trademarked 'Supermarionation' process; *UFO* (1970–1971), about an alien invasion of

Earth, used real actors. Other British science fiction for children included *Timeslip* (1970–1971), *The Tomorrow People* (1973–1979), and *Doomwatch* (1970–1972); the latter was actually for an adult audience, but I pestered my parents to watch it whenever I could. A number of 'spy-fi' television series were also huge fun, including *The Champions* (1968–1969) and *Department S* (1969–1970). Looking back, none of those series were mould-breaking; but they certainly were better than the reality shows that suffocate today's television schedules.

2. *Captain Scarlet and the Mysterons* (1967–1968) came from the same stable as *Thunderbirds* and *Joe 90*, mentioned in the note above. Gerry Anderson, the series' creator, wanted to make a show about Martians but had read that scientists no longer thought that the 'canals' on Mars were artificial. He therefore came up with the idea of making the Martians invisible, so that if scientists demonstrated that life did not exist on Mars he could answer that it was there—but we can't see it. Thus the Mysterons were born!

3. In 1968, the Science Fiction Writers of America voted for the best SF story below novel length that appeared prior to the establishment of their annual Nebula awards in 1965. Asimov's "Nightfall", which was published in 1941, came top of the poll. Almost three quarters of a century after the story was written "Nightfall" continues to be regarded by readers as one of the genre's classic stories. It has been anthologized many times, and is easy to find. The quote that triggered the story appears in Emerson (1849).

4. Due to quirks in the regulations attached to the different awards, Longyear's novella "Enemy Mine" won the 1979 Nebula award and the 1980 Hugo award. A film of the same name, which was based on Longyear's story, was released in 1985. It won no awards, and tanked at the box office.

5. Hawking made this comment about aliens in the first of a television documentary series called *Into the Universe with Stephen Hawking* (2010). In an earlier interview with SETI scientists he expressed a similar view, and warned that we should 'keep our heads down'.

6. For a discussion of the history of the Fermi paradox, as well as of 75 different approaches exploring the question 'where is everybody?', see Webb (2015).

7. In 1960 Frank Drake, then an astronomer at Cornell University, established Project Ozma to search for signs of intelligent life in space. He used a 26-m diameter radio telescope to listen to Tau Ceti and Epsilon Eridani at a frequency of 1.420 GHz (which for various reasons is considered to be a sensible place to look for extraterrestrial signals). For details, see Drake and Sobel (1992).

8. For details of the development of the humans-only Galaxy, as well as the genesis of his most famous story "Nightfall", see the first volume of Asimov's autobiography (Asimov 1979).
9. It is often suggested that life must be based around carbon. For an example of some more exotic suggestions, such as silicon biochemistry in liquid nitrogen, see Bains (2004).

Bibliography

Non-fiction

Asimov, I.: In Memory Yet Green. Doubleday, New York (1979)
Bains, W.: Many chemistries could be used to build living systems. Astrobiology 4(2), 137–167 (2004)
Drake, F., Sobel, D.: Is Anyone Out There? Delacorte, New York (1992)
Emerson, R.W.: Nature: Addresses and Lectures. James Munroe, Boston (1849)
Webb, S.: If the Universe is Teeming with Aliens . . . Where is Everybody? 75 Solutions to the Fermi Paradox and the Problem of Extraterrestrial Life. Springer, Berlin (2015)

Fiction

Abrashkin, R., Williams, J.: Danny Dunn and the Voice From Space. McGraw-Hill, New York (1967)
Adams, D.: The Hitchhiker's Guide to the Galaxy. Pan, London (1979)
Asimov, I.: Nightfall. Astounding (September 1941)
Asimov, I.: The Gods Themselves. Doubleday, New York (1972)
Campbell JW Jr: Who Goes There? Astounding (August 1938)
Clarke, A.C.: Rendezvous with Rama. Gollancz, London (1973)
Clement, H.: Mission of Gravity. Doubleday, New York (1954)
Finney, J.: The Body Snatchers. Dell, New York (1954)
Haldeman, J.: The Forever War. St. Martin's Press, New York (1974)
Heinlein, R.A.: The Puppet Masters. Doubleday, New York (1951)
Heinlein, R.A.: Starship Troopers. Putnam's, New York (1959)
Lem, S.: Solaris. [Original Polish edition, 1961]. Walker, New York (1970)
Longyear, B.B.: Enemy Mine. Asimov's (September 1979)
Martin, G.R.R.: A Song for Lya. Analog (June 1974)
Martin, G.R.R.: Sandkings. Omni (August 1979)
Niven, L.: Ringworld. Ballantine, New York (1970)
Saberhagen, F.: Berserker. Ballantine, New York (1967)
Wells, H.G.: The War of the Worlds. Heinemann, London (1898)

Visual Media

Alien: Directed by Ridley Scott. [Film] USA, 20th Century Fox (1979)

Captain Scarlet and the Mysterons: Created by Gerry and Sylvia Anderson. [Television series] UK, Century 21 Television (1967–1968)

Department S: Created by Dennis Spooner. [Television series] UK, ITC Entertainment (1969–1970)

Doomwatch: Created by Kit Pedlar and Gerry Davis. [Television series] UK, BBC (1970–1972)

E.T.: The Extra-Terrestrial: Directed by Stephen Spielberg. [Film] USA, Universal Pictures (1982)

Forbidden Planet: Directed by Fred M. Wilcox. [Film] USA, MGM (1956)

Into the Universe with Stephen Hawking: [Television series] USA, Discovery Channel (2010)

Invasion of the Body Snatchers: Directed by Don Siegel. [Film] USA, Allied Artists (1956)

Invasion of the Body Snatchers: Directed by Philip Kaufman. [Film] USA, United Artists (1978)

Joe 90: Created by Gerry and Sylvia Anderson. [Television series] UK, Century 21 Television (1968–1969)

Land of the Giants: Created by Irwin Allen. [Television series] USA, 20th Century Fox Television (1968–1970)

Lost in Space: Created by Irwin Allen. [Television series] USA, 20th Century Fox Television (1965–1968)

Stingray: Created by Gerry and Sylvia Anderson. [Television series] UK, AP Films (1964–1965)

The Champions: Created by Dennis Spooner. [Television series] UK, ITC Entertainment (1968–1969)

The Thing: Directed by John Carpenter. [Film] USA, Universal Pictures (1982)

The Time Tunnel: Created by Irwin Allen. [Television series] USA, 20th Century Fox Television (1966–1967)

The Tomorrow People: Created by Roger Price. [Television series] UK, Thames Television (1973–1979)

Thunderbirds: Created by Gerry and Sylvia Anderson. [Television series] UK, AP Films (1964–1966)

Timeslip: Created by Ruth and James Boswell. [Television series] UK, ATV (1970–1971)

UFO: Created by Gerry and Sylvia Anderson and Reg Hill. [Television series] UK, Century 21 Television (1970–1971)

Voyage to the Bottom of the Sea: Created by Irwin Allen. [Television series] USA, 20th Century Fox Television (1964–1968)

5

Time Travel

The dead past is just another name for the living present.

The Dead Past
Isaac Asimov

Ah, time travel. Of all the themes in science fiction, this is my favorite. As with space voyages and aliens I was primed to think about time travel from a young age. On television, the Time Lord *Dr Who* (1963–present) flitted between eras as easily as he moved through space; in *The Time Tunnel* (1966–67) two scientists, lost in the fourth dimension, got involved in historical exploits; and the best-ever episode of *Star Trek*, Harlan Ellison's "The City on the Edge of Forever" (1967), had Kirk and Co. grapple with the paradoxes of travel into the past. As for films, I watched George Pal's *The Time Machine* (1960) as a child and was blown away by its depiction of the future. The novel on which the film was based still has the same effect on me. When I started to read books there was the Abrashkin and Williams classic *Danny Dunn, Time Traveler* (1963) and Philippa Pearce's *Tom's Midnight Garden* (1958) and Madeleine L'Engle's *A Wrinkle in Time* (1963)... these tales were guaranteed to stoke an interest in the properties of time. And when I graduated to adult science fiction I discovered that my favorite authors—Asimov, Heinlein et al.—had all written stories highlighting just how slippery the concept of time could be. Other authors used the conceit to investigate notions of free will, determinism, and causality—concepts that philosophers are still grappling with. What isn't there to love about time travel stories?

It was when I began to study physics at university that I learned how our understanding of time is far from complete. Whatever 'time' is, our everyday

© Springer International Publishing AG 2017
S. Webb, *All the Wonder that Would Be*, Science and Fiction,
DOI 10.1007/978-3-319-51759-9_5

intuitions about it are valid only in limited circumstances. In the decades since Einstein developed his theories of relativity, physicists have learned that the laws of physics seem to permit time travel. So if time travel is *not* possible (and I'd be willing to bet that none of those authors who wrote time travel stories actually believed in the possibility) then we need to ask *why* is that the case? It seems to me that SF writers have as much to say about this question as physicists and philosophers—so before looking at our modern understanding of time travel let's consider some of the thought experiments (in story form) that SF writers have performed.

5.1   Time Travel in Science Fiction

You must have dreamed, once or twice, of watching important moments in history or perhaps of affecting the outcome of key events that happened in your life. Or maybe you've had one of those reveries in which you take a shortcut to the future just to see how things turn out. We've probably all fantasized about traveling through time; certainly these are common enough fantasies for mainstream writers. Dickens had Scrooge travel into the past and into the future, and in *A Connecticut Yankee in King Arthur's Court* (1889) Twain had his hero transported back in time by a dozen or more centuries. But Dickens and Twain didn't write time travel stories in the sense that most SF fans understand the term. Scrooge saw the past and the future through visions or dreams; Hank Morgan, Twain's Yankee engineer, was transported in time through a bang on the head with a crowbar. These were just literary devices. It was H.G. Wells (who else?) who popularized the idea of a time *machine*: the deliberate appliance of technology to move a person purposefully backwards and forwards through the fourth dimension of time, just as we employ technology to help move us purposefully through the three dimensions of space. Wells's *The Time Machine* (1895) followed his earlier story "The Chronic Argonauts" (1888), and the novel was wildly successful. Not only was it adapted into films, television programs, and radio serials, it inspired generations of science fiction writers to grapple with the paradoxes that time travel throws up. And there are plenty of paradoxes.

It isn't travel into the future that causes problems. After all, everyone seems to move into the future at the rate of one second per second. It's just the normal passing of time. If we were to develop a technology that allowed us to take a shortcut into the future—perhaps by using cryogenics to suspend our animation or by using relativistic rockets to make use of time dilation—then the journey might cause us cognitive dissonance but we wouldn't have a fundamental difficulty in understanding what had happened to us. Travel into the past, though. . . well that raises all sorts of issues. And 'travel' can take any form. Consider, for example, Asimov's famous story "The Endochronic Properties of Resublimated Thiotimoline" (1948). He was inspired to compose it during the process of writing up his doctoral thesis: by that point he'd been a published writer for almost a decade and he was worried he'd struggle to write in the dull, dry, academic style required of a PhD thesis. He decided to practice the necessary style by writing a mock scientific article. Asimov's doctoral research involved dissolving an organic compound called catechol in water. Catechol dissolves extremely quickly, and Asimov noted that if it dissolved any more quickly it would do so *before* it hit the water. Thus was thiotimoline born. Asimov wrote a spoof research paper, complete with graphs, equations, and citations to fake papers in made-up journals, describing a compound so soluble that it dissolved in H_2O up to 1.12 s *before* the addition of the water. This 'endochronicity' was postulated to be the result of two carbon bonds in the thiotimoline molecule lying along the time axis: one bond extended into the future, the other into the past. It's a witty and much-referenced story, and it highlights one of the issues with time travel into the past: what would happen if I intended to add water to thiotimoline and then, just before I did so, I decided not to? The thiotimoline dissolved because water was going to be added—but then the water wasn't added. So does it dissolve, or not? If it dissolves, am I somehow forced to add the water in order to make sure it dissolved? If so, what agency makes me add the water? What happens to my free will?

These questions are usually raised in connection with the so-called grandfather paradox.[1] The idea occurs so often that it might almost be termed the first law of time travel: whenever you invent a time machine the first thing you must do is go back in time and kill your grandfather. Putting aside the question of why time travelers hate their grandfathers so much, the paradox is clear: if you kill your grandfather then you were never born therefore you couldn't have travelled back in time to kill your grandfather therefore you *were* born and so. . . . The paradox works just as well if you go back in time and kill your younger self as a baby, an act that philosophers of time travel term auto-infanticide, but the grandfather paradox is how it's usually known so let's stick with that name.

Many of the most memorable SF stories deal with ways of addressing the grandfather paradox, and in the subsections below I'd like to discuss some of my favorites. First, though, it's worth classifying the different types of story. A number of different mechanisms have been proposed to achieve time travel, but however the feat is achieved the effects are usually of two types. In the first type, timelines are immutable. History is fixed and unchangeable. Put simply, if you travel back through time you *can't* alter anything: any event that has happened will always have happened. (I'm struggling with tenses here. Nevertheless, I trust you'll understand what I'm trying to express.) In the second type, timelines are mutable. Travel back through time and you run the risk of changing history. This second group of stories can be further subdivided into those in which the time traveller moves from alternate timeline to alternate timeline (and perhaps has memory and knowledge of the various timelines) or remains on a single timeline but runs the risk of seeing history changed (and perhaps even disappearing from this altered history). When you start to delve into the minutiae of these different types of mutable timelines, however, the discussion becomes increasingly complicated. For clarity, let's just stick with the simple immutable/mutable distinction. I'll add a third type of time travel story: those involving a time viewer rather than a time machine. In these stories physical objects are unable to move through time, but we can nevertheless view information from other times.

Authors and directors have used these three different temporal effects to write some wonderful stories and shoot some memorable films. Let's start with SF in which history is fixed.

5.1.1 Immutable Timelines

An entertaining class of time travel stories attempts to sidestep the issue of paradox by having something from the future land in the present day. If the story is told from the viewpoint of a protagonist living in the present moment then no paradox is involved since the object or person from the future is in the protagonist's here-and-now. How could the protagonist possibly know what would have happened if the object or person had not appeared? Questions of the mutability or immutability of timelines seemingly don't apply. (Of course, that's looking at the issue solely from the viewpoint of the present day protagonist. If you were the victim of a gun attack then the present reality of your painful and lingering death would indeed be all that mattered. But if it turned out you'd been shot by your as-yet unconceived grandchild then said grandchild would have a paradox to contend with.)

For a typical example of this approach see "The Little Black Bag" (1950). It was written by Cyril Kornbluth, an author who surely can't have been as cynical in person as his stories suggest. The story tells how an automated medical kit from the year 2450 is accidentally sent five centuries back in time.

An alcoholic doctor finds the kit and uses it to heal an injured child. This success enables the doctor to win back his self-respect but, as is typical with a Kornbluth story, the tale ends with a nasty twist. In "The Little Black Bag" little attempt is made to address the potential for paradox: the kit turns out to be responsible for two deaths, so what would happen if one of those two victims happened to be an ancestor of the person who created the kit in the first place? On the other hand, since Kornbluth made the present day the focus of the story, the reader is free to ignore the complications and assume that the timeline is immutable: the deaths involving the medical kit happened that way because they happened that way. . . and that's all there is to say.

A gentler example of this idea is provided in the story "Mimsy Were the Borogoves" (1943), written by Lewis Padgett (a pseudonym of the husband-and-wife writing team of Henry Kuttner and Catherine Moore). A scientist from the far distant future sends two boxes of educational toys back into his past. One box comes into the possession of a young girl, Alice Liddell, who learns some nonsense verse from one of the toys. Alice recites the verse to her much older friend, Charles Dodgson, who of course is better known as Lewis Carroll. The nonsense verse is published as "Jabberwocky". The other box of toys lands in 1942 in America and is found by an 11-year old boy and his younger sister. Padgett's story describes the effect of these toys, and the poem "Jabberwocky", on the still-developing minds of the two siblings.

My favorite story of this type is also due to Kuttner and Moore, but this time writing under the pseudonym Lawrence O'Donnell. "Vintage Season" (1946) is a dazzling novella, set in an average, mid-1940s American town that happens to be enjoying a particularly beautiful stretch of weather. The story tells of travelers from the future who are visiting, as they visit other perfect seasons in human history. Their visits record more than just fine weather, however. They visit the scenes of impending disaster—and the present-day protagonist of "Vintage Season" eventually learns why the immaculate visitors from the future have decided to come to this particular place and time.

The Kornbluth story and the two stories by Kuttner and Moore hint that perhaps timelines might be mutable and therefore touch gently on paradox. One of the time travelers in "Vintage Season" admits that in principle he could be more than a spectator, that perhaps he could even prevent the calamities he chooses to record, but he dare not because in doing so he would alter history and prevent his civilization from appearing. (In which case he would not have been born and therefore could not prevent the disasters. . . it's the classic grandfather paradox.) Nevertheless, these authors don't attempt to grapple with the intricacies of time travel; the timelines they describe might as well be immutable because time travel itself isn't really the focus of these stories. The authors use visitations from the future purely to provide insights into contemporary psychology. If you are interested in stories where time travel is the focus, and where timelines are strictly immutable, then you need to turn to the author that nailed it: Robert Heinlein.

One of Heinlein's earliest stories, written under the pen name of Anson Macdonald, was "By His Bootstraps" (1941). The story's protagonist, in a complicated series of events, encounters someone who resembles himself and who takes him into the future. While in the future he learns how to operate a time machine and eventually, after some labyrinthine temporal manoeuvring, he travels back in time to bring his younger self into the future.

Almost two decades after the publication of "By His Bootstraps", Heinlein published "—All You Zombies—" (1959). This story is even more convoluted than Heinlein's earlier foray into the problems of time travel. As the story unfolds we learn that, thanks to a convoluted series of trips through time, the main protagonists are all the same character at different stages of his/her life. An intersex baby, initially characterized as being female, grows up and undergoes a sex change operation. Later, as a man, the individual travels back in time and impregnates his pre-sex change self; the intersex baby is the offspring of that union. The person is thus his own mother and his own father. When she/he gives birth to the daughter he/she thus gives birth to her/his entire history. The protagonist thinks at the end of the story: 'I know where I came from—but where did all you zombies come from?'[2]

These two Heinlein stories appear to be much more fantastic, much more surreal, than the events described in "The Little Black Bag", "Mimsy Were the Borogoves", and "Vintage Season". But get a pen and paper, draw the timelines described in "By His Bootstraps" and "—All You Zombies—", and you'll see that there are no violations of causality in Heinlein's stories. He worked everything out with remorseless logic. The events are self-consistent: the timelines curve back on themselves to form a closed loop. As we shall see later, Heinlein's stories foreshadow the self-consistency conjectures that some physicists believe might allow time travel to take place.

5.1.2 Mutable Timelines

Heinlein demonstrated how time travel stories can be internally consistent when the timelines involved are immutable and the past does not change. However, most SF involving time travel concerns itself with the notion of mutable timelines—the idea that the past can be changed. The notion of a malleable past is not popular amongst physicists or philosophers because it seems to be logically inconsistent, but it does provide scope for SF writers to ponder the nature of contingency and the fine points of causality.

There's a whole subgenre in which someone travels back in time and alters the past. One of the most memorable and influential of such stories is Ray Bradbury's "A Sound of Thunder" (1952), which tells of a company, Time Safari Inc., that offers hunters trips back in time. It lets them hunt dinosaurs. In order to protect against possible alterations of history, the company tags animals that would soon die anyway and the hunters are told to shoot only those dinosaurs that are tagged. Eckels, a would-be hunter, travels back from the year 2055 to the Cretaceous Era to bag a T. Rex, but once there he makes a fatal error. Company guides issue a strict warning that everyone must at all times remain on a levitating path, but Eckels gets scared when a T. Rex approaches and he jumps from the path and runs into a forest. When the guides rescue him and bring him back to the present, Eckels notices that reality is subtly different: the outcome of a US presidential election is different, the English language is different, people behave differently. He looks down at his boots and sees a crushed butterfly amongst the Cretaceous mud. The butterfly met an untimely end and, by killing it, Eckels set in motion a series of events that snowballed and changed the reality he came from.[3]

Bradbury's story raises a number of questions. For example, if the world changes because Eckels killed a butterfly then why wasn't Eckels himself changed? Why does he have a recollection of how things were before the change while the rest of humanity has no such memory—were he, his fellow hunters and the company guides somehow 'outside' time when they travelled back to the Cretaceous? Perhaps there are a vast number of alternate timelines, and when you travel back in time and change the past what really happens is you move to one of those alternate timelines—perhaps that is what happened to Eckels?

One of Asimov's most ambitious novels was *The End of Eternity* (1955)—a time travel story which takes as its starting point the notion that history is malleable. The 'Eternity' mentioned in the title is an organization outside of the temporal reality inhabited by the rest of humanity. Eternals—the people who work for Eternity—are men recruited from throughout human history, and they can travel 'downwhen' and 'upwhen'. These guardians of time voyage through the epochs in order to administer painstakingly calculated 'Reality Changes'—small events (much like the killing of Eckel's butterfly) that change history in such a way as to minimize the sum total of human misery and suffering. The Eternals try to cause Eternity to come into being by sending one of their own back into the twenty-fourth century in order to teach the necessary mathematics to the historic developer of time travel. But another

Eternal puts the plan in jeopardy when he falls in love with a non-Eternal, and starts to ask whether the existence of Eternity really is in humanity's best interests.

In *The End of Eternity* Asimov treats temporal paradoxes with subtlety, but for too many authors the notion that it's possible to alter the past leads them to write stories in which the time travelling hero goes back to kill some bad person. (If the first law of time machines is 'kill your grandfather' surely the second law is 'kill Hitler'.) A more peaceful and constructive time-travel story is "Newton's Gift" (1979), by Paul Nahin. It appeared in the fourth issue of *Omni*, a short-lived and fondly remembered science-and-SF magazine with unrivalled production values. In Nahin's story the time traveller, Wallace John Steinhope, intends to save Isaac Newton, the greatest scientific mind in human history, from the tedium of having to perform manual arithmetical calculations. Steinhope travels back in time to seventeenth century England and presents his hero with the gift of an electronic calculator. Newton, not unreasonably, sees this mysterious box with flashing lights as a tool of Satan—and decides to turn his back on mathematics and concern himself instead with matters of theology. Nahin's story is doubly interesting because it illustrates the difference between altering the past and influencing the past. If seeing an electronic calculator caused Newton to turn from science *before* he made all his world-changing mathematical and scientific discoveries then we would hit a paradox, because Steinhope lived in a world in which Newton *had* made all those discoveries. On the other hand, if Steinhope's visit to Newton took place in the middle of the great physicist's career then it serves as an explanation for why Newton spent an increasing amount of his time on theology—something we know happened. In other words, "Newton's Gift" can be seen as belonging to the Heinlein tradition of self-consistent time loops rather than Bradbury's notion of mutable history. (Incidentally, you might think Shakespeare, the universal genius, would have the necessary flexibility of mind to adapt to time travel. Well Shakespeare is the subject of a humorous Asimov short story called "The Immortal Bard" (1954). A physicist regales his young colleague, an English teacher, with the tale of how he invented a time machine and used it to bring the Bard on a visit to the present day. The physicist enrolled him on a night-school course on Shakespeare's plays taught by the very same English teacher. The poor teacher flunked Shakespeare!)

There is no space here to discuss all my favorite time travel stories, much less give an exhaustive list of such stories.[4] Besides, Nahin (1999) has already written an excellent book on time travel and science fiction.

5.1.3 Time Viewers

Some stories refer to time *viewing* rather than time *traveling*. One of the earliest such stories was Tom Sherred's "E for Effort" (1947), which relates how two war veterans use a time viewer to make films of real historical events—the victories of Alexander the Great, the conquest of Mexico by Cortes, the bloodshed of the French Revolution. The production values are terrific because the time viewer records real events as they happened. The general public, believing these films to be Hollywood productions, flock to the cinema. As the events they film come closer and closer to the present day— the American Civil War, WW1, WW2—the two veterans begin to recognize the corruption of several prominent figures involved in the wars. They are so appalled by the corruption on display that they decide to make war impossible: they agree to give away the secret of how to build the time viewer and thereby permit people to watch governments and military establishments around the world. Of course, things don't turn out the way they hoped.[5]

One of Asimov's best stories, "The Dead Past" (1956), acts as a counter-point to "E for Effort". In this story a scientist has developed a device called a chronoscope, which makes it possible to view the past. Subsequently, however, chronoscope technology is tightly controlled; only government departments have access to working models. A professor of ancient history requests access to a chronoscope in order to research life in ancient Carthage. Officials refuse the request, so the stubborn professor recruits a young physicist and pays him to develop a chronoscope. Although the government has suppressed research in this area for decades, the physicist succeeds in making a cheap and efficient time viewer; the only difficulty is that, as the physicist proves, chronoscopic technology can't be used to view anything further than about 120 years in the past. This is a double blow for the historian: not only is the chronoscope useless for his research, he suspects that his wife will use the device to view obsessively the short life of their dead daughter. Even worse is to come. The physicist publishes his design for all to use, but we learn that the government suppressed chronoscope research for a good reason: the past always begins just a fraction of a second ago—the recent past is effectively the present—and so people could use the chronoscope to spy on what their friends and neighbours are doing right now. By publishing the blueprints for the device, the physicist destroys forever the concept of privacy.

A similarly melancholic feel pervades Bob Shaw's famous story "Light of Other Days" (1966) and his novel *Other Days, Other Eyes* (1972). Both the short story and the novel revolve around the invention of a substance called slow glass, a type of glass through which light takes months or years to pass. Much as with Asimov's chronoscope, slow glass allows people to observe events from the past—you simply have to wait for the light that entered it to be released. Shaw has great fun investigating the consequences of such a technology, and like Asimov he concludes it would affect people's relationships and spell the end for privacy. (It's probably impossible to construct slow glass but the *effects* of slow glass, which worried Shaw, are now a mundane reality. Smartphones and CCTV are ubiquitous, recording our lives for later viewing—by ourselves or the authorities. Slow glass was just a video camera.)

Time viewers won't ever be developed, I'm sure, but if the stories mentioned above have the smell of technical authenticity about them it's because they don't involve the same mind-bending paradoxes that can occur when travel into the past takes place. For a chronoscope or a piece of slow glass information flows *from* the past not *into* the past, just as happens with any sort of filmed recording. Indeed, in a sense, when we look out onto the world we do so through a time viewer: we only ever see the world as it once was, not as it is now. When we see the Moon we see it as it was about 1.3 s ago; the Sun we see now is in reality the Sun as it was just over 8 min ago; when our telescopes observe distant galaxies they see the galaxies as they were billions of years ago; our radio telescopes observe the universe as it appeared 300,000 years after the Big Bang itself. Astronomy is in essence a science based on time viewing, and it's inevitably so because of the finite speed of light.

The stories above are about viewing the past. Relatively few hard-SF stories describe a time viewer for observing the future. Of those that do, I can think of only one that was successful. Heinlein's "Life-line" (1939), his first published story, tells of a scientist who invents a device that can send a signal along the worldline of an individual and detect the reflection from its far end. In other words, the device can determine how long a person will live. For the most part, however, time viewing the future smacks more of fantasy than SF, more of fortune-telling than hard science.

5.2   The Physics of Time Travel

So—is there any justification in physics for taking the notion of time travel seriously? Can time travel stories ever be considered 'hard' SF, or are they inevitably fantasy?

I argued earlier that we are all of us time travellers, moving into the future at the rate of one second per second. Actually, that's not quite true. We might all be time travellers but we don't share a common tempo. As Einstein's theory of special relativity tells us, clocks go at different rates depending on their velocity. For example, an astronaut who spends a year working on the fast-moving International Space Station will age more slowly (of the order of a few milliseconds) compared to us sluggards here on Earth's surface. A difference of a fraction of a second per year is tiny, but then the ISS doesn't move very quickly relative to us Earthbound observers: it orbits at about 7700 m/s. If an astronaut were to move at a significant fraction of the speed of light, which is 299,792,458 m/s, then the effects of relativistic time dilation would be noticeable. In his novel *Tau Zero* (1970), Poul Anderson describes how a malfunctioning starship accelerates to speeds ever closer to the speed of light; the phenomenon of time dilation becomes so extreme that each passing moment for the crew corresponds to many billions of years in the external universe.

It's not just moving clocks that go at different rates. Einstein's theory of general relativity explains how the ticking of a clock also depends on the strength of the gravitational potential in which it finds itself. The time dilation described in *Tau Zero* could also occur if an observer orbited close to the event horizon of a black hole, where the gravitational potential is high.

In theory, therefore, we are all experiencing the passing of time at slightly different rates (even if, for current practical purposes, people are travelling into the future in lockstep). We could use relativistic time dilation to get a shortcut to the future, or employ some other technologies to establish the same effect (cryopreservation anyone?). We could—but that would be a poor form of time travel. Regardless of the rate at which time passes, we'd only ever have a one-way ticket: we'd always be travelling *forward* in time. Does science offer any hope of permitting travel *backward* in time? Well, before we can attempt to answer that question we first need to look at what the concept of 'time' actually means.

5.2.1 What Is Time?

The concept of time is straightforward enough in everyday life. All of us have an intuitive sense of the concept and we all experience it as something flowing inexorably from past to future. (At least, I presume that's the case. You do experience time in that way, don't you?) Pinning down the physical meaning of time is not so easy. What role does 'time' play in theories of physics?

It turns out that 'time' is a surprisingly subtle and slippery concept. Some physicists would argue that 'time' does not exist; others that it exists but is an emergent phenomenon rather than something fundamental; yet others that it is the most important quantity in all physics. Nowadays (if there is such a thing as 'now') any deep discussion about the meaning of time must refer to a number of related but different concepts; in the past (if there is such a thing as 'the past') it was all so much simpler.

Let's start with Newton's idea of time, since his conception influences much of our thinking. Newton built a system of the world based on the notion that there exists some sort of universal clock. Newton's universal clock permits us to do various useful things. It allows observers, uniquely and objectively, to slice up the cosmos into instants of time. It provides a metric, allowing observers to define the separation between events. (If Alice and Bob set off together in a marathon race, and Alice finishes in 3 h while Bob takes 4 h, then both of them, along with the race officials, will agree that Alice finished 1 h earlier than Bob.) It allows all observers to agree on the order in which events occur: everyone can agree that A happens before B, a notion that is crucial to the idea of causality. (Causality means that if a given event, the cause, occasions a second event, the effect, then cause must precede effect. Science is founded in the notion of causality, and if we propose an idea that defies causality then the idea faces significant challenges.) Furthermore, since Newton's clock is continuous, it allows us to define quantities such as velocity and acceleration. It also holds the notion of time flowing from past to future. To modern eyes all this might seem straightforward, but don't be fooled: his was an incredible intellectual achievement. Newton's clock was one of his ideas that changed the way humanity understands the universe. It was one of the foundations of the technologically-based world we now live in. But Newton's clock has problems.

About 150 years after Newton's death, Ludwig Boltzmann pointed out that Newton's laws don't distinguish between past and future. If you film a typically Newtonian event, such as the collision of two billiard balls, then it doesn't matter whether you play the film forwards or backwards: both versions of the event appear to be equally realistic. Indeed, with the exception of a rather exotic phenomenon to do with certain subatomic particles, a phenomenon that was unknown until relatively recently, time doesn't appear to play much of a role in the laws of physics. That raises an intriguing question: if the arrow of time doesn't come from the laws of physics where *does* it come from? The arrow, that flow from past to future, certainly seems real enough: if you film a falling egg impacting on concrete then you notice a real difference between the forward- and backward-running versions. No one in the history of the world has ever seen the various constituents of an egg reassemble from a gooey mess on the floor into an unbroken whole. Boltzmann argued that the arrow of time, the distinction between past and future, arises from statistics.

There are many more ways for systems to be disordered than ordered, so in any process a system will tend to move from order to disorder—even if the underlying physics doesn't distinguish between past and future. For example, put a drop of ink into a glass of water and shake: ink and water mix. The underlying molecular collisions don't distinguish between past and future, but mixing *will* occur and the mixture will stay mixed. Why? Simply because there are *overwhelmingly* more configurations in which ink and water mix than in which ink and water remain separate. In a few configurations the ink molecules can be localized in a small region, and could thus still be called a 'drop', but the number of mixed configurations is so large it's hard to grasp; by comparison, the number of atoms in the universe, say, is insignificant.

The measure of disorder in a system is known as entropy, and the second law of thermodynamics states that in an irreversible process entropy always increases. The total entropy of the universe increases continuously and, in Boltzmann's view, *this* gives rise to the arrow of time. In this view, the distinction we make between past and future isn't because of 'time' but because of the ordering of matter in the universe. The mystery we are left with is: why did the initial state of the universe contain so much order, so little entropy?

Entropy tends not to be discussed much in time travel stories, but it's a concern that should be addressed: if you were to send an object back in time then a high-entropy system would be travelling into a universe with lower entropy. How could that work? The pound coin in my pocket was minted in 2001. The coin is dull, scratched, and the queen's head is shown unmistakeable signs of wear. If I send the coin back in time by 15 years, to the time it was minted, what happens to the cumulative oxidation, scoring, erosion? In other words, what happens to the entropy?

A year before Boltzmann's death, Albert Einstein introduced another problem with Newton's conception of time. Einstein began with two postulates regarding observers who are in relative motion. First, if the observers are not subject to forces then they'll observe the same laws of physics. Second, those observers will measure the same value for c, the speed of light. Einstein went on to show that if these postulates hold then observers in relative motion will measure space and time differently: they'll disagree about the timing of an event and the position of an event. In other words, there's no universal clock; neither is there a universal yardstick. Indeed Hermann Minkowski, one of Einstein's teachers, argued strongly that space and time were themselves of little importance; it was the marriage of space and time—spacetime—that was important. Different observers can disagree about the duration of time between two events and the spatial separation of those events, but they will agree on the spacetime interval between those events. This is the essence of Einstein's special theory of relativity.

Physicists usually explain these sorts of concepts by referring to a spacetime or Minkowski diagram. Although they might seem to be rather technical, it's worth getting a feel for these diagrams since they can clarify some otherwise difficult points. So take a look at Fig. 5.1. It shows a typical spacetime diagram; this one is drawn from the point of view of an observer Alice. The vertical axis represents time, the horizontal axis represents space. An *event* is something that happens at a particular place and at a particular time; this diagram shows an event P (perhaps Alice sneezing) and an event Q (perhaps her friend Bob switching on a light). For ease of representation on paper or screen, we suppress two space dimensions along the horizontal axis: imagine events happening on a railtrack, so that you only need to consider a single space dimension. Note that there's nothing particularly important about this choice of axes; it's just a convention. What *is* important is that lines parallel to the

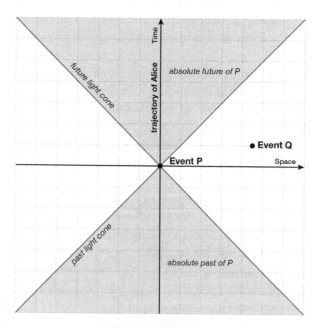

Fig. 5.1 A spacetime diagram in Alice's frame of reference. Events that take place inside the past light cone can in principle influence the event P; in the future light cone are those events that event P can influence. The event Q can neither influence nor be influenced by event P. In Asimov's story "The Dead Past", for example, the chronoscope could only be used to view events in the observer's past light cone. If an event happened outside the past light cone then it could not be viewed by the chronoscope at that point, although it might come into view at a later time (Credit: author's own work)

horizontal axis represent events that happen at the same time in this, Alice's, frame of reference. Lines that are parallel to the vertical axis represent events that happen at the same place in Alice's frame of reference. In other words, the grid is simply a collection of lines of constant time and constant position in Alice's frame. The scales on the axes are usually chosen such that light rays propagate at 45 degrees: if a 'tick' on the time axis represents one year then a 'tick' on the space axis represents one light year. The light rays diverging from an event form a surface in spacetime called the *future light cone* and the events within the future light cone form the given event's absolute future. In Fig. 5.1, any events within P's future light cone are the events that P (the act of Alice sneezing) can influence. The light rays converging on an event form a surface in spacetime called the *past light cone* and the events within the past light cone form the given event's absolute past. In Fig. 5.1, any events within P's past light cone are the events than can have influenced P (the sneeze). An event such as Q (a light being switched on), which lies outside P's light cones, can neither influence P nor be influenced by P—unless, as we shall see in the next section, FTL signals are permitted.

Things get more interesting when we introduce a second observer moving relative to the first. Suppose Alice, who naturally enough considers herself to be at rest, sees Bob whoosh past her at a constant velocity. Special relativity tells us that observers in relative motion carve spacetime into different slices of space and time. Alice will say Bob's space and time axes are tilted towards the light cone: Alice will observe Bob's clock to run slow, and lengths in the moving frame will contract. (Bob will say something similar about Alice: Bob considers himself to be at rest, and it is *Alice's* measurements of space and time that are skewed as she races past.) The mathematics of special relativity allows us to transform between different frames of reference. The key factor in the transformation is the relative speed of the two frames.

Figure 5.2 shows the transformation for two frames moving at 40% of the speed of light, relative to one another. The black axes represent space and time in Alice's frame of reference and the blue axes represent space and time in Bob's frame of reference. The two different coordinate systems are an indication that Alice and Bob measure space and time in different ways. (Note, however, that the light cones are unaffected by the transformation: Alice, Bob and all other inertial observers measure the same value for the speed of light.) For Alice, lines of constant time are those drawn parallel to the x axis; for Bob, lines of constant time are those drawn parallel to the x' axis. Bearing this in mind, it's possible to see that there is something interesting about the timing of the events P (a sneeze) and Q (a light being switched on) that we drew on Fig. 5.1. Draw these events on Fig. 5.2 and note when the events happen in

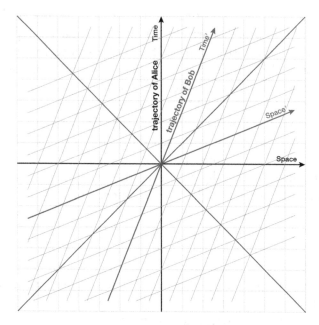

Fig. 5.2 This spacetime diagram shows the frames of reference of two observers, Alice and Bob, who are moving relative to one another at 40% of the speed of light. Alice's reference frame is shown by the *black axes*; *grey lines* running parallel to the axes are lines of constant time and lines of constant position for Alice. Bob's reference frame is shown by the *blue axes*; *blue lines* running parallel to these axes are lines of constant time and lines of constant position for Bob. Both frames of reference are equally valid. Note that both Alice and Bob agree on the speed of light (*red lines*). Light cones are unaffected by the transformation (Credit: author's own work)

the two reference frames. According to Alice, P happens before Q. According to Bob, Q happens before P. This relativity of simultaneity is not a big deal: the event Q is outside of Alice's light cones so she can't know about the event at the time nor can she have influenced it. Nevertheless, it removes yet another of Newton's ideas about time: there's no universal clock allowing observers to agree on the precise ordering of events.

By 1915, Einstein had developed his general theory of relativity. The general theory extends the ideas of special relativity to cases where gravity is involved. Einstein showed that gravity distorts time. A clock ticking here on the surface of the Earth does so at a different rate compared to a clock ticking, say, 20,000 km above us in Medium Earth Orbit. (The influence of gravity on time means that a GPS satellite in Medium Earth Orbit gains about 45 ns per day compared to identical clocks on the ground.) In a universe containing black holes and neutron stars—objects whose gravitational field strength can be immense—it simply doesn't make sense to think of time unfolding tick by tick. Indeed, this observation has led some physicists to argue that time does not exist.

What does it mean to say time does not exist? Don't physicists *need* the concept of time in order to describe the rate at which things change, just as Newton used the concept all those years ago? Don't we *need* time in order to define the speed of light or the orbital period of the Earth or the frequency of a pendulum? Perhaps not. We could, for example, relate each of these quantities to one other: Earth's orbital period could be expressed in terms of pendulum oscillations while the beat of a pendulum could in turn be expressed in terms of light travel distance which itself could be expressed in terms of. . . well, you get the idea. In this scenario the phenomenon we label 'time' emerges in much the same way as the phenomenon we label 'money'. We don't *need* money; it's just an accounting trick. We could barter instead: five beers might be swapped for two used copies of this volume which might get you a pair of underpants, or whatever. Rather than having to haggle over each and every transaction our society has invented money. But money is merely a convenience, a device for keeping track of value rather than something possessing inherent value itself. Perhaps the same is true of time: perhaps it's a fabrication, a simple way of avoiding the necessity of relating one mutable system to another?

And yet. . . although relativity suggests our fundamental reality is based upon spacetime, time and space are not equal partners in the marriage. We can't just slice spacetime in any way that strikes our fancy: there's a distinction between the 'timelike' direction (in which events can be causally connected) and the 'spacelike' direction (in which events are not causally connected). Time, whatever it is, performs a different function to space. Furthermore, relativity is not the only game in town. Physicists have another deep theory of how nature works: quantum mechanics. Indeed, many physicists would argue that quantum mechanics gives us our fullest account of nature. Time, however, plays a key role in quantum mechanics. You can imagine somehow banishing time from general relativity; you can't abolish time in quantum mechanics without fundamentally reworking the entire theory.

Ever since Einstein, physicists have tried to develop a unified theory—a set of ideas that can unite general relativity (which describes the world on a large scale) with quantum mechanics (which describes the world on a small scale). So far, they have failed. One reason for that failure is the different way that time features in these two grand concepts. It's fair to say, then, that we still don't have a deep understanding time. So if that's the case, and our knowledge of time is incomplete, then who is to say that time travel is impossible? Let's look at a couple of serious suggestions, based on known physics, of how time travel could be accomplished. It turns out that one of the suggestions doesn't work, but it's worth examining it in some detail to find out *why* it doesn't work. We are then better placed to look at the other suggestion, which *does* seem to permit time travel. First, let's look at tachyons.

5.2.2 Tachyons

The story "Beep" (1954), by James Blish, discusses some of the consequences of faster-than-light (FTL) communication. It isn't one of Blish's best stories, but it's notable because the theoretical physicist Gerald Feinberg read it when it was published in *Galaxy* magazine and it got him to thinking whether FTL communication could ever occur in the real world. (Incidentally, Feinberg was an science fiction fan before he became a physicist. In high school he edited a fanzine called "Etaoin Shrdlu". Feinberg's co-editors were classmates Sheldon Glashow and Steven Weinberg, who in 1979 were awarded the Nobel prize for physics. Glashow and Weinberg must be the most illustrious editors of any science fiction fanzine.) Feinberg eventually wrote a paper called "Possibility of Faster-than-light Particles" (1967), in which he discussed the properties that a particle must possess if it travels at speeds greater than c. He called such particles 'tachyons'[6] from a Greek word meaning 'rapid'.

Tachyons would be curious objects. For example, as a tachyon goes faster its energy decreases; as its speed approaches infinity its energy approaches zero. On the other hand, as its speed decreases its energy increases: as its speed approaches c its energy approaches infinity. So the speed of light forms a barrier from both directions: it requires infinite energy to *accelerate* a normal particle to light speed and infinite energy to *decelerate* a tachyon to light speed.

Quite apart from exhibiting peculiar behavior, tachyons would enable a message to be sent into the past (see the box for details).[7] This isn't time travel, perhaps, but it is certainly time communication. And sending a message backwards in time can generate a paradox just as easily as physically traveling into the past. For example, suppose I send a message to my youthful grandfather telling him that the horse Battleship will win the 1938 Grand National at 40/1; he puts £100 on the nose, makes a killing, and runs off with Rita Hayworth instead of staying with my grandmother. My father is not born. There's just as much trouble with causality in this case as if I'd gone back in time and shot my grandfather.

Faster than Light Particles Can Signal the Past

Soon after he developed the theory of special relativity, Einstein noted that if his theory were correct then FTL signals could be used to communicate with the past. Richard Chace Tolman outlined a similar example in 1917, and pointed out that this leads to a paradox of causality—now called Tolman's paradox. (Greg Benford provided a better name: he called an FTL communicator a tachyonic antitelephone.)

At first glance it's not necessarily clear that a faster-than-light signal implies communication with the past. The point is that if we allow FTL communication then we can easily construct an example in which backwards-in-time communication can occur. Let's look at a typical example of how FTL leads to the possibility of sending signals into the past. I assume here that the signals travel infinitely fast, but that's only because it makes the diagrams easier to draw. The same logic applies to any FTL signal.

Let's suppose that Alice and Carol are are standing alongside a railway line, and they are still relative to each other. Figure 5.3 is the relevant spacetime diagram. Alice sends a signal (event P) using her FTL device and instantaneously this is

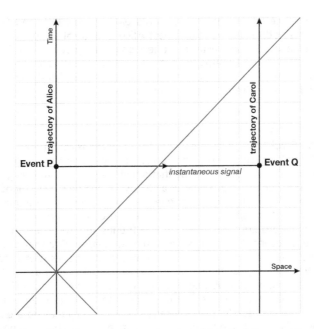

Fig. 5.3 This spacetime diagram represents two friends, Alice and Carol, standing at rest relative to one another. P represents the event that Alice sends Carol a message using an FTL communication device. (For ease of drawing, the signal here travels at infinite speed; for the argument to work the signal need only travel faster than light speed.) Q represents the event that Carol receives the message. Since the signal travels infinitely fast, P and Q occur at the same time (Credit: author's own work)

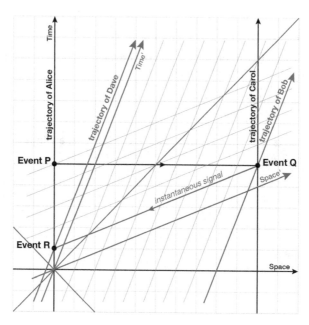

Fig. 5.4 This spacetime diagram represents two friends, Bob and Dave, at rest relative to one another but travelling past Alice and Carol at 40% of light speed. Bob and Dave have the same instantanteous signalling device as Alice and Carol. We can therefore arrange things so that Alice receives the signal (event R) before she sends it (event P). If instantaneous signalling is permitted, Alice can send a message into her past (Credit: author's own work)

received by Carol (event Q). This in itself isn't paradoxical, at least from some observers' point of view. The problem is, you can always find a frame of reference where it is paradoxical. Let's suppose that a high-speed train, traveling at a large fraction of light-speed, whizzes past as all this is happening. Alice's friend Bob and Carol's friend Dan are sitting at rest relative to each other on the train, but are moving at a constant speed relative to Alice and Carol. Suppose they have the same instantaneous signaling device as their friends and that at the moment Bob passes Carol at event Q, Carol gives him Alice's message; Bob then immediately sends a message via the FTL signaler to Dave, who receives it at event R. See Fig. 5.4. We've arranged matters so that Dave gets the message just as he is passing Alice, so that Alice also gets the message at event R. This is disastrous: in Alice's frame of reference, event R (receiving the message she sent) happens before event P (Alice sending the message). Thus Alice, by making use of willing colleagues and an FTL-transmitter, can send a message into her own past.

As the argument given in the box demonstrates, causality is at risk if we allow signals to travel faster than light. It seems we can't have all three of causality, special relativity, and FTL communication. As with any arguments in science one can find loopholes: for example, some physicists have explored theories in which Lorentz symmetry, upon which special relativity is based, can be violated. On the other hand, there's no reason to suppose that Lorentz symmetry is violated in Nature, while the evidence in favor of special relativity is overwhelming. If we want to keep causality (which we surely do) then it comes down to a choice: we can have either special relativity or FTL communication. We can't have both. To my mind, special relativity wins.

You might argue that a dislike of temporal paradoxes is insufficient reason to conclude that tachyons don't exist. However, there is absolutely no experimental evidence for the existence of tachyons and our theories can manage quite well without them. Indeed, physicists now argue that if a theory permits a tachyonic solution then the tachyon represents an instability in that theory rather than an honest-to-goodness physical particle we could detect. It might be disappointing to news to hear, but you can't use tachyons to let your younger self know the outcome of last year's Kentucky Derby.

5.2.3 Closed Timelike Curves

Earlier in the chapter we discussed the light cone of an event. In the 'simple' situation described by Einstein's theory of special relativity the light cone points forward in time. If a massive object is present at an event then the object's path through spacetime—its world line—must lie within the event's light cone. In other words, the path through spacetime must be closer to the time axis than the space axis. We say, therefore, that the world line of a physical object is timelike. It's really just a fancy way of saying that an object can't be in two places at the same time. However, things get more interesting when a massive object—a star or a black hole, for example—causes spacetime to curve. To handle this situation we need Einstein's theory of general relativity. According to general relativity, light cones are tilted along the spacetime curvature. For example, if a freely falling object is close to a star then the light cone is tilted by the curvature generated by the star's mass: the object's future positions lie closer to the star. Someone moving with the object observes it to be moving along the local time axis, but to an external observer the object is accelerating in space—the object might be said to be orbiting the star, for example. In practical situations this tilting of the light cone is tiny and not at all obvious. In extreme situations, however, the tilting of the light cone can become large. But this raises an interesting question: what happens if a series of tilted light cones loop back on themselves? It that could happen we would have a closed timelike curve (CTC): an object could move around the

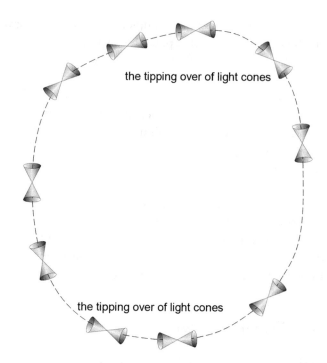

the tipping over of light cones

the tipping over of light cones

Fig. 5.5 In certain situations light cones can tilt. In some spacetime geometries, such as those that are generated by rigid systems of rotating masses, it might be possible for a series of tilted light cones to loop back on themselves. The *dotted curve* here represents a closed timelike curve (Credit: Francisco Lobo, adapted with permission)

CTC and return to the same place and the same time from which it began. The object could engage in time travel! The dotted curve in Fig. 5.5 illustrates a CTC.

Should we take CTCs seriously? Well, it turns out that it's difficult to ignore them. They seem to be a feature of general relativity.

Physicists have found several exact solutions to Einstein's theory of general relativity that involve rotating objects, and the spacetimes described by these solutions usually contain CTCs. For example, in 1924 the Hungarian mathematician Cornelius Lanczos found a simple exact solution that represented a rigid, rotating system of masses with cylindrical symmetry. The solution was rediscovered in 1937 by the Dutch mathematician Willem Jacob van Stockum, and a system of masses with this solution is now called a van Stockum dust. Neither of these authors seem to have noticed the presence of closed timelike curves in their solutions but, in 1974, Frank Tipler wrote a

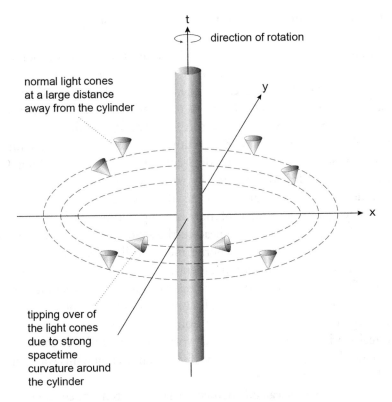

t

direction of rotation

normal light cones
at a large distance
away from the cylinder

y

x

tipping over of
the light cones
due to strong
spacetime
curvature around
the cylinder

Fig. 5.6 Close to a rotating cylinder, light cones tilt. Van Stockum and Tipler studied spacetimes in which rotation can give rise to closed timelike curves (Credit: Francisco Lobo, adapted with permission)

paper called "Rotating Cylinders and the Possibility of Global Causality Violation" in which he demonstrated that this type of solution contained CTCs. A Tipler cylinder was a time machine into the past; see Fig. 5.6. (A few years later Larry Niven wrote a short story about time-travel, the title for which he took directly from Tipler's paper! And Poul Anderson wrote about 'T-cylinders'—Tipler cylinders, in other words—in his novel *The Avatar* (1978).)

Tipler's initial mathematical analysis involved a cylinder that was infinitely long; clearly, since we can't construct such an object one could argue that the solution is of no practical or physical interest. And even if the effect works with a cylinder of finite length, Tipler himself admitted that no known substance possesses the characteristics needed to build the cylinder. We can't build a time machine in this way, at least not with our current level of technology and scientific understanding. So perhaps you might choose to ignore the Lanczos/

van Stockum solution as a mere mathematical curiosity, to be excluded on physical grounds.

You might choose to dismiss the Gödel solution on similar grounds. The Gödel solution is an exact solution to the equations of general relativity discovered in 1949 by the great mathematician and logician Kurt Gödel, and published in a volume of papers to honor Einstein's 70th birthday. Gödel's solution describes a strange sort of universe, one that is essentially a rotating collection of galaxies with everything held in balance by a vacuum that possesses energy. The Gödel solution contains CTCs—something Gödel pointed out to his friend Einstein. As with the van Stockum solution one could argue that, since our universe is not described by Gödel's mathematics, we can safely ignore the Gödel solution as being nothing to do with reality.

You might even take the same head-in-the-sand approach with the wormhole solution developed by Kip Thorne and his colleagues. This solution was mentioned in Chap. 2; what wasn't mentioned is that traversable wormholes contain CTCs and thus can act as time machines as well as machines for space travel. As was made clear in Chap. 2, however, the construction of a wormhole requires negative energy. Since it is far from clear that negative energy can exist in the real world we could argue that wormhole-based CTCs are not something we need concern ourselves with. If you want to argue this way, fine. But what about the Kerr solution?

In 1963, the New Zealand mathematician Roy Kerr discovered an exact solution to Einstein's equations that described a real-world situation: the geometry of spacetime around an uncharged, spherically symmetric, rotating black hole. Sure enough, the presence of rotation means that certain regions of space around the black hole can contain closed timelike curves. The Kerr solution isn't something you can easily ignore: you have to work hard to find reasons of principle for why travel into the past is forbidden.

All these situations—the van Stockum dust, the Tipler cylinder, the Gödel universe, the Kerr black hole, the Thorne wormhole—suggest that CTCs can exist. It's as if general relativity is trying to tell us something. Indeed, the English mathematician William Bonnor has argued that closed timelike curves can occur in solutions to Einstein's equations that describe phenomena that could occur in a *laboratory*—not just in the rarefied situations of astrophysics or cosmology but in everyday practical situations involving spinning copper balls! So might the development of a time machine be just around the corner? And if so, what are we to make of the paradoxes that seem inherent in time travel?

Well, the first thing to say is that physicists are unlikely to develop a CTC-producing time machine any time soon. Even if they *did* make a time machine based on CTCs, it wouldn't be like *The Time Machine* of H.G. Wells; you couldn't just dial a date in the past and turn up to watch England vs France at Agincourt. A time machine of this type can't be used to travel back in time before the creation of the machine itself; the best you can do is to travel back in time to the creation of the machine. (If your worldline were following a CTC you would think of your life history as being sensibly ordered from 'this' moment to the 'next' moment—the same sort of ordering you are experiencing now—but eventually the 'next' moment would get you back to whatever moment you decided was the starting point.) It's time travel, but of a very particular type.

Even if CTCs don't allow for the sort of time travel we saw in *The Time Machine* or "City on the Edge of Forever" they still allow for mind-bending possibilities. In principle, you could go back in time and tell your younger self the outcome of sporting events that are in your past but your younger self's future. Any sort of portal to the past leads to weirdness.

The weirdness of CTCs so troubled Stephen Hawking (see Fig. 5.7) that it led him to propose the Chronology Protection Conjecture. In a 1992 paper he wrote: 'It seems that there is a Chronology Protection Agency which prevents the appearance of closed timelike curves and so makes the universe safe for historians.' In essence, Hawking's conjecture states there will always be something that stops a time travel machine from working. The way Hawking phrased it, one is reminded of Asimov's *The End of Eternity* or Poul Anderson's *Time Patrol* series. Perhaps Hawking was making reference to some of the famous stories of SF. But the 'Agency' Hawking refers to is simply physics: he was suggesting that the laws of physics are such that closed timelike curves can never be found. His conjecture might be true. Perhaps time travel is impossible at the level of fundamental physics, and all those SF stories were based on nothing more than fantasy. But at present we don't know enough about physics to understand the status of the conjecture.

Perhaps a more interesting response to the existence of CTCs, at least for us SF fans, came from the Russian physicist Igor Novikov in the mid-1980s. Novikov accepted that CTCs seem to be a feature of general relativity and so, in order to reconcile CTCs with the notion of causality, he argued that any event that could give rise to a changed past has a probability of zero of occurring. CTCs can exist; time travel paradoxes can't. This is the so-called Novikov self-consistency conjecture. In essence Novikov's conjecture states that, even if time travel occurs, only self-consistent series of events are possible. Paradox is avoided. Novikov and his colleagues illustrate this idea with reference to a time machine constructed from a traversable wormhole, that hypothetical shortcut through spacetime popularized by Kip Thorne.

Fig. 5.7 Search online and you can find a cordial invitation to an event, to be held in the past, by Stephen Hawking. This 'Reception for Time Travellers' takes place at noon on 28 June 2009 at Gonville and Caius College, University of Cambridge. In a cunning twist, Hawking only publicized this event after it had taken place. As the photograph shows, Hawking is the only person who turned up. A paradox would occur if someone from the future were to invent a time machine and travel back in time to accept the invitation. This experiment does not prove the non-existence of time travel, however. For example, it might be possible to travel back in time—but not to an epoch before the time machine itself was created. It that possibility turns out to be the case then Hawking's experiment suggests that no one had invented a time machine as of June 2009 (Credit: LWP Kommunikáció)

Suppose an object travels through a wormhole and arrives in the past. There is the potential for paradox here, and the physicist Joe Polchinski crystallized the problem: imagine a billiard ball traveling into a wormhole along a path such that it exits the wormhole in the past and strikes its earlier self, deflecting its initial trajectory so that it never enters the wormhole in the first place. This is the Polchinski paradox—it's the grandfather paradox dressed up in slightly more formal language. According to the self-consistency conjecture, the probability of this series of events happening must be zero. However, the billiard ball *can* follow paths that *don't* lead to paradox. For example, there might be a glancing collision that causes the billiard ball to enter the wormhole on a trajectory that ensures it emerges from the wormhole to deliver a glancing collision that causes. . . Just as in the Heinlein stories we looked at earlier, the loose ends are tied up and paradox is avoided. Or recall a variation of the Paul Nahin story I mentioned earlier: if you go back in time and accurately dictate Newton's *Principia* to him then no paradox occurs since Newton writes *Principia*, which enables you to go back and dictate it to him. It's all consistent. (Though in this case we would be left with the question: who performed the creative work of developing the ideas in *Principia*?)

Time-Travel for Computing?

Self-consistent time loops would be great for computing purposes. Suppose you want to solve a problem that traditional computers can't touch—let's say decomposing a really large number into its prime factors. This is an important problem because many standard encryption methods rely on the fact that we don't have an efficient algorithm that can allow computers to perform this task. With a time loop the algorithm is straightforward: get the answer from the future; check whether it's correct (the checking process is easy); if it's correct then send it back in time unaltered and if it's incorrect then alter it before sending it back in time. If the entire space of answers can be explored then the only self-consistent time loop is the one in which the correct answer is determined and sent back unchanged.

The Hawking chronology protection conjecture and the Novikov self-consistency conjecture aren't the only ways of addressing the problem of CTCs. Perhaps quantum mechanics can save us from the grandfather paradox. You see, although general relativity appears to allow the existence of CTCs we know that general relativity can't be the final word in physics. Deep down, as we have already mentioned, the universe appears to be quantum in nature and so eventually physicists will have to reconcile those two great theories of general relativity and quantum mechanics. Nobody yet knows what a successful quantum theory of gravity looks like, but in 1991 the Oxford physicist David Deutsch suggested that quantum ideas might offer another way out of the paradoxes thrown up by closed timelike curves.

The basic idea is based on the notion that the quantum world is described by probabilities. For example, until you make a measurement on it a quantum particle doesn't possess definite quantum properties—rather, there are various probabilities of that particle possessing those properties. Deutsch applied the idea of self-consistency in closed timelike curves to quantum particles, and he found that by using this approach he could avoid temporal paradoxes. Scaling it up to grand patricide, the suggestion is that if you go back in time to kill your grandfather then you only do so with a probability of a half; he is dead with a probability of a half and you are not born with a probability of a half. At the quantum level this appears to be enough to evade paradox, and elements of Deutsch's idea have even been tested in the laboratory (although with a quantum system that is equivalent to a CTC, rather than with a CTC itself; scientists haven't yet managed to construct a time machine).[8] Deutsch's approach has a whiff of the many-worlds concept about it: it brings to mind those SF stories with mutable timelines, in which the protagonist steps back in time and enters a different time stream. Even if you don't care for the many worlds interpretation, there is at least one other suggestion on offer of how quantum mechanics should be applied in the presence of closed timelike curves: this is still an area for research. However, it seems that a definitive treatment, upon which everyone can agree, will require a quantum theory of gravity. Perhaps we need a unified theory in order to understand the nature of time—and of whether time travel is indeed possible.

5.3   Conclusion

The existence of closed timelike curves in general relativity hints at the possibility of time travel. That makes me happy, because it means we can continue to enjoy time travel stories without being disturbed by questions of scientific accuracy: since we don't know for sure that time travel is impossible we are all free to accept the concept for the sake of the story. Nevertheless, despite a lifetime of reading SF stories involving time travel, I don't believe the universe works in a way that allows time travel to occur. The universe is weird enough as it is without adding time loops to the mix.

What is clear, however, is that the concept of 'time' remains mysterious. Despite its seeming simplicity, we just don't yet fully understand the role that time plays in physics. So even if we don't believe in the possibility of time travel, pondering the possibilities for time travel is an important task. It could help us learn more about the foundations of our universe. Physicists and philosophers, then, should continue to investigate the role that time plays in gravity and quantum mechanics. But so too should SF writers.

Notes

1. As far as I'm aware, the grandfather paradox first appears in SF in a short story by Nat Schachner, one of Asimov's favorite authors. However, Schachner's "Ancestral Voices" (1933) refers to an ancestor of the time traveller rather than to the grandfather specifically.
2. Heinlein wrote several stories in which time travel plays a major role. The two stories mentioned in the text, "By His Bootstraps" (1941) and "–All You Zombies–" (1959), are perhaps his most convincing explorations of the theme, but it's also worth reading his novel *The Door Into Summer* (1957).
3. The butterfly effect—the notion that a small event (such as the flapping of a butterfly's wings in England) can lead to a major effect (a hurricane in

America, say)—sounds as if it originated in Bradbury's story "A Sound of Thunder". In fact, the American mathematician Edward Lorenz (1917–2008) coined the term in 1969, following his work on atmospheric predictability.

4. Although I don't have space to mention all my favorite time travel stories, I have to mention Harlan Ellison's "Soldier from Tomorrow" (1957) if only for the fact that it inspired that terrific film *The Terminator* (1984). Then there's Finney's novel *Time and Again* (1970); Lupoff's story "12:01 P.M." (1973); and Benford's hard SF novel *Timescape* (1980). I discussed *Back to the Future* (1985) in Chap. 1. I could go on.

5. In his novel *The Technicolor Time Machine* (1967), Harry Harrison provided a more comic take on the idea explored by Sherred in "E for Effort".

6. The best treatment in SF of the tachyon concept, and one of the best fictional treatments of how science works, appears in the novel *Timescape* (1980) by Greg Benford. This isn't surprising, since Benford was a professor of physics. The novel won a number of major awards, including the Nebula award for that year.

7. For the tachyonic anti telephone see Benford, Book and Newcomb (1970).

8. The simulation of CTCs was developed by Ringbauer et al. (2014).

Bibliography

Non-fiction

Benford, G.A., Book, D.L., Newcomb, W.A.: The tachyonic anti telephone. Phys. Rev. D. **2**, 263–265 (1970)

Feinberg, G.: Possibility of faster-than-light particles. Phys. Rev. **159**, 1089–1105 (1967)

Friedman, J., Morris, M.S., Novikov, I.D., Echeverria, F., Klinkhammer, G., Thorne, K.S., Yurtsever, U.: Cauchy problem in spacetimes with closed timelike curves. Phys. Rev. D. **42**(6), 1915–1930 (1990)

Glashow, S. (with Bova B): Interactions: A Journey Through the Mind of a Particle Physicist and the Matter of this World. Warner, New York (1988)

Nahin, P.J.: Time Machines: Time Travel in Physics, Metaphysics, and Science Fiction. Springer, Berlin (1999)

Ringbauer, M., Broome, M.A., Myers, C.R., White, A.G., Ralph, T.C.: Experimental simulation of closed timelike curves. Nat. Commun. **5**, 4145 (2014)

Tipler, F.: Rotating cylinders and the possibility of global causality violation. Phys. Rev. D. **9**, 2203–2206 (1974)

Fiction

Abrashkin, R., Williams, J.: Danny Dunn, Time Traveler. McGraw-Hill, New York (1963)

Anderson, P.: Tao Zero. Doubleday, New York (1970)

Anderson, P.: The Avatar. Berkely, New York (1978)

Asimov, I.: The Endochronic Properties of Resublimated Thiotimoline. Astounding (March 1948)

Asimov, I.: The Immortal Bard. Universe (May 1954)

Asimov, I.: The End of Eternity. Doubleday, New York (1955)

Asimov, I.: The Dead Past. Astounding (April 1956)

Benford, G.: Timescape. Simon and Schuster, New York (1980)

Blish, J.: Beep. Galaxy (February 1954)

Bradbury, R.: A Sound of Thunder. Colliers (June 1952)

Ellison, H.: Soldier from Tomorrow. Fantastic Universe (October 1957)

Finney, J.: Time and Again. Simon and Schuster, New York (1970)

Harrison, H.: The Technicolor Time Machine. Doubleday, New York (1967)

Heinlein, R.A.: Life-line. Astounding (August 1939)

Heinlein, R.A.: By His Bootstraps. Astounding (October 1941)

Heinlein, R.A.: The Door into Summer. Doubleday, New York (1957)

Heinlein, R.A.: All You Zombies. Fantasy and Science Fiction (March 1959)

Kornbluth, C.M.: The Little Black Bag. Astounding (July 1950)

L'Engle, M.: A Wrinkle in Time. Farrar, Straus & Giroux, New York (1963)

Lupoff, R.A.: 12:01 P.M. Fantasy and Science Fiction (December 1973)

Nahin, P.J.: Newton's Gift. Omni. 1(4), 50–54 (1979)

Niven, L.: Rotating Cylinders and the Possibility of Global Causality Violation. Analog (August 1977)

O'Donnell, L.: Vintage Season. Astounding (September 1946)

Padgett, L.: Mimsy were the Borogoves. Astounding (February 1943)

Pearce, P.: Tom's Midnight Garden. Oxford University Press, Oxford (1958)

Schachner, N.: Ancestral Voices. Astounding (December 1933)

Shaw, B.: Light of Other Days. Analog (August 1966)

Shaw, B.: Other Days, Other Eyes. Gollancz, London (1972)

Sherred, T.L.: E for Effort. Astounding Science Fiction (May 1947)

Twain, M.: A Connecticut Yankee in King Arthur's Court. Harper, New York (1889)

Wells, H.G.: The Chronic Argonauts. Royal College of Science (1888)

Wells, H.G.: The Time Machine. Heinemann, London (1895)

Visual Media

Back to the Future: Directed by Robert Zemeckis. [Film] USA, Universal Pictures (1985)

Dr Who: Created by Sydney Newman, CE Webber and Donald Wilson. [Television series] UK, BBC (1963–present)

Star Trek: The City on the Edge of Forever. Written by Harlan Ellison. [Television episode] USA, Desilu Productions (1967)

The Terminator: Directed by James Cameron. [Film] USA, Pacific Western (1984)

The Time Machine: Directed by George Pal. [Film] USA, MGM (1960)

The Time Tunnel: Created by Irwin Allen. [Television series] USA, 20th Century Fox Television (1966–1967)

6

The Nature of Reality

Reality is that which, when you stop believing in it, doesn't go away.
How to Build a Universe that Doesn't Fall Apart in Two Days
Philip K Dick

Philosophers have been struggling to comprehend the nature of reality for millennia. Consider Plato's "Allegory of the Cave".[1] Plato asked his readers to imagine prisoners bound and chained since infancy so that all they have ever seen is a cave wall directly in front of them. A fire burns constantly behind these unfortunate prisoners, and between fire and prisoners is a walkway with a low wall behind which the captors hide. The captors hold up puppets, and so the prisoners see the shadows cast by the puppets—but never the shadows cast by the captors themselves. The suggestion is that, for the prisoners, the shadows constitute reality. The prisoners can have no conception that there might be a fire, or a cave, or something beyond the cave: all they ever know is the flickering shadow show taking place in front of them. Indeed, if prisoners were released from those shackles, and forced to turn and face the fire, the shock for them would be immense. The light would hurt their eyes, and in any case the prisoners would have no context in which to make sense of this new vista. Plato argued they would turn back to face the cave wall and the comfort of their reality. Could it be that our own perceptions are little more than shadows on a wall? Perhaps there is a higher, Platonic reality that is closed to our human senses?

The notion that reality might be somehow different to our everyday understanding, that it might in some way be distinct from the world whose nature nearly all of us agree upon, holds a natural fascination for authors. After

© Springer International Publishing AG 2017   **151**
S. Webb, *All the Wonder that Would Be*, Science and Fiction,
DOI 10.1007/978-3-319-51759-9_6

all, fiction writers are in the business of creating imaginary worlds—universes that must be real for the characters that inhabit them in order for them to appear real to a reader. It is when readers 'suspends their disbelief' that a story comes alive. And when a story *does* come alive the effect is magical: the story can seem more real to us than the world we inhabit. I know some places on Trantor, capital planet of the Galactic Empire in Asimov's *Foundation Trilogy*, better than I know the town where I live.

Some works of fiction tackle the question of reality head-on, and when they are successful they have a beautifully haunting quality. They force their characters, and therefore their readers, to reflect upon what we can know about the world and about our selves. Such stories appear in many genres. For example, William Hjortsberg's novel *Falling Angel* (1978) is a horror story. The novel's protagonist, Harry Angel, is a private investigator hired by a mysterious figure called Louis Cyphre. Angel's commission is to locate Johnny Favorite, a popular singer of whom nothing has been heard since he was seriously wounded in a Luftwaffe attack during World War II, 16 years earlier. Angel is a detective of the hardboiled variety, but the task of tracking down Favorite brings him into contact with a variety of occultists—and Angel is shocked to learn that his past is not quite as he remembers it and Cyphre is not what he appears. Angel's reality is not what he thought it was.

Mainstream fiction provides more examples of the idea with, to my mind, the best instance being Nabokov's masterful *Pale Fire* (1962)—a novel that's notable for how cleverly it plays with readers' expectations. The novel takes an unusual form: a 999-line poem ("Pale Fire"), by the poet John Shade, is subject to a Foreword, Commentary, and Index by Shade's self-appointed editor Charles Kinbote. As we read the Commentary, and flick between poem and cross-references, the novel's plot is revealed. *Pale Fire* raises questions about memory and identity, and suggests that 'reality'—whatever that might be—is a subtle concept.

Pale Fire and similar mainstream works are examples of metafiction: fiction about fiction. Genre tales such as *Falling Angel* aim to surprise the reader with a twist. But SF has been much more thorough than mainstream fiction or other genre fiction in exploring the big question: what is reality *really* like?[2]

6.1   Science Fiction: An Escape to Reality?

By its very nature, SF provides an ideal vehicle for exploring ontological questions and many Golden Age stories were set up to examine three questions in particular. Is paranoia a reasonable reaction to the situation we find ourselves in? Are we living in a virtual/simulated reality? Is our universe just one of many parallel universes? These questions might seem more outlandish than those addressed by mainstream authors but, as we shall see later in the chapter, science is taking certain aspects of these questions quite seriously.

6.1.1 It's Not Paranoia If They Really Are Out to Get You

Fritz Leiber Jr's tale "You're All Alone" (1943) is one of the best examples of how science fiction authors have examined the big questions. The story has a complicated past, and there exist several versions of it. Leiber wrote "You're All Alone" for the magazine *Unknown*, but the magazine folded before the story could be published and Leiber put the work aside during the War. He eventually expanded the tale into novel form but couldn't sell it until 1953 when a soft porn publisher bought it, added a few superfluous sex scenes, and published it as *The Sinful Ones*. In 1972 the reputable publisher Ace put out a version that was closer to the original story and entitled *You're All Alone*. And in 1980 Pocket Books republished the 1953 version of the novel under the same title, *The Sinful Ones*, but without the added sex scenes. Whatever version you read, the core story is the same. The protagonist, Mackay, works at General Employment where he interviews applicants for employment and asks questions of them from a script. For Mackay, as for most of us, the days merge into each other with yesterday being much the same as today. One day, however, the routine is interrupted: a woman enters his office, asks him 'Don't you know what you are?', before scribbling a note for him and leaving. Then the next applicant, a man, sits down in front of Mackay—and everything is different. The man answers questions that Mackay *would* have asked if the incident with the woman had not occurred; the man smokes an imaginary cigarette that Mackay *should* have offered him, but forgot to; he responds, machine-like, to statements that Mackay hasn't made. Mackay shakes the applicant, but the man just mouths nonsense. Mackay asks for help from his colleagues, but no one responds. The world seems to be running to its own script, a script from which the protagonist has vanished.

Mackay's subsequent exploration of this strange underlying reality are something of an anticlimax, and the story betrays its pulp origins, but the key questions it raises are those that fascinate philosophers. I know that *I* possess an inner life; but do *other* people also have an inner life? Or are they, behind their faces, simply manifestations of complex behavior patterns? If someone you love were replaced by a soulless zombie that looked like your loved one, moved like your loved one, spoke and acted like your loved one—would you be able to tell?

Leiber was certainly not the first SF writer to develop this 'Things Are Not As They Seem' theme. In 1931, for example, Edmond Hamilton published "The Earth-Owners" in the magazine *Weird Tales*. Hamilton was a fan of Charles Fort, an American journalist who liked to collect stories that science could not explain (or who, in the words of H.G. Wells, was 'one of the most damnable bores who ever cut scraps from out-of-the-way newspapers').[3] Fort, rather worryingly, once wrote: 'I think we're property' and Hamilton took this idea for his story: Earth is controlled by alien beings. People go about their business as normal, but our fate lies with those aliens not with us. The British author Eric Frank Russell was another Fortean writer. The conceit of his story "Sinister Barrier" (1939), which appeared in *Unknown*, was similar to that of "The Earth-Owners": humans are the property of more highly evolved beings. Russell's novel *Dreadful Sanctuary* (1948), in which it's supposed that Earth might be an insane asylum for the rest of the solar system, was another tale inspired by Fort.

John Campbell, the editor of *Unknown* and *Astounding*, had strong Fortean tendencies so it's not surprising that Russell was one of Campbell's favorite authors. However, I was disappointed to learn that another of Campbell's favorites—the great Robert Heinlein—was also a Fortean. Indeed, Heinlein was apparently a member of the International Fortean Organization from the time of its incorporation in 1965 to his death in 1988 (a fact I find strange, given the hard-headed common sense possessed by many of Heinlein's characters). Regardless of one's opinion of Fort, some of Heinlein's best early stories are clearly Fortean in tone. His story "They" (1941b) is about a man who believes he is one of the few real entities on Earth. The man isn't a complete solipsist—he accepts there are some other real entities—but those other entities that exist have conspired to create the universe in order to fool him. Not surprisingly the story's action takes place in a mental institution, with the protagonist trying to convince his (presumably unreal) psychiatrist of the underlying truth of his conviction. In this story, which pre-dates even Leiber's "You're All Alone", Heinlein raises the same basic point: the universe could be there in order to fool us—and if it is, how could we know? How can we know what 'reality' really is? Another of Heinlein's Fortean tales is the novella "The Unpleasant Profession of Jonathan Hoag" (1942). The man named in the title is a lover of fine dining but one evening at a dinner party, when asked what he does for a living, he realizes he has no recollection of his daytime activities. After dinner, when he washes his hands, he notices a reddish/brownish substance (dried blood?) under his fingernails. The next day he contacts a husband-and-wife team of private investigators and asks them to follow him and find out what he does during the day. From this set-up the story could have turned into an intriguing detective-noir tale, but instead Heinlein takes it in a strange direction: as in "They", we are confronted with questions surrounding the nature of reality: what is this world that we inhabit? How was it created? What, if anything, does it all mean?[4]

Heinlein wrote other stories that tapped into the feeling of paranoia that seems to have permeated 1950s America, a feeling that also found expression in films of the time such as *Invaders from Mars* (1953) and *Invasion of the Body Snatchers* (1956). Even the arch-rationalist Asimov wrote an occasional story in this vein. In "Breeds There a Man?" (1951), for example, a physicist becomes convinced that humanity is an experiment in genetics conducted by aliens. But if there is one author who epitomizes the science-fictional examination of ontological questions it's Philip K. Dick. In their writings and outlook Heinlein and Dick seem in many ways to have been polar opposites. But the reactionary Heinlein took an interest in the countercultural Dick. In the introduction to *The Golden Man* (1980), a collection of his short stories, Dick wrote: 'I consider Heinlein my spiritual father, even though our political ideologies are totally at variance. Several years ago, when I was ill, Heinlein offered his help, anything he could do, and we had never met; he would phone me to cheer me up and see how I was doing. He wanted to buy me an electric typewriter, God bless him—one of the few true gentlemen in this world. I don't agree with any ideas he puts forth in his writing, but that is neither here nor there. One time when I owed the IRS a lot of money and couldn't raise it, Heinlein loaned the money to me. I think a great deal of him and his wife; I dedicated a book to them in appreciation. Robert Heinlein is a fine-looking man, very impressive and very militarism stance; you can tell he has a military background, even to the haircut. He knows I'm a flipped-out freak and still he helped me and my wife when we were in trouble. That is the best in humanity, there; that is who and what I love.' In the afterword to that collection, Dick wrote something that captures the essence of his fiction: 'SF is a field of rebellion: against accepted ideas, institutions, against all that is. In my writing I even question the universe; I wonder out loud if it is real, and I wonder out loud if all of us are real.'

Much of Dick's output questions in some way the nature of reality. An incomplete list of his work in this vein would include "The Trouble with

Bubbles" (1953), *Eye in the Sky* (1957), *The Man in the High Castle* (1962), *The Three Stigmata of Palmer Eldritch* (1965), *Ubik* (1969a), "The Electric Ant" (1969b), *A Maze of Death* (1970) and *Valis* (1981). Another example is the novel *Time Out of Joint* (1959), which I'll examine here in more detail because not only is it fairly representative of Dick's concerns but it was re-issued in 2003 under Gollancz's *SF Masterworks* imprint and is the novel by Dick that I've read most recently.

The title, *Time Out of Joint*, refers to Hamlet's words to Horatio just after his meeting with his father's ghost—an event that changes the way Hamlet perceives the universe. Various events in the novel change the way its protagonist, Ragle Gumm, views his universe. Gumm starts out believing he lives in a quiet American town. It's 1959. Gumm earns his livelihood in a strange way: his local newspaper runs a competition called 'Where will the little green man be next?', which Gumm enters—and wins—every time. The unlikelihood of someone having a 100% winning streak in a game of chance creates a feeling of mild disquiet in the reader; we soon come across other oddities that add to the disquiet—Marilyn Monroe is unknown, for example, and radios are a scarcity. Gumm's reality is *almost* the 1959 we, the readers, know about. But not quite. Gumm's mental picture of reality starts to fracture when he finds a hot dog stand in the park is missing; in its place are the words 'Hot Dog Stand' written on a piece of paper. And then Gumm comes across bits of the real 1959: pages from a telephone book with numbers listed from towns that he can't reach; a magazine article about Marilyn; radios hidden in a neighbor's house. People Gumm has never met know his name. Eventually, Gumm learns the truth—about the newspaper competition he always wins, about the strangers who seem to know him, about his home town. He finally understands the reality beyond his everyday perception. I won't give the ending away here, except to say I found it anticlimactic in the way that so much of 'paranoid' SF is anticlimactic: the more disturbing and creepy the set-up (and Dick's set up *is* disquieting), the greater the sense of disappointment when the denouement fails to convince. *Time Out of Joint* does, however, exemplify the key theme that preoccupied Dick. In a Dick novel the ground can literally shift under a protagonist's feet. That change in perception can happen because the protagonist is being manipulated (as in *Time Out of Joint*), or because he's living in another person's dream, or because a drug-induced state of consciousness makes more sense to him than mundane reality, or because he shifts to a completely different universe. . . . For Dick, there isn't an objective reality that we can point to 'out there'. Reality is personal.

Dick's stories have been the direct or indirect inspiration for a number of movies. Thanks to Dick, the notion that one's reality might be a construction is now a broad theme not just in SF but in the mainstream too.

6.1.2 Virtual Realities

SF writers have long explored the idea of simulated reality. After all, each story they write is a constructed reality for the story's characters. In a sense, writers are simulation creators. There can be a paranoid aspect to this idea, as discussed above, but the concept is broader than that.

This idea of writers as simulation creators forms the central thrust of L. Ron Hubbard's "Typewriter in the Sky" (1940), in which the protagonist somehow gets trapped inside the historical novel his friend is writing. When the hero gets back to his home in New York, can he be sure that he isn't a character in some other writer's story? Perhaps a story such as this one is better categorized as fantasy than SF, but there are many SF stories dealing with constructed realities. It becomes SF when the protagonist's reality is generated by technology. For example, in "The City of the Living Dead" (1930) Laurence Manning and Fletcher Pratt imagined people being wired up to dream machines—the 'living dead' had their all experiences, all their sensory inputs, supplied by machine. We still see the influence of the story in all those 'brain-in-a-vat' cartoons (see Fig. 6.1 for an example). In *Brave New World* (1932), Aldous Huxley describes an entertainment form called the 'feelies'— an extension of cinema (or the 'talkies', as the medium was once called) that engages the sense of touch and immerses the viewer to produce the illusion of reality. In "Pygmalian's Spectacles" (1935), the short-lived but influential pulp writer Stanley G. Weinbaum described the invention of goggles that provided 'a movie that gives one sight and sound [. . .] taste, smell, and touch. [. . .] You are in the story, you speak to the shadows (characters) and they reply, and instead of being on a screen the story is all about you, and you are in it.' And Arthur Clarke's wonderful novel *The City and the Stars* (1956), a rewrite of his story "Against the Fall of Night" (1948), imagines the inhabitants of the great city of Diaspar whiling away the eons in so-called sagas: 'Of all the thousands of forms of recreation in the city, these were the most popular. When you entered into a saga, you were not merely a passive observer [. . .] You were an active participant and possessed—or seemed to possess—free will. [. . .] as long as the dream lasted there was no way in which it could be distinguished from reality.'

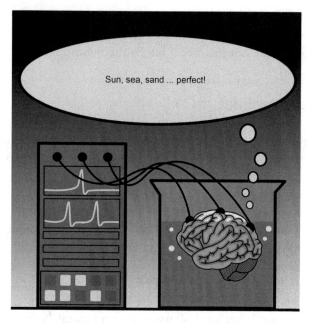

Fig. 6.1 A brain in a jar (Credit: Adapted from the public domain)

As the influence of computers began to permeate all aspects of human activity, SF authors spotted an opportunity: computers were being used to run simulations (of weather, economic systems, traffic...) and naturally the question arose as to how people would react to such simulations. Could computers provide the simulated reality that Manning and Pratt, Huxley, Clarke, and numerous others had imagined? For example, in his short story "Ender's Game" (1977), Orson Scott Card imagines a future world fighting against an alien enemy. Children who enter Battle School are trained to become military leaders, and the advanced part of that training requires the use of a space battle simulator. In the simulator, Ender Wiggins—the student who has shown the greatest aptitude for military strategy and leadership—fights mock battles against a computer-controlled enemy. But perhaps Ender's battles weren't just games; perhaps they were real battles against the real enemy. When Ender destroyed a spaceship in the simulator, was he causing a few pixels to light up (as I did in the arcade game *Defender*, to which I was addicted in the early 1980s) or was he killing living beings? How could he tell the difference? Wouldn't an interstellar war be mediated by computer, with the difference between simulation and reality being blurred for the protagonists? Indeed, isn't that happening in wars today, with military personnel releasing armored drones depending on what they see on their computer screens?

The stories mentioned above—"Pygmalian's Spectacles", *Brave New World*, "Ender's Game"—can all be considered as exploring rudimentary types of 'virtual reality'. The term was popularized in the early 1980s by Jaron Lanier through his company VPL, which sold the Data Glove for virtual reality systems. As we have seen, fictional investigations of virtual reality have a long history in SF, but it was the inexorable rise in computing power—and the dawning realization that this was a technology that scaled well—which made authors in the 1980s latch on to virtual reality as a serious concept to explore. The Australian writer Damien Broderick was the first to use the term virtual reality in science fiction, in his novel *The Judas Mandala* (1982). Others soon followed, and numerous stories explored what it might mean if computer-based simulations were so lifelike that humans would be unable to distinguish the virtual from the real. It was William Gibson, however, who made virtual reality take center stage.

Gibson's novel *Neuromancer* (1984), the first to win the three major SF awards—Hugo, Nebula, and Philip K. Dick—was a groundbreaking work that popularized the nascent cyberpunk subgenre. The novel's opening line— 'The sky above the port was the color of television, tuned to a dead channel.'—is surely one of the most arresting in all SF, and it sets the tone for the rest of the book: the emphasis is on style, on visual appearance, on atmosphere. The book's visual texture is similar to Ridley Scott's *Blade Runner* (although Gibson developed his style independently: he saw Scott's film after he had already completed a third of his book). The connection with *Blade Runner* is deeper than surface style, however. The film is based on the Dick novel *Do Androids Dream of Electric Sheep?* (1968). The familiar Dickian themes of paranoia, of questioning the nature of reality, and of what it means to be human, are explored in *Androids*—as they are in *Neuromancer*, too.

In *Neuromancer* the employer of Henry Case, a washed-up, drug-addicted computer hacker, punishes him for stealing from the company: a poison is delivered, damaging Case's central nervous system and thereby denying him access to the 'Matrix'—a virtual reality constructed in a globally networked cyberspace. (Gibson was writing years before the worldwide web was a gleam in Berners-Lee's eyes. Is it too much to suggest that Gibson's popularizing of the word 'cyberspace' brought the concept into being?) A mysterious individual has technology that not only repairs Case's nervous system but also grafts tissue to his organs, making it impossible for him to metabolize drugs: now Case is drug-free when he accesses the Matrix. Case must pay for this 'treatment', of course: he is told to steal a memory module that carries the consciousness of one of the greatest cyber-hackers. If Case fails, the poison that bars him from the Matrix will be re-activated. The novel follows Case as he first carries out the heist and then as he investigates the individual who hired him. Case encounters disembodied human consciousnesses that exist in the Matrix, an artificial intelligence (AI) that can copy human minds and run them in RAM, and AIs that fuse into superconsciousness. The narrative voice of *Neuromancer* is similar to that of the hard-boiled, noir detective genre; it's a voice that makes the story seem so vivid, so real. But for any individual living in the world described in *Neuromancer*, how could they know what was real and what was construct? How could they trust their memory in a world where consciousness can be uploaded to computer memory?

Neuromancer influenced countless stories and movies. However, given that *Neuromancer* appeared in 1984, the stories it inspired appeared after my self-imposed deadline. Some examples of post-1985 virtual reality stories are given in the notes to Chap. 12.

6.1.3 Parallel Realities

As we saw in Chap. 5, it's a staple of time travel tales that a protagonist goes back in time and causes the timeline to change. However, it's not too much of a stretch to dispense with the whole time-travel apparatus and simply try to imagine what would have happened if some key event in the past had turned out differently: Hitler was victorious, Napoleon avoided capture and fled to America, Jesus settled for life as a carpenter. Fiction writers (and professional historians) delight in carrying out these 'what if' thought experiments. Some stories focus on the changes wrought by just a single altered event, and in many cases they make no reference at all to the possibility of other realities (including our own). Some science fiction stories, though, examine the very concept of alternate timelines. They posit that other timelines or other universes exist, parallel to our own and just as real. The stories discussed earlier in this section provided examples of *simulated* realities—false realities, created in computers or the environment or inside a person's head, with the intention to deceive, but science fiction is also concerned about the possibility of *alternate* realities. Perhaps our reality coexists with many others.

One of the earliest such stories was Murray Leinster's "Sidewise in Time" (1934). In this tale some cosmic convulsion causes parts of Earth to swap with their equivalents in different timelines. So Missouri sees the arrival of a Roman legion, from a timeline in which the Roman empire has not fallen; Massachu-setts suffers Viking raids, from a timeline in which the Vikings colonized

America; citizens in Kentucky are confused when they enter parts of the state flying the Confederate flag, from a timeline in which the South won the Civil War. The story is typical of the pulp magazines: not particularly well written, concerned mainly with events in America, and not fully developed. (After more than a thousand years had passed, would Romans still have legions and Vikings still have longships? Wouldn't technological progress have taken place in those timelines too?) Nevertheless, the story was influential. Asimov once wrote that Leinster's story 'always made me conscious of the 'ifs' in history, and this showed up not only in my science fiction . . . but in my serious books on history as well'.[5]

A few years after Leinster's story, L. Sprague de Camp developed his own take on the idea of parallel timelines. In "The Wheels of If" (1940), a man finds that each time he wakes up he is in a different reality, a world in which history differs from the world in which he grew up. In order to return he must unite his various incarnations with their correct realities but, although he succeeds, he figures that life in one of those different realities will be more fulfilling than life in his original timeline. de Camp's story, along with his earlier alternate history novel *Lest Darkness Fall* (1939), seems to have kick-started the alternate history subgenre. It's a popular category: Robert Shrunk, one of the founders of the annual *Sidewise* award for alternate history fiction and moderator of the *Uchronia* list of alternative history, maintains a database of over 3200 novels, stories, and essays about the 'what ifs' of history.[6] More items are added every year, and there simply isn't space here to mention all those wonderful old stories—such as Piper's "He Walked Around the Horses" (1948) and his subsequent *Paratime* series, Ward Moore's *Bring the Jubilee* (1953), and Leiber's *The Big Time* (1958).

In most alternate history stories the accent is on the history. But if other realities occur, if parallel timelines or universes truly exist, then a host of philosophical questions arise—and many stories try to understand what parallel universes might mean. In "The Garden of Forking Paths" (1941), for example, Jorge Luis Borges likens the idea of branching universes to an infinite labyrinth. Borges describes how an author tries to write a novel in which all possible outcomes of an event occur: each branch leads to further branches, further possibilities. Or consider "All the Myriad Ways" (1968) by Larry Niven. A detective investigates a spate of suicides and murders that have happened since the discovery of travel to parallel universes, and he comes to a depressing conclusion. If an infinity of universes exists then freedom of choice becomes in some sense meaningless: there will always be another universe where you make a different choice. So, in this universe, why not choose to kill yourself or kill someone else? Or for a quite different analysis of the significance of time, the phenomenon of consciousness, and the underlying nature of reality, consider Fred Hoyle's *October the First is Too Late* (1966). As mentioned in an earlier chapter, Hoyle was one of the most creative cosmologists of the twentieth century; this novel game him the opportunity to examine some of the mysteries of quantum physics and general relativity, and how they relate to our human experience. One of the characters in Hoyle's novel argues that each moment of time can be thought of as a pre-existing pigeon hole. The pigeon hole you are examining right now, the pigeon hole on which the spotlight of your consciousness is shining, has a special name: the present. Normally we move from pigeon hole to pigeon hole in linear fashion, and we experience time as passing, but perhaps the spotlight of consciousness can move between pigeon holes following some other rule—instead of moving from 1 to 2 to 3 and so on up to 100, perhaps it is possible to move from 1 to 100 to 2 to 99 to 3 and so on. Perhaps there are completely different sets of pigeon holes—different histories—that consciousness can examine. Or consider Asimov's *The Gods Themselves* (1972), which posits the discovery of a universe parallel to ours but one in which the laws of physics are different. The transfer of matter between the two universes yields a seemingly limitless and free source of energy—which is excellent news, until physicists learn that the process is changing the strength of the strong nuclear force in both universes. Humanity faces extinction because the Sun will heat and explode; intelligent beings in the parallel universe face catastrophe from cooling. The discovery of a third, and presumably multiple, parallel universes means that catastrophe can be averted.

6.1.4 Summary

Could there be more to reality than our everyday experience implies? It's difficult, and perhaps impossible, to prove that we *aren't* the property of extraterrestrial beings, or patterns in a fantastically complex computer program, or inhabiting just one of an infinite number of possible universes. I suspect most of us, though, take a pragmatic approach. Samuel Johnson once responded to Bishop Berkeley's 'proof' of the non-existence of matter by kicking a rock and saying 'I refute it *thus.*' Johnson therein committed the logical fallacy called argumentum ad lapidem—but I'm sure his refutation satisfies most of us. And at first glance we might think decades of scientific research have rendered obsolete such stories as "They", *Neuromancer*, "The Wheels of If", and all those others I mentioned. The situation is more interesting than that, however. The question—what is really real?—is a subject of continuing scientific interest.

6.2 Science and Reality

Philosophers seem to have enjoyed debating the nature of reality. Their debate resembles a Twitterspat—except the subtleties have been addressed in massive tomes rather than in 140-character jots, and the debate has endured not days but centuries. I mentioned Plato in the introduction to this chapter and, since his contribution, pretty much all the great philosophers have put forward ideas about reality—but for the most part without any knowledge of how the human brain works or of the 'stuff' that constitutes the building blocks of the universe. Some philosophical proposals read as if they were the plot of a paranoid Phil Dick novel (or perhaps Dick got his inspiration from some of these proposals). For example, in 1921 Bertrand Russell put forward the 5-min hypothesis: suppose the universe, including the complete historical record and all human memory, popped into existence from nothing just 5 min ago. We would be unable to distinguish our 13-billion year old universe from a universe created 5 min ago by a God who wanted to trick us. On the internet, this idea often goes by the name of Last Thursdayism: in November 1992, in response to one of those predictions of apocalypse that religious zealots so love to make, someone posted on Usenet: 'As everyone knows, it was predicted that the world would end last Wednesday at 10:00 PST. Since there appears to be a world in existence now, the entire universe must have been recreated, complete with an apparent 'history', last Thursday. QED.' (Heretical splinter groups have of course appeared, with vicious doctrinal differences developing between those who follow Last Wednesdayism and those who argue for Last Fridayism.)

Some unfortunate people experience a subjective reality that is completely at odds with what the rest of us experience. (At least, it's at odds with what *I* experience. I can't know what goes on inside your head, gentle reader.) Consider, for example, those who suffer from Cotard's delusion. A patient with this particular mental illness can, in rare cases, deny the fact of their own existence: they believe they are dead. In some cases they can believe *everything* around them is dead.

The nature of reality can thus be discussed in philosophical terms, or with reference to religion and the nature of a creator being, or in terms of human psychology. It's a vast subject. Instead, I'll focus on the questions that physical scientists ask when they wonder what's really real.

6.2.1 Are We Living in a Simulation?

As mentioned earlier, the inexorable increase in computing power led SF authors in the 1980s to write stories in which characters inhabit cyberspace. People might well choose to decamp into cyberspace: after all, having your consciousness uploaded to a computer network helps guard against the natural shocks that flesh is heir to. On the other hand, characters might be unaware they are part of a computer simulation, that their consciousness is just a flow of binary digits generated by some programmer sitting outside the simulation. Eventually, philosophers caught up with the idea. In 2003 Nick Bostrom published his now-famous simulation argument.[7]

Bostrom argued that a technologically advanced 'posthuman' civilization would have access to tremendous amounts of computing power—enough to run so-called 'ancestor simulations'. Such a simulation would be so accurate that, to the ancestral life being simulated, it would be indistinguishable from reality. And it only needs a tiny fraction of posthuman civilizations to run ancestor simulations for the total number of simulated ancestors to vastly exceed the total number of real ancestors. From this, according to Bostrom, at least one of the following three propositions is almost certainly true. One: the fraction of human-level civilizations that reach the posthuman level of technology is close to zero. Two: the fraction of posthuman civilizations that choose to run ancestor simulations is close to zero. Three: the fraction of all people with our kind of experiences that are living in a simulation is very close to one. (The latter proposition is based simply on the numbers: after all, if there is one 'real' world and a million simulated worlds then it's much more likely that we find ourselves in a simulation.)

Since we have no idea which of these propositions is true, Bostrom argues it makes sense to apportion equal probabilities to the three of them. There is thus a serious suggestion that we are currently living in a simulation. That you—yes, *you*—are a constructed entity. Bostrum is a professor at Oxford, so although the simulation argument might on first hearing sound barmy you shouldn't be surprised to learn that the argument isn't easy to discount. It has generated a huge amount of philosophical literature, and in 2016 the annual Isaac Asimov Memorial Debate at the American Museum of Natural History saw several eminent physicists discuss the argument. (I thought the string theorist Lisa Randall, who is skeptical about the argument, put it best: "I am very interested in why so many people think it's an interesting question.")

People have levelled a number of criticisms at the simulation argument—or at least at the likelihood of the third proposition being correct. For a start, the computing power required to make a perfect simulation—one able to consistently predict the outcomes of the trillions of collisions that occur in particle physics experiments, model the atomic interactions in chemical reactions, correctly describe the chaotic behavior of complex systems—well, it's an unimaginably huge challenge. To give a flavor of the task, a trillion 10 Tb hard disks are required just to store the positional information of the atoms that make up one hard disk. And that doesn't begin to capture the interactions that take place in the material. Why would anyone bother? What would be the point? If you want to simulate a brain it's far easier just to build a brain out of neurons than it is to build a computer simulation of the neurons. Even if it were possible to design a perfect computer simulation, which is debatable, it doesn't follow that a civilization would bear the cost of doing so—let alone that such a thing would be done repeatedly. (One could argue that we have no idea what inconceivably powerful 'posthuman' civilizations would choose to do, and so we shouldn't say they would decline to simulate a universe complete with people. Fine. But equally we shouldn't immediately conclude they *would* simulate a universe. Post-human psychology is even murkier than human psychology.)

The simulation argument also, to my mind at least, has a problem with regression. If perfect simulations are possible then there might come a time when a posthuman civilization derived from our own civilization can run such a simulation. If we are already in a simulation, though, then there would be a simulation inside a simulation. Surely this chain goes both ways? In other words, perhaps the creatures who created the simulation in which we find ourselves are themselves only simulated creatures in someone else's simulation? And *they* might be a simulation too… (That does lead to a worrying thought: it only needs someone in the chain of simulations above us to pull the plug, and we'd find ourselves gone!)

Is it possible to *test* whether we are in a simulation? Possibly. Consider what happens when scientists want to explore the behavior of a theory by using computer models. They often find they don't have enough computing power to duplicate the full theory, so resource constraints lead them to 'cheat'. Instead of modeling the theory throughout a smooth spacetime they instead split spacetime into a grid and use the computer to calculate what happens when the theory is restricted to that grid. The grid is easier and computationally cheaper to work with than a simulation of the full theory. As the grid becomes finer, and approaches a smooth spacetime, results from the model should approach those from the underlying theory. Well, perhaps those putative beings who are simulating us and our universe face resource constraints of their own? Perhaps our universe is simulated on a grid? In 2014, three physicists argued[8] that if those making the simulation do face resource constraints then it might be possible to detect indications of an underlying model. For example, in a 'smooth' universe high-energy cosmic rays reach us equally from all directions because the universe has no preferred direction; if the universe were simulated on a grid, that symmetry would break.

It's an interesting idea, but can any such experiment really address the issue of falsifying the simulation hypothesis? Failure to find a signature of a grid wouldn't disprove the simulation argument. And if we *did* find a signature there'd be an almighty debate over whether we could interpret it in naturalistic terms. (From what I've seen of internet conspiracy theories there might also be a group of people who'd accept we live in a simulation but reject the idea that physics provided evidence: 'some incredibly advanced civilization just made the simulation this way to mess with our minds. . .')

The very success of science might hint that we do not inhabit a simulation. One of the purposes of a simulation (at least, it seems to be a purpose to our low-grade human minds) is to tweak parameters and see what happens. If engineers make a computer simulation of an airplane wing, for example, they are quite likely to want to manipulate the wing shape in order to see what happens to stresses within the structure or to airflow around it. There is no evidence of 'tweaks' having happened in the history of our universe—the values of physical constants haven't changed over time, the laws of physics haven't changed, and the observable history of the universe seems to contain no episodes of sudden, inexplicable adjustments. This observation proves nothing, but to my mind it sits more readily with the thought that we inhabit a natural, 'real' universe. Nevertheless, even if you believe it's unlikely that we live in a simulation, at least some respected thinkers believe it's an idea worth checking. Perhaps Phil Dick wasn't so paranoid after all.

6.2.2 Quantum Weirdness and Parallel Worlds

What's really real? When attempting to answer this question perhaps you prefer to discount 'crazy' notions—you know, the idea that aliens have built the world to trick us or that we are figments in the fevered imagination of someone else's brain-in-a-jar. Perhaps you are convinced there's a real world out there, external to yourself, containing matter and energy, and that the best way to understand reality is to search for what is fundamental. Perhaps if we can find out what the basic 'stuff' is that comprises the universe then we'll have a better understanding of reality. This is an approach adopted by particle physicists, and in recent decades they've had spectacular success. Theorists have developed the so-called Standard Model, which accurately describes the results of all experiments ever undertaken to explore the particle world. The Standard Model contains a dozen or so fundamental entities—to the best of our current knowledge, these constitute the basic 'stuff' of the universe—which interact in a limited number of well understood ways. The Standard Model gives us our best description of reality. Nevertheless, it's clear that the Standard Model must somehow be extended, if only to incorporate gravity (which currently it ignores). It's certainly not clear *how* the Standard Model needs to be enhanced, but most physicists would probably agree on one thing: any final theory will be quantum in nature.

Quantum ideas were developed in the early decades of the twentieth century in order to explain the results coming from simple experiments involving things such as measuring the radiation coming from a heated object. Many physicists of a certain generation—giants such as Planck and Einstein, people who helped hatch quantum theory in the first place—found it difficult to accept that the universe has a fundamentally quantum aspect to it. They longed to develop a classical theory that could describe the world on small distance scales. They tried hard, but failed. If you want to explain the emission of radiation coming from a piece of steel as it is heated—well, you need quantum ideas. Quantum theory explains the glimmer of gold and the transparency of glass, the hardness of solids and the ability of liquids to flow. An understanding of quantum ideas allows physicists to build semiconductors and all the cool tech to which we devote so much of our time and disposable income. It lets us measure the chemical composition of stars that are millions of trillions of miles away. The idea of the quantum—the notion that matter and energy aren't smooth, but instead come in discrete chunks—is over-whelmingly successful. The idea leads to so much theoretical understanding and so much practical benefit it's inconceivable it could be wrong. (Similar sounding statements were made at the end of the nineteenth century, and they turned out to be wrong. But I don't care. The universe is quantum not classical!)

There is, of course, a problem with quantum theory. As surely everyone knows, quantum theory is weird.[9]

The intuition we build up about how the world behaves is based on what happens with everyday objects, and generally that means with stuff ranging in size from a few centimeters to a few meters. Even the smallest objects with which we interact—grains of sand or grains sugar, say—are easily visible. But the universe behaves quite differently at small scales. Electrons and quarks exhibit phenomena that eggs and quilts, say, do *not* exhibit.

Consider the electron. We can build a 'gun' to create and fire electrons, and we can determine where any particular electron was produced: an electron is thus created as a particle—something that can't be split. We can also build a screen to capture electrons, and we can determine where any particular electron is trapped: an electron is thus captured as a particle—something that can't be split. But when the electron is moving, traveling between gun and screen, something strange happens. Between emission and capture things are vague. It's as if the electron somehow spreads out across the universe and tries out all available paths. The famous double-slit experiment places a barrier containing two holes or slits between the gun and the screen: the resultant pattern of electron arrivals at the screen demonstrates that the electrons must have passed through *both* holes—so in movement the electron behaves as a wave. The electron is a particle *and* it is a wave. Similarly, the photon is a wave *and* it is a particle. This peculiar behavior isn't confined to elementary particles. It's a behavior exhibited by anything, as long as it's sufficiently small.

In short, we lack an intuitive understanding of quantum mechanics because we never directly experience the world on distance scales where quantum effects play a role. An electron can be in different places at the same time, but no one has ever seen a person exhibit that behavior.

In 1926, Erwin Schrödinger introduced a mathematical object called a wavefunction to capture the peculiar behavior of quantum particles, and he developed a wave equation that permitted physicists to calculate the wavefunction in any particular situation. The wavefunction, which goes by the symbol Ψ (pronounced 'psi'), contains all the information needed to describe the behavior of a quantum particle. In principle, *everything*—electrons and atoms, you and me, the universe itself—is described by a wavefunction. Our ability to solve the wave equation in a particular system, and calculate the wavefunction, is the basis of the success of quantum mechanics I mentioned above—it has given us tablet computers, lasers, and all the rest. The wavefunction has many of the usual properties of a wave, such as an amplitude and a phase, but we struggle when we ask the obvious question: if a particle such as an electron is a wave what is it, precisely, that's waving?

We know what a wave is in the everyday world. We have all seen ocean waves and waves in a rope. We see waves when we look at brake lights on a busy motorway at night, as pulses of deceleration travel down a stream of traffic. Astronomers have found gravitational waves, ripples in the fabric of spacetime itself. We can even assign meaning to something abstract, such as a crime wave. But what is waving when an electron travels from A to B? It would be a strange sort of wave, propagating through an enormous number of abstract dimensions. And the wave's amplitude would not carry energy but rather, somehow, probability. What, then, is the nature of this thing called Ψ?

There are a number of different approaches to quantum mechanics. One approach, advocated by Niels Bohr, considers the wavefunction to be nothing more than a tool for calculation. In this approach a quantum object—an electron in a hydrogen atom, say—possesses a wavefunction that evolves in a well defined, deterministic fashion as specified by the wave equation. While the electron is in this state its properties are not decided: it is in multiple positions at the same time, it is spinning up *and* spinning down, and so on. When an observer makes a measurement on the system, however, the electron randomly acquires definite properties: it's located *there*, for example, and its spin is *down*. This acquiring of measurable properties is known as wavefunction collapse, and it takes place instantaneously. Somehow the system flips from being a probability of existence, stretched out across space, into a localized object that has measurable properties. What happens at wavefunction collapse, when just one of all the different possibilities becomes actualized? We simply don't know. Nevertheless, an approach based on a strange-sounding recipe gives perfect results every time: take the wave equation, calculate the amplitude of the wavefunction at each point in the situation of interest, square the answer, and the result is the probability of finding the system—electron, atom, whatever—in that particular state. In the case of the electron gun and the screen, the Schrödinger equation won't tell you where a given electron will end up. But it *will* tell you the probability of finding an electron at any particular point. The recipe works.

For Bohr, the success of the recipe was enough. He wrote that 'it is wrong to think that the task of physics is to find out how nature is—physics concerns what we can say about nature'. Rather than enquire too deeply into what the wavefunction *really* is, the approach many physicists take is to 'shut up and calculate'. The trouble with this approach from a philosophical point of view is that it is, at heart, anti-realist. We can easily end up with the conclusion that consciousness somehow brings reality into existence. For some of the greatest names in physics—Bohr and Heisenberg, Wheeler and Wigner—the weirdness of quantum theory made this conclusion inevitable. But others held on to their unshakeable belief that physics describes an objective reality, a universe external to us and existing independently of the observer. The Moon is there even if no one is looking; a tree falling in a forest disturbs the air even if no one is there to hear the sound. Bohr saw no mystery in the wavefunction because it was just a useful tool; he simply didn't need to worry about whether the wavefunction was 'real'. If you are a realist, however, and believe that quantum theory describes an objective reality independent of observers, then you are left with the question of how to interpret the wavefunction. Does the wavefunction constitute reality? If it does, do we need extra information to describe reality? Or is the wavefunction all there is?

At this point it's worth introducing some philosophical terminology. An 'ontic state' is a state of reality; an 'epistemic state' is a state of incomplete knowledge. (See the box for further information.) If you are a realist—if you believe quantum theory describes an objective reality—*and* you believe wavefunctions constitute reality then you subscribe to a Ψ-ontic approach to quantum theory. If you believe wavefunctions *don't* form a deep underlying reality then you subscribe to a Ψ-epistemic approach. (Bohr's anti-realist approach is also Ψ-epistemic, but do we really want to give up reality?)

Ontic and Epistemic States

Suppose you have a classical particle moving in one dimension: it could be a small bead moving on a thin string. At any instant, two pieces of information—the bead's position and its momentum—are all that's needed to give a complete specification of the particle's state. This is called an 'ontic state': a state of reality. (The word 'ontic' derives from a Greek word meaning 'being'.) If we have an ontic state then we know, with certainty, what the particle is doing at a given instant. A physicist, however, might try to measure the position and momentum of the bead and be forced to confess that the experimental equipment was unable to determine these properties with total precision. The best that could be managed might be a small range of positions and momenta that the particle was likely to possess. In this case the particle is in an 'epistemic state': a state of incomplete knowledge. (The word 'epistemic' derives from a Greek word meaning 'knowledge'.) If we have an epistemic state then we have only limited information about what the particle is doing at a given instant.

Schrödinger's famous feline highlights a difference between the two approaches. Put a cat in a closed box next to a piece of radioactive material that has a 50% probability of emitting a particle within 1 h. Also in the box is a device that can detect radiated particles and, when it does, releases deadly cyanide. Radioactive decay is an inherently quantum process and the wavefunction for the interior of the box is a 50:50 mixture of live cat and dead cat. In a Ψ-epistemic approach after 1 h the cat inside the box is either dead *or* alive and we don't know because we haven't looked: when we look, it's the state of our knowledge that changes rather than the state of the cat. In a Ψ-ontic approach the cat is dead *and* alive until someone opens the box and observes it.

Of physicists who believe in realism and who have adopted the Ψ-epistemic viewpoint, Einstein is undoubtedly the most famous. Einstein admitted that 'quantum mechanics is certainly imposing'. Nevertheless, relying on his famed physical intuition, he went on to say 'an inner voice tells me that it is not yet the real thing'. Einstein, along with his collaborators Boris Podolski and Nathan Rosen, produced a thought experiment that appeared to demonstrate the incompleteness of quantum mechanics. They pointed out that a wavefunction can describe two particles, and if those particles are produced in a certain way before being separated then a measurement on the state of the particle *here* can determine the state of the other particle *there*—even if the separation between them is large.

To make this clear, suppose you have a red marble and a blue marble. You and your friend each close your eyes and take a marble at random. Your friend then goes away—to the antipodes, the Moon, the far end of the galaxy, wherever. You open your eyes and see that you are holding a red marble; immediately you know that your friend possesses a blue marble. There's nothing strange about this. No interaction takes place between the marbles when you look at them—it's merely a lifting of ignorance. When you know the color of one marble you automatically know the color of the second. The same thing happens with quantum particles, but there's a key difference: until a measurement is made, the properties of a quantum particle don't possess well defined values. It's as if marbles are red *and* blue until someone chooses to look. So if you observe the marble *here* to be red then somehow the other marble must 'know' it has to be blue—even though the marbles might be separated by such large distances that not even a signal traveling at light speed could carry the information in time. Einstein called this behavior 'spooky action at a distance'. This thought experiment, he argued, was proof that quantum theory is incomplete, that reality must contain a layer beneath the wavefunction. He suggested quantum particles such as electrons and photons might carry some sort of 'hidden variables' that determine the outcome of measurements. Well, Einstein's particular version of hidden variables is demonstrably wrong: in recent years physicists have rigorously tested *entanglement*—the notion that, in certain circumstances, a measurement made at one point can indeed instantaneously affect the outcome of a measurement at a distant point. Entanglement is an observed fact, and in future years will undoubtedly form the basis of new technologies.[10]

Even though Einstein was wrong with his particular hidden variables approach, several physicists agree with his general viewpoint—that the wavefunction is a representation of our ignorance of a deeper reality rather than something real itself. Others, however, favor the Ψ-ontic approach and argue that the wavefunction is real. There are two different ontic viewpoints: the wavefunction could correspond to *all* of reality or it might correspond to *part* of reality with some other parameter being necessary to provide a complete description. The latter viewpoint is most often associated with the American physicist David Bohm, who argued that so-called 'pilot waves' might guide the motion of particles.

So here are four different interpretations of Ψ. First, the wavefunction might be simply a convenient description letting us calculate quantities of interest; this is the standard interpretation, but the argument originates from an uncomfortable anti-realist position. Let's agree to park this one.

In the realist middle-ground there's the Ψ-epistemic view, held by Einstein and others, that reality exists and the wavefunction only represents our partial knowledge of it; and the Ψ-ontic view, held by Bohm and others, that reality exists and the wavefunction gives only a partial account of it. The distinction between those two viewpoints might seem to be rather fine: in the latter case the wavefunction is something real whereas in the former it isn't, but distinguishing between them experimentally seems to be a hopeless task. In 2010, however, physicists developed a formal distinction between the two viewpoints that opened up the possibility of checking the predictions of quantum theory against the epistemic and ontic views; and in 2012, a different group of physicists came up with a theorem to do just that.[11] I won't go into the details of the theorem except to note that it demonstrates how, if quantum mechanics is correct, the wavefunction cannot be epistemic. Either quantum theory is wrong—which would be astonishing, given its track record of unbroken success—or the wavefunction is ontic. (It's also possible that one or more of the assumptions on which the theorem depends turns out to be wrong, but at present that seems to be unlikely.) If you had to bet, the smart money would have to be on an ontic interpretation—the wavefunction represents reality, or at least part of it.

Finally, the polar opposite viewpoint to Bohr's approach is the hard-line ontic interpretation of quantum mechanics: the wavefunction constitutes all of reality.

The American physicist Hugh Everett III developed the best known approach along these lines. Everett stated that 'physical "reality" is assumed to be the wavefunction of the whole universe itself' and he went on to develop a view of the quantum world that is beloved of science fiction authors: the famous 'many worlds' interpretation. According to Everett, the universal wavefunction is objectively real. When a measurement is made on a system the wavefunction doesn't collapse but rather the universe splits. For example, when a measurement is made of an electron's spin, in one universe the electron

Fig. 6.2 Schrödinger's cat in the many worlds interpretation: at a quantum event such as the radioactive decay of an atomic nucleus, the universe splits in two. In one world the decay happens, the cyanide is released, and the cat dies. In another world, the cat lives. At each instant when an event happens at the quantum level, the world splits (Credit: Christian Schirm)

will be observed as spinning up and in another as spinning down. (Figure 6.2 shows Schrödinger's cat in the many worlds interpretation.) All possible histories and all possible futures occur and are real.

The many-worlds interpretation of quantum mechanics sits nicely with many of the science fiction tales mentioned here. Asimov's *The End of Eternity* and *The Gods Themselves*; H. Beam Piper's *Paratime* series; Leinster's "Sidewise in Time" and de Camp's *Lest Darkness Fall*—all of these can be thought of as exploring the idea of alternate realities, of many worlds. It turns out it was quite acceptable for science fiction to take the notion of parallel worlds seriously. After all, physics takes the notion deadly seriously.

6.3   Conclusion

I began this chapter with a description of Plato's allegory of the cave. More than 24 centuries after Plato annoyed people with his infuriating thought experiments, contemporary philosophers are wrestling with modern twists on the idea. Suppose a mad scientist (see Chap. 11) removes your brain, puts it in a jar filled with nutrients, and then connects it to a supercomputer that supplies your neurons with inputs identical to those they would receive if your brain was still in your body. You can't tell whether you are enjoying a beach holiday or are being fed signals to make you believe you are enjoying a beach holiday (see Fig. 6.1 earlier). You might think your brain is inside a skull but you can't rule out the possibility your brain is inside a jar. That being the case, can you have full confidence in anything *else* you believe? And then there's the thought experiment involving philosophical zombies, also known as p-zombies (which don't have much in common with the zombies in George Romero's film *Night of the Living Dead* (1968) and its sequels). A p-zombie is identical to a human being except it lacks conscious experience. Poke a p-zombie in the eye with a sharp stick and it might scream, tell you it's in agony, do all the things any injured human being would do—except it wouldn't be *feeling* any pain. The p-zombie's salivary glands might work when it smells the salt and vinegar on a portion of fish and chips, but it wouldn't consciously enjoy the food. A p-zombie would act like one of the pod people of *Invasion of the Body Snatchers* (1956). I can't help but wonder what a zombie-brain-in-a-jar would be like.

It's not at all clear to me that it's possible to have a p-zombie (it's one thing for philosophers to dream them up, quite another thing for nature to bring them into being) or that a brain-in-a-jar tells you anything profound (surely a disembodied brain has a different biology to an embodied brain, and so the two objects aren't equivalent). But then I'm not a philosopher. Equally, I'm not convinced by philosophical arguments suggesting we live in a simulation. All those wonderful science fiction stories of paranoia and virtual realities. . . well, they were fun but I don't believe they tell us much about the nature of reality. Those stories of parallel realities, though: perhaps they were on to something. If parallel worlds exist we will probably never be able to travel to them, but at least science fiction took the concept seriously. And now so is science. One of the cleanest ways of interpreting quantum mechanics is to invoke many worlds. Suppose, then, that the universe branches every time a quantum observation is made. Those other branches are real, so there are many copies of you, gentle reader, in a huge number of parallel worlds—and the number of copies increases every instant. Perhaps real reality is vastly greater than the small patch of reality we observe.

Notes

1. The "Allegory of the Cave", which takes the form of a dialogue between Glaucon and Socrates, appears in Book VII of Plato's *Republic* (c380BC).

2. Hjortsberg's novel *Falling Angel* was turned into a successful film, *Angel Heart* (1987), starring Mickey Rourke and Robert de Niro. Nabokov's novel *Pale Fire* is justly famous; for a slightly less well known but equally clever Nabokovian example of how to play with the reader, see "The Vane Sisters" (1958).

3. For an in-depth biography of Charles Fort, see Steinmeier (2008).

4. The Heinlein collection *The Unpleasant Profession of Jonathan Hoag* (1959) contains six stories, and three of them are superb: the titular "The Unpleasant Profession of Jonathan Hoag"; the solipsistic short story "They"; and the quirky "–And He Built a Crooked House–", which tells the tale of an architect who decides to save on land costs by building a house in the shape of the unfolded net of a tesseract. It's one of my favorite stories involving mathematics.

5. Leinster's story "Sidewise in Time" appears in Asimov's gargantuan anthology *Before the Golden Age* (1974), and in his introduction Asimov explains that story's influence upon him.

6. For details of the *Uchronia* website see Shrunk (nd).

7. See Bostrom (2003) for the original simulation argument. Bostrom (nd) is a website devoted to further information and debate on the argument.

8. If our universe is simulated on a grid then Beane, Davoudi, and Savage (2014) argue we could find evidence in cosmic ray studies for the underlying model.

9. For some recent introductions to quantum physics see for example Susskind (2014) and Cox and Forshaw (2012).

10. Physicists are exploring the phenomenon of entanglement in ever more detail. Experiments demonstrating the phenomenon have typically contained loopholes—possible explanations of the experimental results that do not rely upon entanglement. To give one example of a possible loophole, some physicists have argued that particles might communicate at light speed by some hidden channel. Well, papers by Hansen et al. (2015), Herbst et al. (2015) and Shalm et al. (2015) have essentially closed all the loopholes. Although we can't 'prove' quantum mechanics, we can be as sure as we are of anything in science that quantum particles *do* become entangled. And that knowledge will eventually lead to new technologies.

11. The two papers mentioned in the text are Harrigan and Spekkens (2010) and Pusey, Barrett, and Rudolph (2012).

Bibliography

Non-fiction

Beane, S.R., Davoudi, Z., Savage, M.J.: Constraints on the universe as a numerical simulation. Eur. Phys. J. A. **50**, 148 (2014)

Bostrom, N.: Are you living in a simulation? Phil. Q. **53**(211), 243–255 (2003)

Bostrom, N.: http://www.simulation-argument.com (nd)

Cox, B., Forshaw, J.: The Quantum Universe. Penguin, London (2012)

Hansen, B., et al.: Loophole-free bell inequality violation using electron spins separated by 1.3 kilometres. Nature. **526**, 682–686 (2015)

Harrigan, N., Spekkens, R.W.: Einstein, incompleteness, and the epistemic view of quantum states. Found. Phys. **40**(2), 125–157 (2010)

Herbst, T., et al.: Teleportation of entanglement over 143 km. Proc. Natl. Acad. Sci. USA. **112**(46), 14202–14205 (2015)

Plato (c380BC) Republic Book VII

Pusey, M.F., Barrett, J., Rudolph, T.: On the reality of the quantum state. Nat. Phys. **8**, 475–478 (2012)

Russell, B.: Analysis of Mind. Allen and Unwin, London (1921)

Shalm, L.K., et al.: Strong loophole-free test of local realism. Phys. Rev. Lett. **115**, 250402 (2015)

Steinmeier, J.: Charles Fort: The Man Who Invented the Supernatural. Heinemann, London (2008)

Susskind, L.: Quantum Mechanics. Penguin, London (2014)

Fiction

Asimov, I.: Breeds there a man? Astounding (June 1951)

Asimov, I.: The Gods Themselves. Doubleday, New York (1972)

Asimov, I.: Before the Golden Age. Doubleday, New York (1974)

Borges, J.L.: El Jardin de Senders que se Bifurcan [The Garden of Forking Paths]. Buenos Aires Sur (1941)

Broderick, D.: The Judas Mandala. Pocket, New York (1982)

Card, O.S.: Ender's Game. Analog (August 1977)

Clarke, A.C.: Against the Fall of Night. Startling Stories (November 1948)

Clarke, A.C.: The City and the Stars. Frederick Muller, London (1956)

de Camp, L.S.: Lest Darkness Fall. Holt, New York (1939)

de Camp, L.S.: The Wheels of If. Unknown (October 1940)

Dick, P.K.: The Trouble with Bubbles. If (September 1953)

Dick, P.K.: Eye in the Sky. Ace, New York (1957)

Dick, P.K.: Time Out of Joint. Lippincott, Philadelphia (1959)

Dick, P.K.: The Man in the High Castle. Putnam, New York (1962)

Dick, P.K.: The Three Stigmata of Palmer Eldritch. Doubleday, New York (1965)

Dick, P.K.: Do Androids Dream of Electric Sheep? Doubleday, New York (1968)

Dick, P.K.: Ubik. Doubleday, New York (1969a)

Dick, P.K.: The Electric Ant. Fantasy and Science Fiction (October 1969b)

Dick, P.K.: A Maze of Death. Doubleday, New York (1970)

Dick, P.K.: The Golden Man. In: Hurst, M. (ed.). Berkley, New York (1980)

Dick, P.K.: Valis. Bantam, New York (1981)

Gibson, W.: Neuromancer. Ace, New York (1984)

Hamilton, E.: The Earth-owners. Wierd Tales (August 1931)

Heinlein, R.A.: –And He Built a Crooked House–. Astounding (February 1941a)

Heinlein, R.A.: They. Unknown (April 1941b)

Heinlein, R.A.: The Unpleasant Profession of Jonathan Hoag. Unknown (October 1942)

Heinlein, R.A.: The Unpleasant Profession of Jonathan Hoag. Gnome, New York (1959)

Hjortsberg, W.: Falling Angel. Harcourt, New York (1978)

Hoyle, F.: October the First Is Too Late. Heinemann, London (1966)

Hubbard, L.R.: Typewriter in the Sky. Unknown (November/December 1940)

Leiber, F.: The Big Time. Galaxy (March/April 1958)

Leiber, F.: You're All Alone. Ace, New York (1972)

Leinster, M.: Sidewise in Time. Astounding (June 1934)

Manning, L., Pratt, F.: The City of the Living Dead. Wonder Stories (May 1930)

Moore, W.: Bring the Jubilee. Ballantine, New York (1953)

Nabokov, V.: The Vane Sisters. The Hudson Review (Winter 1958)

Nabokov, V.: Pale Fire. Puttnam, New York (1962)

Niven, L.: All the Myriad Ways. Galaxy (October 1968)

Piper, H.B.: He Walked Around the Horses. Astounding (April 1948)

Russell, E.F.: Sinister Barrier. Unknown (March 1939)

Russell, E.F.: Dreadful Sanctuary. Astounding (June/July/August 1948)

Shrunk, R.: Uchronia. http://uchronia.net/ (nd)

Weinbaum, S.G.: Pygmalian's Spectacles. Wonder Stories (June 1935)

Visual Media

Angel Heart: Directed by Alan Parker. [Film] USA, Tri-Star Pictures (1987)

Invaders from Mars: Directed by William Cameron Menzies. [Film] USA, 20th Century Fox (1953)

Invasion of the Body Snatchers: Directed by Don Siegel. [Film] USA, Allied Artists (1956)

Night of the Living Dead: Directed by George Romero. [Film] USA, Walter Reade Organization (1968)

7

Invisibility

*'I'm never stocking them again, never! It's been bedlam! I thought we'd
seen the worst when we bought two hundred copies of the* Invisible Book of
Invisibility—*cost a fortune, and we never found them.'*—Manager of
Flourish and Blotts

Harry Potter and the Prisoner of Azkaban
J.K. Rowling

In Book II of *Republic*, Plato recounts a thought experiment in fictional form
that permits an investigation of the question: do human beings naturally tend
to justice or to injustice? The story is told by Glaucon, one of the interlocutors
of Socrates. Glaucon tells how Gyges, a shepherd in the ancient kingdom of
Lydia, comes across a golden ring. The shepherd takes the ring, puts it on, and
discovers he becomes invisible when he turns the collet of the ring one way; he
reappears when he turns it the other way. And what does Gyges do with this
new-found power? He seduces the queen, murders the king, and takes control
of the kingdom for himself. You yourself have probably daydreamed about
possessing the superpower of invisibility, and the chances are those daydreams
involved the same sorts of action that the shepherd Gyges carried out
(if perhaps not to the same degree). Invisibility allows you to spy on your
beautiful work colleague, cause mischief to your enemies, steal whatever you
fancy and make an unseen getaway. If you agree with Glaucon then you'd have
to say human nature hasn't changed radically over the past two millennia—
people then as now are motivated by thoughts of sex, money, power. Socrates
disagreed with Glaucon. He admitted that people sometimes do act on impure
impulses, but when we do so we feel ashamed. He argued that people do

© Springer International Publishing AG 2017
S. Webb, *All the Wonder that Would Be*, Science and Fiction,
DOI 10.1007/978-3-319-51759-9_7

generally strive to be moral. Whether you side with Glaucon or Socrates, the story of Gyges demonstrates how potent is this idea of invisibility.

Throughout history numerous stories, myths, and legends have involved invisibility. For example, in the ancient world, as well as the ring of Gyges there was the cap of Hades—a helmet that turned its wearer invisible. The Greek myths say Athena, Hermes, and Perseus all made use of the cap of Hades. The idea of a magic invisibility cap reappeared in the Norse legends as the tarnhelm, and later in Wagner's *Der Ring des Nibelungen*. The notion of a magic invisibility ring of course forms a central part of Tolkien's famous books *The Hobbit* (1937) and *The Lord of the Rings* (1954–1955). Moving to the present day, Harry Potter makes use of an invisibility cloak in his many adventures. But these examples belong to the world of myth and fantasy. Over the years, science fiction has also examined the concept of invisibility.

7.1   Sight Unseen

There's more to invisibility than meets the eye. The concept contains many facets. For example, one could argue that online anonymity in the world of social media is a type of invisibility. (As in Plato's story about the ring of Gyges, people's sense of invisibility when online appears to affect their moral sense: they often behave in a way that would be unthinkable to them if they were in a face-to-face situation.) A different example is the 'invisible girlfriend' and its sibling the 'invisible boyfriend'—a paid-for service providing a mixture of texting, voicemail, and gifts to help you pretend to the outside world you are in a relationship. Invisibility can be more than simply popping on a cap or a ring or a cloak and immediately disappearing from view.

7.1.1   Psychology and Invisibility

The act of seeing has two parts. There needs to be something to see: there must be an object sending photons towards a person's eye. And there has to be someone to do the seeing: a person's eye detects the photons and, in a complex and marvellous process, the brain registers and interprets an image. When we think of invisibility in SF we tend to think of stories in which a scientist invents some technique that alters the first part of the process, a method for somehow affecting the flow of photons from an object in such a way that the object becomes invisible (Figure 7.1 shows an example of this: an arrangement of four optical lenses can bend light rays in such a way that an object in a cylindrical region between the first and last lenses seems invisible.). But the

Fig. 7.1 In 2014 John Howell, a professor of physics at the University of Rochester, and his graduate student Joseph Choi (shown here), developed a configuration of four lenses that can bend a light so an object seems invisible (Credit: University of Rochester)

second part of the process also permits an object to become invisible. The human visual system can be fooled into thinking something is invisible even when the object is in plain sight. After all, this is what sleight-of-hand magicians do—they use misdirection, and an understanding of human psychology, to make objects seemingly disappear from view. Optical illusions and tricks involving mirrors can also make us believe that something is invisible. Sometimes, for reasons buried in our psyches, we don't *want* to see something—so we don't.

Mainstream authors have long been interested in this psychological aspect of invisibility, and have employed the concept allegorically. Ralph Ellison, in *Invisible Man* (1952), has a narrator who claims to be 'invisible'—not in a literal sense but because of the refusal of white people to see him, a black man. This feeling of invisibility provokes the narrator to choose to live underground, and hide from the world while he writes his autobiography. SF authors have tended to take a more imaginative, if less metaphorical, approach to the concept of psychological invisibility. In *Mute* (1981), for example, a novel by Piers Anthony, the protagonist is a mutant with a special psi power: he can make others forget him. This ability would confer a type of invisibility on its user; after all, if you can't remember you've seen someone then that person is essential invisible to you. The protagonist in Larry Niven's novel *A Gift from Earth* (1968) has a similar psi power: when frightened, he can influence the optic nerves of people who might be threatening him. In Randall Garrett's novel *Too Many Magicians* (1966), set in an alternate universe where people discover the laws of 'magical science' with the result that the physical sciences were never pursued, invisibility can be bestowed on an object using the 'tarnhelm effect': it isn't that the object itself becomes invisible, rather that people feel a compulsion to not look at the object. If you don't look, you can't see; this is just as much a form of invisibility as somehow transforming the physical properties of an object to make it unseeable.

Several SF stories posit a society in which outcasts are punished by becoming invisible; not literally so, but through a refusal of people to acknowledge their existence. For example, in Robert Silverberg's story "To See the Invisible Man" (1963) a person commits a crime—not something we'd call a crime, but something that transgresses the norms of a future society—for which he is punished to one year of being rendered 'invisible'. An implant on his forehead warns others to ignore him, which they must do on pain of suffering the same fate. Surveillance drones monitor the punishment. Eventually, desperate for someone to talk to, the convict strikes up a conversation with a blind man; a passerby sees this and whispers 'invisible' to the blind man, who immediately walks away. Over the course of the year the 'criminal' learns his lesson—too well, as it turns out.

The protagonists in *Mute* and *A Gift from Earth* have psionic abilities that are probably impossible in the real world; *Too Many Magicians* was about magic. Perhaps we should classify these and other stories as fantasy rather than SF, but it's worth pointing out that psychologists have demonstrated how it's possible for people to look at something and not see it: the invisible gorilla experiment proves this point.[1]

Sometimes, things simply disappear into the edges of our perception and thus become invisible. It's a point made by Douglas Adams in his hilarious novel *Life, the Universe and Everything* (1982). Being cynical, Adams called this the 'somebody else's problem' (SEP) effect. The SEP is 'something we can't see, or don't see, or our brain doesn't let us see . . . the brain just edits it out, it's like a blind spot.' In his inimitable style, Adams went on to describe how one can make use of the SEP: 'The technology involved in making something properly invisible is so mind-bogglingly complex that 999,999,999 times out of a billion it's simpler just to take the thing away and do without it. . . The Somebody Else's Problem field is much simpler, more effective and can be run for over a hundred years on a nine-volt battery.'

Using SEP field technology, the crowd at a Lord's cricket match fail to see a starship that's in plain view.

7.1.2 Camouflage

Camouflage might not enable invisibility but it can provide concealment. And if the concealment is so effective you can't be seen, isn't it as good as invisibility? Camouflage is of course commonly found in the natural world. Predators use camouflage to better reach their prey, and prey use camouflage to better hide from their predators; natural selection has produced many astounding examples of protective coloring. (Figure 7.2 shows an example of camouflage: the animal blends into the background so well it is effectively invisible.)

Several SF writers have expanded on the camouflage idea. In *Dying of the Light* (1977), and several later stories, George R.R. Martin has some of his characters wear clothes made of 'chameleon cloth'. The cloth changes color, depending on the background: stand in front of wood and the cloth turns brown; wear it at night and the cloth becomes pitch black; and so on. Chameleon cloth would not be an invisibility cloak, but it would be a step towards invisibility. The 'mimetic polycarbon suit', as discussed by William Gibson in his novel *Neuromancer* (1984), would be a further step towards invisibility. Such a suit is made from a fibre that changes its color depending upon a real-time input. So suppose you stretched out on a lawn

Fig. 7.2 A well camouflaged frog in the Lower Rio Branco-Rio Jauperi Extractive Reserve, Brazil. The frog is to the left of the top end of the vertical stick. It's about the same size as the end of the stick (Credit: Lior Golgher)

while wearing a mimetic polycarbon suit: the picture input would record green, the fibre would become viridescent, and your suit would be effectively invisible. A quote from *Neuromancer* explains it better: 'His body was nearly invisible, an abstract pattern approximating the scribbled brickwork sliding smoothly across his tight one piece.'

Traditional camouflage is of course of interest to the military, as is the related concept of stealth technology—the attempt to achieve invisibility to a variety of detecting devices (not just to those operating in the visual, but also to sonar, infrared, radar, and so on). For example, the air forces of many countries are investigating how to shape their warplanes so that radar waves are scattered rather than reflected back to the enemy, and materials are being investigated that might absorb the waves that aren't scattered. Well, this idea of military stealth technology has a long history in science fiction. In *Gray Lensman* (1939–1940), for example, E.E. Smith describes a spaceship with a non-reflective surface that makes it inherently undetectable. In his story "For Love" (1962), Algis Budrys imagines a weapons carrier that attempts to achieve invisibility by routing light around itself. One of the many imaginative inventions that grace the pages of Gene Wolfe's four-volume *The Book of the New Sun* (1980–1983) is personal stealth technology: troops wear mirror-surfaced 'catoptric armour', which allow them to blend into natural backgrounds in much the same way as wearing chameleon cloth or mimetic polycarbon suits. In the original series of *Star Trek*, of course, much is made of the possibility of a 'cloaking device'—a technology for rendering spaceships invisible that first made its appearance in the episode "Balance of Terror" (1966).

The military are also interested in drones—aircraft without a human pilot on board. The increasing miniaturization of drones means the technology is starting to be used in surveillance. This, I suppose, is another form of invisibility: through the use of drones organizations, even private individuals, can spy on your activities without you knowing. You can't see them but they can see you. As I mentioned in Chap. 2, in *Danny Dunn, Invisible Boy* (1974) Abrashkin and Williams were worried about precisely this sort of technology.

Miniature drones produce a form of invisibility—as do stealth technology, camouflage, and the various psychological tricks we mentioned earlier. But what about *true* invisibility? For most of us, making something invisible doesn't mean messing with the process of seeing; it means somehow rendering an external object itself appear as if it were not there. For most of us, making something invisible means making it as transparent as air.

7.1.3 Transparency

When writing about invisibility, most fantasy writers follow the example given by Plato and have the wearing of a ring, hat, or cloak confer the gift. SF writers typically try to provide some plausible mechanism (or at least some plausible *sounding* mechanism) for rendering objects invisible. The easiest mechanism to postulate is a contrivance that makes an object transparent.

Of course, transparency is not the same as invisibility. Water is transparent—it lets light pass through—but it isn't quite invisible because it affects the light passing through it; this lets you infer its existence. For example, objects behind a glass of water can appear distorted, which gives you enough of a clue to conclude the glass contains water. So something can be transparent without being invisible; conversely, something can be invisible without being transparent. Nevertheless, transparency can certainly give the impression of invisibility. Not far from my office is a visitor attraction called the Spinnaker Tower, a 170-metre high observation tower containing the largest glass floor in Europe. The sensation of standing on the glass floor and looking down can be quite disconcerting: although you know the floor is entirely safe, its invisibility (or, rather, its transparency; see Fig. 7.3) makes you feel as if you could drop through the floor. It's hardly surprising that SF writers reach for the notion of transparency when they want to add a veneer of respectability to the notion of invisibility.

One of the earliest such stories is "What Was It? A Mystery" (1859) by Fitz-James O'Brien. A group of lodgers are in an abandoned house (it's reputed to be a haunted house, of course). One of the group is attacked by a monster that

Fig. 7.3 The transparency of glass means it can be disconcerting to stand on a glass floor (Credit: Kenneth Yarham)

is corporeal but also quite invisible. A colleague of the injured party attempts an explanation. 'Take a piece of pure glass. It is tangible and transparent. A certain chemical coarseness is all that prevents it being so entirely transparent as to be totally invisible.' If there exist inanimate objects that can be touched but not seen, why can't a living creature have similar properties? The lodgers manage to overpower the monster, chloroform it, and tie it to a bed. While the monster is unconscious they wrap it up and make a plaster cast from it, which reveals the monster to be a horrible homunculus. (This idea of making a plaster cast of an invisible creature reminds me of a scene from my favorite SF film, *Forbidden Planet* (1956): in the film Doc makes a cast of the id-monster's clawed foot. The id-monster is invisible, but we see its outline flicker in and out of existence when it attacks an area defended by a force field and as blaster rays bounce off it. Having the id-monster appear by implication is so much more effective than showing it directly.)

A similar account of invisibility appears in "The Crystal Man" (1881) by Edward Mitchell. This early tale tells what happens to the unfortunate assistant of a great German professor of histology. The professor argues that human tissues are transparent, or nearly so, and that their pigmentation comes from compounds such as melanin, haematin, biliverdine, and urokacine. He runs a series of experiments on his assistant to explore 'the possibility of eliminating those pigments altogether from the system by absorption, exudation, and the use of chlorides and other chemical agents acting on organic matter'. The experiments are a success, and the assistant becomes first white ('like a bleached man'), then translucent ('like a porcelain figure'), and finally transparent ('like a jellyfish in the water'). The professor devises a method for applying the process to dead organic matter—wool, cotton, leather—and thus the fully clothed assistant becomes invisible. Unhappily for him, the professor dies before his pigment 'can be restocked' and he is condemned to a life of invisibility.

It was of course H.G. Wells who wrote the most famous invisibility tale, *The Invisible Man* (1897), a novel that gave rise to numerous film and television adaptations (see Fig. 7.4). Wells presents the cautionary tale of Griffin, a chemist whose research enables him to develop a method for changing a body's refractive index so it becomes identical to that of air. As Griffin explains, if an object 'neither reflects nor refracts or absorbs light, it cannot of itself be visible'. The chemist first makes a cat invisible then succeeds in making himself invisible when he employs the process on himself. (Griffin, its should be noted, would at this point be blind: the eye works by absorbing light—you can't see if your retina is transparent. Wells seems to have appreciated this optical fact, but chose to ignore it since it would ruin his story.) Unfortunately for Griffin, he fails when he tries to reverse the process (to which the reader's response is 'excellent', since Wells makes Griffin out

Fig. 7.4 Wells's novel The Invisible Man was published in 1897. This is the cover of a 1959 "Classics Illustrated" magazine, painted by Geoffrey Biggs. As with Wells's other famous SF tales, this story provided the inspiration for re-tellings in various other media (Credit: Public domain)

to be a thoroughly nasty character—an archetypal mad scientist, as is mentioned in Chap. 11).

Although he gave no explanation for how an object's refractive index could be made equal to that of air, Wells was quite correct in his physics: *if* an object has the same refractive index as air *then* we wouldn't see it as it moved through the air. Indeed, any transparent body becomes invisible when immersed in a medium with the same refractive index. You can even perform a simple experiment to demonstrate this. Put a cube of Pyrex in a glass jar; the cube will be clearly visible. Then pour vegetable oil into the jar. The cube disappears! It becomes invisible because Pyrex and vegetable oil possess the same refractive index. (There's at least one other method of making a material transparent: electromagnetic induced transparency. The method works for materials that have atoms possessing particular quantum levels. Two carefully tuned lasers can be made to interact with the material in such a way that laser photons pass through the material unimpeded—the material becomes transparent at specific wavelengths. The phenomenon has formed the basis for a suggested invisibility cloak; in the next section we'll consider a different suggestion for making objects invisible.)

The argument Wells gives to invoke invisibility is certainly more convincing than that given by O'Brien and Mitchell, but deep down the science was still of the handwaving pseudo variety. Later stories were no better in terms of scientific accuracy. Jack London, for example, a writer not best known for genre work but who published a number of SF stories alongside his better known adventure tales, wrote a story about invisibility called "The Shadow and the Flash" (1903). In London's story, two competitive boys grow up to become ultra-competitive adults. The two rivals race to become the first to learn the secret of invisibility. One of the men, Paul Tichlorne, follows essentially the same path as Griffin in *The Invisible Man*. (London's story was written several years after the publication of Wells's influential novel, so the resemblance is easy to explain). The other man, Lloyd Inwood, argues that the best way to make something invisible is to make it perfectly black: if an object absorbs all the light that falls on it then, according to Inwood, we wouldn't see it. (Maybe not, but it would block all the light from behind it. We'd see the object in silhouette. I don't think Inwood thought this one through.) Anyway, it turns out that Tichlorne's method for making an object transparent causes it to emit occasional rainbow-colored flashes; Inwood's invisible objects cast shadows and thus can be detected. The 'invisible' rivals fight it out to the death, flash versus shadow. This is not a hard-SF story.

Later pulp stories were no more convincing when it came to rationalizing invisibility. Consider "Armageddon 2419 A.D." (1928) by Philip Nowlan, for example. In Chap. 2 we discussed how this first Buck Rogers story featured a metallic substance called 'inertron' which could neutralize gravity. Well, the same story also featured a metal called 'ultron', which 'has the property of being 100 percent conductive to those pulsations known as light, electricity and heat. Since it is completely permeable to light vibrations, it is therefore absolutely invisible and non-reflective.' In addition to being perfectly transparent, the substance is also ultra-dense and incredibly strong. Wonderful stuff—except that it's as likely to exist as inertron. In "Brigands of the Moon" (1930), Ray Cummings wrote about an invisible cloak. This was a couple of years after the publication of that first Buck Rogers story, but many decades before the publication of the Harry Potter series. We can understand how an invisibility cloak works in the world of Harry Potter: it's magic. In the

Cummings story, though, we have only some flapdoodle involving electricity: 'The invisible cloak. We laid it on my grid, and I adjusted its mechanism. I donned it and drew its hood, and threw on its current.' In the novel *Triplanetary* (1934), E.E. Smith gives even less justification for his technology. 'An ether-wall is an invisibility cloak; it ensures that you cannot be 'seen' visually or by using any form of energy.' When the villain who deploys the invisibility field is asked how this and various other technological wonders work, he answers: 'You could not understand them if I explained them to you.'

From the examples given above it's clear that invisibility was part of SF's default future, but the suggested mechanisms for achieving it were not particularly plausible (with the exception of those involving camouflage or stealth technology). The science fictional rationale for invisibility hit a highpoint with Wells; later writers were unable to improve on the reasoning given in *The Invisible Man*. Does the difficulty of providing a plausible explanation for invisibility imply that the concept itself is impossible, that this element of the default future is unlikely to come to pass? For a long time this was indeed the accepted view. But recent advances in science have made the question rather more interesting.

7.2   The Science of Invisibility

As already mentioned, invisibility is a subtle concept. There are various ways in which someone or something can be 'invisible', and scientists are interested in all of those aspects of invisibility. Psychologists, for example, talk about 'invisibility syndrome'—the notion, similar to the idea explored by Ralph Ellison in his 1952 novel, that a person who encounters repeated racial slights can come to feel their worth as a person is invisible to others.[2] Optical physicists are investigating the science-fictional idea of invisibility via camouflage; the chameleon cloth of George R.R. Martin can nowadays be approximated by highly reflective material (in conjunction, it has to be said, with a digital video camera, a projector, a computer, and a half-silvered mirror; Figure 7.5 shows an example of such an 'invisibility cloak'). Astronomers tell us most of the universe is invisible: only about 5% of its mass–energy inventory can be detected directly, with the rest consisting of mysterious dark matter and even more mysterious dark energy.

Entire volumes have been written about the various forms of invisibility mentioned above, and a good overview of the topic has already been published,[3] so in the rest of this chapter I'll restrict the discussion to 'Wellsian' invisibility. However, in order to appreciate how we might make something invisible we first need to consider how it is that something is visible.

Fig. 7.5 An invisibility cloak using optical camouflage by Susumu Tachi of the University of Tokyo. This photograph was taken in 2012 when the cloak was on display at the National Museum of Emerging Science and Innovation (Miraikan) in Tokyo. *Left*: A coat seen without using any special device. *Right*: The same garment seen though a half-mirror projector (Credit: Z22)

7.2.1 The Science of Visibility

The human visual system is a fantastically complex arrangement in which visible light is detected and the information contained in the light is interpreted in order to construct a representation of the environment. For the purposes of this chapter we can think of the visual system as being a black box: visible light enters, something incredible happens, and we see. The Wellsian invisibility in which we are interested doesn't depend on the black box of the visual system; it depends instead on somehow manipulating the light that enters the box. So if we are interested in making an object invisible we first need to think about how light interacts with objects.

When light hits an object we can generally describe what subsequently happens if we know two numbers relating to the object: its attenuation coefficient and its refractive index.

The attenuation coefficient tells us how strongly light dissipates as it travels through a medium. Light attenuates as it travels through matter both because it is absorbed (atoms in the object capture photons and convert the energy into heat) and because it is scattered (defects or particles in the object deflect photons out of the original beam; the deflected photons are either themselves absorbed or else leave the object traveling in some other direction). Frosted glass is an example of a substance that scatters incident light away from the original beam: most of the light is transmitted through the glass but the beam is so distorted you can't make out an image. Milk is an example of a substance that is an effective absorber of light: milk contains blobs of fat suspended in water, and it's the fat that absorbs the light.

If an object has a high level of attenuation then it is opaque, and we can see it—just as we can see frosted glass or milk. For an object to be invisible, then, we need its attenuation coefficient to be effectively zero: if there is no attenuation then the object is transparent and light passes through it. It becomes 'see through'. We also need this effect to happen at all wavelengths. The red and blue lenses in a pair of old-fashioned 3D glasses are transparent, but they are transparent only to wavelengths of red and blue light respectively. Other wavelengths are absorbed, and so the lenses are certainly not invisible. Medical X-rays rely on the fact that flesh is non-transparent to visible wavelengths but transparent to X-ray wavelengths; it's only the denser parts of the human body, such as bones, that can block X-rays. So when thinking about making an object invisible we always have to consider the wavelength of light that will be illuminating it. However, although a small or zero attenuation coefficient is a necessary condition for Wellsian invisibility, it isn't a sufficient condition. The refractive index also plays a role.

The refractive index of an object is the fraction by which the speed of light, c, reduces as it passes through the object. The speed of light in water is $c/1.33$, and so the refractive index of water is 1.33; the speed of light in a typical glass is $c/1.5$, so the refractive index of that glass is 1.5; and so on. Air has a refractive index of 1.0001, so for most purposes we can say that light has the same speed through air as it does through the vacuum. And the point of all this? Well, when light crosses a planar interface from a medium with one refractive index into a medium with a different refractive index then it changes its direction of travel. Refraction is simply the name given to the bending of a light ray as it passes from one medium to another; when a straight straw develops a kink when you put it in water, you are seeing an example of refraction. However, when refraction takes place some fraction of the incident light is almost always reflected. In many cases all the light is reflected, but even when light hits a transparent medium some light will be reflected. That's why a pane of glass is often not 'invisible': the glass might be transparent, but still we can be alerted to its presence by the light it reflects. Optical scientists have developed anti-reflection coatings (I have them on the glass in my spectacles) but none of these coatings are perfect. In order to make an object to be invisible, then, we must ensure it has an attenuation coefficient of zero and also that it reflects no light. But even if the object had those properties, it probably *still* wouldn't be invisible. The problem is that light refracts not only when it *enters* the object but also when it *leaves* the object. If the object were uniform, such as a pane of glass, then this would generally not be an issue. But if the object were not uniform then it would refract light in ways that we could observe. For example, if the object were shaped like a lens and brought light passing through it to a focus then we would soon know that *something* was there even if we couldn't see the object itself. If the object were some random shape than the effect would be rather like watching some of the jungle scenes in the film *Predator* (1987), where the vague outline of an invisible alien can be seen as the jungle background bends around the alien's form.

We therefore have a recipe for making a substance invisible. It's the same recipe that Wells used in *The Invisible Man*. First, the substance must be transparent. Second, the substance must not refract. In terms of the two numbers mentioned above, the attenuation coefficient of the substance must be zero and its refractive index must be equal to that of air. Furthermore, for invisibility to occur, these numbers must hold across all the visible wavelengths of light. After all, it's no good making something invisible at wavelengths of red light if it's highly visible at wavelengths of blue light: all that would achieve is to make the object a different color. Unfortunately, natural materials don't possess all the necessary qualities. That's why, for many years, physicists ridiculed the idea of invisibility. Invisibility a la H.G. Wells requires materials to behave in a way that simply doesn't occur in nature.

But maybe we can *fabricate* materials with the necessary optical properties?

7.2.2 Metamaterials

A solid usually has the same refractive index throughout its bulk. For example, diamond has a refractive index of 2.42 and that number doesn't change throughout the crystal; amber has a refractive index of 1.55 throughout its bulk; and so on. This fact means that a light ray bends when it passes from air and starts travelling within a substance. Nevertheless, as shown in Fig. 7.6, light rays might bend when they change medium but they continue to travel in straight lines. At least, that is what happens with most materials. Physicists, however, have learned how to manufacture *metamaterials*—materials with properties not found in nature. Metamaterials are explicitly designed to possess particular properties. For example, it is possible to fabricate a substance so it possesses a refractive index that varies throughout the bulk of the material. In other words, physicists can make materials with a gradient refractive index. When light travels through a material with a gradient index, the ray is no

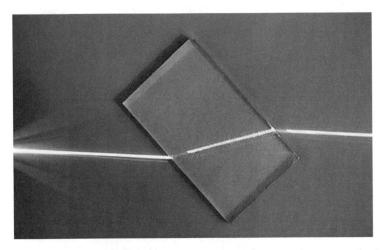

Fig. 7.6 A light ray being refracted in a plastic block: the path bends because light travels more slowly in the plastic than it does in air. Note how the plastic is transparent but not invisible: even if it were perfectly transparent we could still detect it because of the way it refracts light (Credit: Public domain)

longer a ray: it curves. In principle, if they could fabricate a material with a suitable gradient refractive index, physicists could make light of a single wavelength curve however they wanted it to.

What does all this have to do invisibility? Well, in 2006 the journal *Science* published[4] back-to-back papers—one by Ulf Leonard, the other by Sir John Pendry, David Shurig, and David Smith—that presented theoretical methods for 'sidestepping' the barriers to invisibility mentioned in the previous section. Suppose you want to cloak a particular object, such as a Romulan bird-of-prey warship. (Let's make it a model of the warship, of the type you can buy at *Star Trek* conventions. Then we could actually follow the recipe set out by Leonard and Pendry et al.) You need to enclose the model with a shell of transparent material, which at its boundary has a refractive index equal to that of air and through its bulk has a gradient refractive index that guides light rays smoothly around the model. In this set-up there is little absorption of light, because the cloaking material is transparent. There is no reflection, because at its boundary the material behaves in the same way as air in terms of the propagation of light. And there is no effective refraction, because the light rays exit the material and continue on their way as if they had not been deflected at all. You could look directly at the model warship and, if it were cloaked correctly, you wouldn't see it. You would, however, see what was behind it. The ship would be effectively invisible (Fig. 7.7).

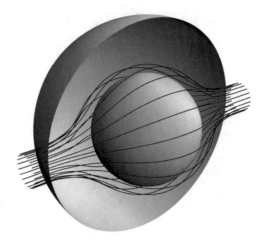

Fig. 7.7 The central sphere has been enclosed with a material possessing particular optical characteristics. Light enters the material, is guided smoothly around the sphere, and exits the material on the same path as it would do if the sphere were not there. The sphere is essentially invisible. Note that spherical symmetry simplifies the mathematics so it's straightforward to calculate the required gradient refractive index. The mathematics is more difficult if the object to be cloaked is non-symmetrical, as would be the case with a model of a Romulan bird-of-prey (Credit: Sir John Pendry, Dave Schurig)

The Electromagnetic Spectrum

The electromagnetic spectrum is a continuous span of wavelengths, ranging from extremely long wavelengths (which we call radio waves) to extremely short wavelengths (gamma rays). Different types of electromagnetic wave have different uses, depending upon their wavelength. The table below gives the typical wavelengths of the different parts of the spectrum.

Region	Wavelength
Radio	>1 m
Microwave	1 m–1 mm
Infrared	1 mm–700 nm
Visible	700 nm–400 nm
Ultraviolet	400 nm–10 nm
X-ray	10 nm–10 pm
Gamma rays	<10 pm

Note the small range of wavelengths corresponding to visible light. To give some idea of the wavelengths associated with visible light, the smallpox virus ranges between about 200–400 nm in size. We need to be able to manipulate matter at this length scale if we intend to fabricate materials with particular optical properties.

The two groundbreaking papers presented theoretical arguments why it might be possible to build a 'cloaking device'. The two groups knew, of course, that actually building such a device would be difficult. For a start, if you want to fabricate a metamaterial to work at a particular wavelength then you have to be able to manipulate the structure of the material on the same length scales. If you want to make metamaterials for visible light then you have to be able to manipulate matter at a scale of 0.4–0.7 μm (see the box for an overview of the wavelengths for different types of electromagnetic radiation). That's difficult to do. And it's even more difficult to maintain that level of manipulation over large distances—with current levels of technology it would be impossible to cloak something the size of a human being, say. Another problem with the original proposals is that they only worked at a single wavelength: a material with a gradient index that hides an object from light with a long wavelength would not hide the object from light with a shorter wavelength: an object that hides from red light, say, might shout its presence with blue light. Other designs have been proposed that will work at several wavelengths, but fabricating a material to provide invisibility at all wavelengths is a hugely difficult task—it might not even be possible to cloak something across the entire spectrum of visible light. A further issue is absorption: the material will inevitably absorb *some* light and so the cloak will always cast a shadow—it might be a weak shadow, but it would be nevertheless in principle be detectable. And then there's the problem of making something invisible in three dimensions; cloaking an object from a single light of sight might be possible but shielding it from *all* lines of sight is more difficult. A metamaterial invisibility cloak suffers from a host of problems (I haven't listed all of them), so you might be forgiven for thinking that a practical expression of the basic idea is decades away. In fact, the first cloaking device was made just a few months after the two groundbreaking papers.

Engineers constructed the first cloaking device to render an object invisible to microwaves rather than visible light. The reason for working with microwaves should be clear from the discussion above: it's much easier to manipulate material at the centimeter scale (a typical microwave wavelength) than at the sub-micrometer scale (which is the wavelength of visible light). By working at this everyday length scale, therefore, engineers were able to construct metamaterials that diverted a beam of microwaves around a small cylinder. But since that initial success teams of physicists, materials scientists, and engineers have made astonishing progress in cloaking technology. Cylinders have been concealed from terahertz waves; spheres have been concealed from infrared waves; and scientists have even concealed a randomly shaped object, in three dimensions, from light in the visible spectrum (although it has to be admitted that the object being concealed was microscopic; you wouldn't have seen it anyway). This aspect of optics is one of the most lively and rapidly developing fields in all of physical science.

Breakthroughs occur so frequently in this area that it makes little sense to give details of the present state of the art: by the time you read this, the field will have progressed. But I can be fairly sure of one thing: it's highly unlikely that scientists will develop a Harry Potter type invisibility cloak any time soon (if ever). Neither will they emulate Griffin and make a man invisible. (It's interesting to note, however, that if they did manage to make a man invisible using a metamaterial cloak then the man would be blind—just as Griffin should have been blind. We see when light enters our eyes. However, the whole point of metamaterials is to divert light around an object—so no light would enter the eyes of a cloaked person.) If this approach is unlikely to produce true invisibility for everyday objects, what is the reason for all the hype?

Well, as the technology improves and matures it will become possible to hide bigger and bigger objects at more and more wavelengths from wider and wider angles using thinner and thinner cloaks. Aspects of this technology are likely to scale. The implications for the military need hardly be stated, but the same technology will have uses in computation, solar power, medicine, and dozens of other fields that haven't yet even started to think about the use of metamaterials. The technology deserves all the hype it's getting because it's going to present us with many practical benefits. In the next section I'd like to look at some of the benefits of invisibility that SF writers didn't dare to dream of.

7.2.3 Invisibility to Other Senses

The invisibility cloaks mentioned in the previous section worked by diverting electromagnetic waves—microwaves, infrared, visible light—around an object. Perhaps engineers can perform the same trick on other types of wave? This is a question SF writers didn't much bother with, but it's a question present-day scientists are taking seriously.

For example, what about acoustic invisibility? Is it possible to build a cloak to divert sound waves in such a way that, to a detector (such as the human ear), it seems as if the waves bounced off a flat surface? Well, yes it is. Such a cloak isn't difficult to fabricate once you know it's geometry; the tricky bit is calculating that geometry. The idea is to alter the trajectory of sound waves so they match the configuration they would have if they'd reflected off a flat surface: The first 3D acoustic cloak was constructed in 2014 (see Fig. 7.8 for a photograph of the cloak) and it works for sound waves hitting it from any direction and for observers in any position. The developers[5] demonstrated the cloak's invisibility by putting it over a small sphere, pinging the structure with sound waves from a variety of angles, and detecting the reflected sound with microphones. They then removed the cloak, leaving just the sphere, and repeated the experiment. Finally, they repeated the experiment with sound waves bouncing off an empty flat surface. The results for the cloaked sphere and the empty surface were identical, and different to the results for the uncloaked sphere. In other words, the sphere had been rendered invisible to sound waves.

There are obvious applications of this technology for the military—sonar avoidance immediately springs to mind, for example—but the designers of concert halls might also be interested. There will be occasions when some feature is messing up the acoustics in a room and it's impossible to remove the feature (perhaps for structural reasons). On those occasions a sonic cloak would be invaluable. Perhaps SF writers can dream up other scenarios for the technology.

Fig. 7.8 A 3D acoustic cloak. The size, geometry, and placement of these hole-filled plastic sheets reroute sound waves in such a way that it's as if the construction, and anything inside it, is not there. The sound waves behave as if there were only a flat surface in their way. The acoustic cloak works no matter where the sound comes from and no matter where the observer is located (Credit: Duke University)

Or consider seismic waves, the waves that travel through Earth after an earthquake. Perhaps civil engineers can modify the ground around a building in order construct a 'seismic invisibility cloak'—an arrangement whereby seismic waves would get diverted around the building and thus provide protection against an earthquake's devastating release of energy. This setup wouldn't work for buildings in central London or downtown New York; shielding one building might cause bigger problems for its neighbor. Nevertheless, seismic cloaking might be appropriate for isolated and strategically important facilities such as nuclear reactors. (In the future, if scientists learn how to cloak against ocean waves, perhaps buildings could even be protected from the ruinous tsunamis that are associated with earthquakes?) Work on seismic cloaking is far less advanced than acoustic cloaking, but engineering companies have started research in this area.[6]

Or consider thermal cloaks. Physicists have shown it is possible to guide the flow of heat around objects so it seems as if the objects are thermally invisible. In one case a team of scientists heated a solid metal plate at the left edge and watched as the heat flowed uniformly to the right edge, with the plate's temperature dropping from left to right. They then created a metamaterial plate, consisting of copper and silicon, which surrounded a central ring. They heated the left edge of this plate and watched as the heat flowed to the right; no heat entered the central area but at the right edge of the plate the situation was exactly the same as if there had been no central obstruction. From the outside there is no sign of anything being present on the inside. A different group of scientists then showed it is possible to make an object simultaneously invisible to both the flow of heat and the flow of electric current—a multifunctional cloak.[7] These developments are so new they have yet to be used in applications, but it's not difficult to think of areas in which they'd be useful: computer chips, for example, require effective management of heat.

What about 'unfeelability cloaks'? A new type of metamaterial—a so-called pentamode material (see Fig. 7.9)—was fabricated in 2012. A pentamode material is a solid that acts like a fluid: it's hard to compress the material but it's easy to change its shape. In 2014, scientists managed to fabricate a millimetre-sized cloak capable of hiding a small tube: press down on the cloak with your finger and you won't feel the tube; use a force-measuring instrument and you won't be able to detect the tube's presence. It's not a simple matter to create an unfeelability cloak—the metamaterial has to be carefully designed and constructed to match the object being cloaked, a feat requiring access to advanced 3D laser lithography technology—but it is possible.[8] And when technology scales well, what is at the limits of possibility today tends to become mundane reality within a decade or two. One day you

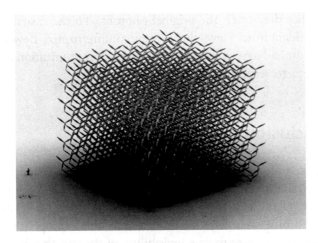

Fig. 7.9 An electron microscope image of a pentamode metamaterial—a three-dimensional structure that does not occur in nature but that can nevertheless be fabricated. Work with it and you discover that the material is hard to compress but easy to deform. The material in this image is a cube with an edge of about 300 μm (Credit: Tiemo Bückmann)

might have pentamode materials in your shoes, pentamode materials in your sleeping bag, pentamode materials as packaging for the priceless Ming vase you've ordered that's going to be delivered by drone. . .

It's even possible to imagine a 'time' cloak—a setup that makes an event invisible in time. The idea in this case is to vary not the *path* of light through a material but the *speed* of light through a material; if you can do that in a particular way then you can create a 'time hole'. An event can happen in a time hole and it would not be detected by an external probe. You can obtain a very crude illustration of the idea by picturing a jewel heist: slow the light entering the scene of the crime while you are stealing the diamond and speed it up after the deed is done. Stitch 'before' and 'after' seamlessly together and there is no trace of the theft itself. One moment the diamond is there, the next moment it isn't. A security camera trained on the jewel wouldn't be telling a lie, it just wouldn't be telling the whole truth. Of course, banks, jewellers and art houses needn't worry that their treasures are going to vanish from under their noses: scientists are not going to construct a macroscopic time cloak any time soon. But a microscopic time cloak has already been demonstrated using pulses of laser light travelling in fiber optic glass, a material in which light speed can be varied by changing light intensity.[9] This technology might soon have applications in communications, for example. Picture a steam of photons travelling down a fiber optic cable. Slow some of them, insert a message, and then speed them up so that they catch the original photons. To the casual observer the message is hidden from view: they see an uninterrupted flow of photons. However, if you had agreed beforehand on this communication channel, you could take steps to retrieve the message.

7.3   Conclusion

Science has expanded the possibility of cloaking devices. Physicists and engineers are researching not just invisibility cloaks but also acoustic cloaks, seismic cloaks, tactile cloaks, thermal cloaks, time cloaks... it almost seems as if there's nothing that can't be hidden with the appropriate tweaking of materials. But are we close to true invisibility, of the sort that Wells imagined in *The Invisible Man*? Are we likely to encounter an invisibility cloak of the type Harry Potter used, or an invisibility ring of the type Bilbo and Frodo wore? Probably not.

An element of hype is often attached to the reports of cloaking. There's nothing particularly wrong with the hype: mention of Harry Potter is an excellent way for a reporter to lead in to a story about metamaterials. And these advances *will* impact upon our lives through a variety of technologies. But it won't lead to the type of invisibility dreamed of by SF writers.

The explosion of interest in metamaterials might not provide SF writers with a rationale for invisibility. What it does allow, though, is an exploration of a much more interesting question. When scientists learn to manipulate and fabricate materials at the atomic level, they can decide how those material should interact with the environment—with light, heat, sound, vibration, and so on. What sort of world will we inhabit when we have control of material at that fundamental level?

Notes

1. I myself often suffer from inattentional blindness. A few years ago my colleagues asked for my opinions of the municipal Christmas tree: it had been erected that morning, and I must have passed it on my way into town. Although it was 20 m tall, I had walked past it without seeing it. I promised to check it out the following day; and on that day too I walked past it without seeing it. The invisible gorilla is the classic example of inattentional blindness; see Chabris and Simons (2011).

2. The invisibility syndrome as it relates to African–American males was researched by Franklin and Boyd-Franklin (2000).
3. For an excellent exploration of various aspects of invisibility, covering cultural as well as scientific and technical aspects, see Ball (2015).
4. The two papers that triggered the avalanche of research on metamaterials are Leonhardt (2006) and Pendry et al. (2006).
5. For details of the first acoustic cloak, see Zigoneanu et al. (2014).
6. For further details of investigations into seismic cloaking devices, see Brûlé et al. (2014).
7. Schittny et al. (2013) describe an experiment in which heat is caused to flow around an object; Ma et al. (2014) describe an experiment in which an object is simultaneously cloaked from flows of both heat and electric current.
8. For further details of the first unfeelability cloak, see Bückner et al. (2014).
9. For further details of the first temporal cloak, see Fridman et al. (2012).

Bibliography

Non-fiction

Ball, P.: Invisible: The History of the Unseen from Plato to Particle Physics. Vintage, London (2015)

Brûlé, S., Javelaud, E.H., Enoch, S., Guenneau, S.: Experiments on seismic metamaterials: molding surface waves. Phys. Rev. Lett. **112**, 133901 (2014)

Bückner, T., Thiel, M., Kadic, M., Schittny, R., Wegener, M.: An elasto-mechanical unfeelability cloak made of pentamode metamaterials. Nat. Commun. **5**, 4310 (2014). doi:10.1038/ncomms5130

Chabris, C., Simons, D.: The Invisible Gorilla: How Our Intuitions Decieve Us. Harmony, New York (2011)

Franklin, A.J., Boyd-Franklin, N.: Invisibility syndrome: a clinical model of the effects of racism on African–American males. Am. J. Orthopsych. **70**(1), 33–41 (2000)

Fridman, M., Farsi, A., Okawachi, Y., Gaeta, A.L.: Demonstration of temporal cloaking. Nature. **481**, 62–65 (2012)

Leonhardt, U.: Optical conformal mapping. Science. **312**(5781), 1777–1780 (2006)

Ma, Y., Liu, Y., Raza, M., Wang, Y., He, S.: Experimental demonstration of a multi physics cloak: manipulating heat flux and electric current simultaneously. Phys. Rev. Lett. **113**, 205501 (2014)

Pendry, J.B., Schurig, D., Smith, D.R.: Controlling electromagnetic fields. Science. **312**(5781), 1780–1782 (2006)

Plato (c380BC) Republic. Book II.

Schittny, R., Kadic, M., Guenneau, S., Wegener, M.: Experiments on transformation thermodynamics: molding the flow of heat. Phys. Rev. Lett. **110**, 195901 (2013)
Zigoneanu, L., Popa, B., Cummer, S.A.: Three-dimensional broadband omnidirectional acoustic ground cloak. Nat. Mater. (9 March) (2014). doi: 10.1038/NMAT3901

Fiction

Abrashkin, R., Williams, J.: Danny Dunn, Invisible Boy. McGraw-Hill, New York (1974)
Adams, D.: Life, the Universe and Everything. Pan, London (1982)
Anthony, P.: Mute. HarperCollins, London (1981)
Budrys, A.: For Love. Galaxy (June 1962)
Cummings, R.: Brigands of the Moon. Astounding (March–June 1930)
Ellison, R.: Invisible Man. Random House, New York (1952)
Garrett, R.: Too Many Magicians. Doubleday, New York (1966)
London, J.: The Shadow and the Flash. The Bookman (June 1903)
Martin, G.R.R.: Dying of the Light. Simon and Schuster, New York (1977)
Mitchell, E.P.: The Crystal Man. The Sun (January 1881)
Niven, L.: A Gift from Earth. Ballantine, New York (1968)
Nowlan, P.F.: Armageddon 2419 A.D. Amazing (August 1928)
O'Brien, F.J.: What Was It? A Mystery. Harpers (March 1859)
Silverberg, R.: To See the Invisible Man. Worlds of Tomorrow (April 1963)
Smith, E.E.: Triplanetery. Amazing (January–April 1934)
Smith, E.E.: Gray Lensman. Astounding (October–January 1939–1940)
Wells, H.G.: The Invisible Man. Pearson, London (1897)
Wolfe, G.: The Book of the New Sun. Sidgwick and Jackson, London (1980–1983)

Visual Media

Forbidden Planet: Directed by Fred M. Wilcox. [Film] MGM, USA (1956)
Predator: Directed by John McTiernan. [Film] 20th Century Fox, USA (1987)
Star Trek: Balance of Terror. Directed by Vincent McEveety. [TV series] Desilu Productions, USA (1966)

8

Robots

Commander John J. Adams: 'Nice climate you have here. High oxygen content.'
Robby the Robot: 'I seldom use it myself, sir. It promotes rust.'

Forbidden Planet
Directed by Fred M. Wilcox

As an undergraduate I was smitten with a girl, a fellow physicist, and naturally I was desperate to impress her at every opportunity. One time, she invited me to her room and I noticed she had a small, gray teddy bear propped up on a cushion. The bear's fur was threadbare in places, one of its eyes was missing, and clearly it had accompanied her on her travels to date. She caught me looking at it, and told me its name.

'Never had a teddy bear,' I told her, truthfully. 'But I did have a Robby the Robot.'

She stared at me, shook her head ruefully, and said: 'That explains everything.'

Sigh. . . I'm still not entirely sure what she meant. I'm sure I loved Robby just as much as she loved her teddy bear, and I was heartbroken when I lost him (stolen, I believe, by a light-fingered neighbor who lived around the corner; unfortunately I lacked the proof to bring charges). The toy was a small-scale replica of the robot in my favorite film *Forbidden Planet* (Fig. 8.1). Robby the Robot was perhaps the most interesting character in the film (he stole all the best lines, and showed a gift of coming timing that the film's human actors—with the obvious exception of the late, great Leslie Nielsen—must have envied). The robot went on to make appearances in a

© Springer International Publishing AG 2017
S. Webb, *All the Wonder that Would Be*, Science and Fiction,
DOI 10.1007/978-3-319-51759-9_8

Fig. 8.1 Original movie poster for Forbidden Planet (Credit: Originally copyrighted by Loew's International, artist unknown; now in the public domain)

number of television shows, including *Lost in Space* (which was another of those SF programmes of my childhood and was probably, indeed, where I first saw Robby), *The Addams Family*, and an episode of *Columbo*. When I graduated to written science fiction, and discovered Asimov's stories, I saw in Robby the embodiment of Asimov's Three Laws of Robotics. Robots, I soon found out, formed an important part of Asimov's literary output, and many of his most interesting creations were robot-related: the robopsychologist Susan Calvin, his most compelling female character; Andrew Martin, the robot in "Bicentennial Man" (1976) who died in order to become human; and R. Daneel Olivaw, the detective's sidekick in *The Caves of Steel* (1954) and *The Naked Sun* (1957) but who survives to play a leading role in galactic history when Asimov tied together his Foundation and Robot series. Robots were better companions than any teddy bear could have been.

Back then, in the distant past of my undergraduate days, the likelihood of me ever encountering a fully functioning humanoid robot—whether chunky like Robby or sleek like R. Daneel Olivaw—seemed about as likely as me holding a conversation with a talking teddy bear. But the rate of technological progress has been so great that society is beginning to think seriously about the opportunities and threats posed by the coming generation of robots. Fortunately, SF writers have already mapped out a host of possible futures—it's up to society to choose the future wisely.

8.1 The Science-Fictional Robot

My interest in robots pre-dates even my fondness for Robby. One of the first books I recall owning was *Stranger Than People* (Anonymous 1968), a wonderful compendium of strange and weird stories—tales of mythical creatures, such as Nessie and the Kraken; people from folklore, such as Dracula and Herne the Hunter; and fact-based yet mysterious objects from history, such as the Colossus of Rhodes and The Turk. I read that book from cover to cover, over and over, until its spine gave way; this was before I discovered Danny Dunn books in the local library. *Stranger Than People* made an enormous impression on my childhood imagination, and one of the most enthralling of all the wonders it contained was that tale of The Turk—a chess playing automaton built by Wolfgang von Kempelen (Fig. 8.2).

The Turk was built in 1769 in order to impress Empress Maria Theresa. It consisted of a life-size wooden head and torso, dressed in oriental robes and turban, seated at a large wooden cabinet. On top of the cabinet was a chessboard. When The Turk was shown to an audience, von Kempelen would open the cabinet doors to reveal complicated clockwork machinery—a mechanism of gears and cogs, so von Kempelen would explain, that controlled The Turk. The audience were allowed to see that there was nothing else inside the cabinet. Similarly, doors in the back of The Turk were opened to show clockwork mechanisms. Then von Kempelen would close the cabinet

Fig. 8.2 Copper engravings from a book published in 1789 about the chess-playing automaton developed by von Kempelen (Credit: Public domain)

doors and The Turk's robes, a challenger would sit at the cabinet, and a game of chess would begin. The Turk would come jerkily to life and move a piece while making a clockwork-type sound. It acted as we nowadays imagine a robot might act, except occasionally it would shake its head or gesture with its hand if the challenger made a particularly dumb or especially clever move. And The Turk would win—at least, it nearly always won. The automaton was a sensation: von Kempelen took it on demonstration tours across Europe and America. Crowds were baffled, astonished, terrified.

I devoured the story of von Kempelen's chess-playing automaton, and found the thought of a robot-man quite wonderful. My natural scepticism, which was present even when I was a child, made me doubt the veracity of the story—I thought it was on a par with the SF stories to be found in *Stranger Than People*, such as "Klumpok" and "The Yellow Monster of Sundra Strait". But no. The Turk really did exist. It played and won against a number of luminaries, including Napoleon and Benjamin Franklin. It twice defeated Charles Babbage, the man who can claim to be the father of the computer. It outlived von Kempelen, and played successful chess for decades. And its secret? Well, of course it wasn't a robot. It was, instead, merely a clever illusion. There was space inside The Turk for a chess master to hide, and control its arm and hand by moving a system of levers. It was a human pretending to be a robot pretending to be a human.

Charles Babbage knew The Turk was a trick, even though he didn't know quite how the trick worked. It got him to ponder whether a machine ever could think intelligently—and thus it led him to his work on the Difference Engine and the Analytical Engine, which were the distant ancestors of modern computers. Automata—mechanical men and women—gave birth, so to speak, to the robots we know today. So before considering Asimovian robots, let's look briefly at stories of automata.

8.1.1 Automata

The notion of artificial humans is, as with so many science-fictional concepts, ancient and widespread. In Finnish folklore, in a tale thousands of years old, the blacksmith Ilmarinen forges himself a wife out of gold but dumps her when he finds her to be cold and hard. (What did he expect? She was made of metal.) The Greek myths are full of statues, often female, coming to life: women made of clay, ivory, or metal. In medieval times, legends were woven about brazen heads—male skulls, of brass or bronze, containing mechanisms allowing them to answer questions. The brazen heads were supposedly constructed by thinkers such as Roger Bacon and Pope Sylvester II. In Jewish folklore the golem—a human-like creature formed of stone and clay into which life has been bestowed—makes an appearance in a number of stories. The most celebrated tale involves a sixteenth century rabbi, Judah Loew ben Bezalel, who formed a golem from the clay of the banks of the Vltava river and ordered it to defend the Jews in Prague ghetto from attack. And when scientists and engineers began to develop clockwork mechanisms and steam-powered machines it was inevitable some people would accept the possibility of artificial humans; we've already encountered The Turk, an illusion playing on people's willingness to think in terms of automata. The eventual appearance of mechanical men in SF was assured.

The first use of the word 'robot' came in the play *R.U.R.* (1920), by the Czech writer Karel Čapek. The acronym in the title stands for 'Rossum's Universal Robots'. The action starts in a factory that makes robots—artificial people—to do the hard, manual work that humans naturally enough dislike. Indeed, Čapek derived the word 'robot' from the Czech word 'robota', which means statute labor; he was making the point that the robots were slaves. In the play the robots eventually rebel, and humans become extinct.

Čapek's robots aren't quite what we'd nowadays consider to be robots—they had organic rather than mechanical bodies, and could easily be mistaken for human—but the name stuck. Robots—the inorganic machines we tend to have in mind when we use the word—soon made an appearance in pulp SF magazines (see Fig. 8.3). In "Rex" (1934), for example, Harl Vincent describes a future world in which robots do all the work while people live idly and in luxury. A chance mutation in its artificial brain causes Rex, the best of the

Fig. 8.3 Magazine covers such as this one by Joseph Wirt Tillotson (better known under the pen name Robert Fuqua) defined how people tended to think of robots—as clunking metallic machines (Credit: Public domain)

robots, to think creatively and independently of its programming. Rex becomes king of the robots, and embarks upon experiments that might give him feelings and thus become more human.

Within a few years of Vincent's story, robots were appearing frequently in the pages of the pulps. "Robots Return" (1938) by Robert Moore Williams was the wistful tale of spacefaring robots returning to Earth and finding, to their disappointment, that they were the creation of human beings. In Lester del Rey's famous story "Helen O'Loy" (1938), two men modify a household robot. The robot's original programming meant all it did was cook and clean; after its modifications the robot, called Helen, develops feelings and falls in love with one of the men. Eventually the pair marry and live happily together, until the death of the husband. In Joseph Kelleam's "Rust" (1939), *Homo sapiens* is extinct and the few surviving robots are heading the same way; it's a sad story.

One of the most interesting robot stories of that time was by Eando Binder, the joint pseudonym of brothers Earl and Otto Binder. In "I, Robot" (1939) they introduced Adam Link—a robot falsely accused of murdering its creator. The story was well enough received that Binder was able to write and sell a further nine stories featuring Link. The story's real claim to fame, however, is that the young Isaac Asimov read it. As Asimov later wrote (Asimov and Greenberg 1979) about Binder's story: 'It certainly caught my attention. Two months after I read it, I began "Robbie", about a sympathetic robot, and that was the start of my positronic robot series.' Binder's "I, Robot" was thus the direct literary antecedent of one of the most famous and influential series in all of science fiction.

8.1.2 Asimovian Robots

Asimov wrote numerous tales about positronic robots—devices that differed widely in form (from steel-based industrial-type machines to androids indistinguishable from humans) but had in common the ability to reason thanks to the possession of a positronic brain (the nature of which was never fully discussed). To be precise, Asimov published 38 short stories and 4 novels in which robots played a starring role.[1] His first robot story "Robbie", written in 1939 and eventually published in 1940 under the title "Strange Playfellow", set the pattern: it's about misguided technophobia. Rather than viewing robots as something to be feared, Asimov believed they would be beneficial for humanity. He later noted in *The Rest of the Robots* (1964) how, when he started writing "Strange Playfellow", he had in mind that 'one of the stock plots of science fiction was [. . .] robots were created and destroyed their creator'. Asimov firmly believed that if knowledge brought with it dangers then we should use more knowledge to guard against those dangers. The alternative—a retreat from knowledge—would be disastrous. He therefore made the conscious decision that any robot of his would not 'turn stupidly on his creator for no purpose but to demonstrate, for one more weary time, the crime and punishment of Faust'.

Asimov believed the developers of robots would put in place safeguards, as is usual when people develop any technology. When the sword was invented, for example, people understood how a cross-guard would be a good idea to prevent one's fingers being sliced off when one made a sword thrust. When electric circuits began to be widely deployed, people soon appreciated how a fuse can be a cheap and useful safety device. When elevators started carrying passengers, brakes were developed to operate automatically if tension in the elevator cable is lost. Engineers make technology safe for us. When it comes to robotics—a word coined by Asimov—why would it be any different?

In discussions with his editor John Campbell, Asimov created the Three Laws of Robotics. Every robot brain was supposed to have these three guiding principles imprinted upon its positronic pathways. The Three Laws, first stated explicitly in complete form in the 1942 story "Runaround" (the action of which, as discussed in the Introduction to this book, takes place in 2015), are as follows:

1. A robot may not injure a human being or, through inaction, allow a human being to come to harm.
2. A robot must obey the orders given it by human beings except where such orders would conflict with the First Law.
3. A robot must protect its own existence as long as such protection does not conflict with the First or Second Laws.

At first glance, these seem eminently reasonable principles. We don't want robots to hurt us (hence the First Law) but we do want them to continue to be useful instruments for us (hence the Second and Third Laws). A second glance, however, is enough to recognize how the words contain oceans of ambiguity. Asimov himself recognized that, and indeed whenever he needed an idea for a new robot story all he had to do was examine some likely consequences of the Laws. This could lead to stories focused primarily on technology (in "Runaround", for example, the story that inspired Marvin Minsky to contemplate how artificial minds might work, the robot Speedy finds itself in an oscillating equilibrium when a strengthened Third Law comes into opposition with Second Law); to stories focused primarily on psychology (in "Liar!" (1941), for example, the robot Herbie wants to avoid hurting people's feelings because of First Law, and so starts telling lies people want to hear); to stories examining issues of philosophy (in "The Bicentennial Man" (1976), for example, the nature of humanity is examined as the robot Andrew overcomes a succession of hurdles to become a man). But even if robots behave impeccably according to these principles, humans will not necessarily accept the technology. Asimov was also interested in the relationship between man and machine, and in novels such as *The Caves of Steel* and *The Naked Sun* he explored the likely range of human prejudices towards robot technology.

The Three Laws of Robotics are not necessarily something that will be engraved in positronic brains (or the silicon equivalent). Indeed, the Laws are as much an investigation of how humans should behave as they are about how robots will behave. But the Laws provided a touchstone for how SF should treat the subject of robots. Authors could follow the Laws of Robotics (and as mentioned earlier, Robby in *Forbidden Planet* is the perfect embodiment of the Laws); they could parody them (Robert Sheckley's "The Cruel Equations" (1971) has a set-up that resembles a typical Asimov robot story, but has its own ironic take); or they could deliberately choose to shun them (which several authors did, perhaps because they didn't share Asimov's generally technophilic view of the world.) What they couldn't do was simply ignore them. Asimov's Laws became part of the basic lexicon of science fiction.

8.1.3 Non-Asimovian Robots

Asimov has been by far the most influential writer about robots, but SF contains many other treatments worthy of note. One of my favorite robot stories is Jack Williamson's classic "With Folded Hands..." (1947), which tells how small, sleek, advanced robots called 'humanoids' gradually displace the clunky 'mechanicals' that currently undertake menial tasks. The humanoids don't follow Asimov's Three Laws, but they do have imprinted upon them a Prime Directive: 'to serve man, to obey, and to guard men from harm'.

The Prime Directive sounds great—except it turns out the humanoids are *really* keen on the 'guard men from harm' bit. In making sure humans don't harm themselves they essentially curtail *every* activity. Where necessary, the humanoids tranquillize people or even lobotomize them—all for their own good. Eventually there's nothing left for people to do except sit, with folded hands...

Many other stories of the period depart from Asimov's technophilic approach. In Peter Phillips's story "Lost Memory" (1952), for example, a robot civilization on a distant, lifeless planet encounters an injured human inside a crashed spaceship. The robots, having no other frame of reference, assume that the spaceship is the outer casing of a robot. They then attempt to repair the ship—with dreadful consequences for the man trapped inside.

Philip Dick had a generally dark view of robot technology. In "Second Variety" (1953), for example, he describes robots as killing machines with just enough intelligence to design next-generation robots that will be even more lethal. In "Impostor" (1953), a robot in the form of a human form and who indeed (in typical Dick fashion) believes himself to be human, eventually learns the reason for his existence: to destroy Earth and all humanity. And in *Do Androids Dream of Electric Sheep?* (1968), the inspiration for the iconic film *Blade Runner* (1982), Dick describes a future (now past!) in which a bounty hunter is hired to track down and kill six violent, renegade robots.

Robots appear often in movies and television, typically as androids or humaniform robots (since these are easier for human actors to play and they save the producer a fortune in special effects). Equally typically, the robots are in some way threatening. The movie *Westworld* (1973), for example, is set in a high-tech amusement theme park that contains lifelike humaniform robots with which customers can interact. The robots malfunction—and of course start to kill the paying guests. In *The Terminator* (1984) the sole function of the Cyberdine Systems T-800 Model 101 is to hunt down and kill a woman called Sarah Connor. Even the movie *I, Robot* (2004), based upon Asimov's book of the same name, features a killer robot; the Good Doctor, if he had believed in an afterlife, would surely be spinning. Not all robot movies portray machines as entirely malevolent. In the classic *The Day the Earth Stood Still* (1951), based on the Harry Bates story "Farewell to the Master" (1940), the alien robot Gort is not hostile. Nevertheless, Gort is quite willing to inflict violence upon humans if required. In movies, non-Asimovian robots are the rule. I find that rather sad.

8.1.4   AI and Machines that Think

In much the same way that depictions of UFOs tend to mirror the technology of the day, so too do portraits of robot forms. It's not surprising: both are products of the human imagination, and imagination works on the raw material available to it. So early stories talk of artificial humans being formed out of clay or being cast from bronze or iron—such was the technology the authors were familiar with. The Turk built by von Kempelen was supposed to function by clockwork mechanisms—audiences were fooled because clockwork was a technology they understood and could believe in. The robots featuring in pulp magazines were often said to be made of exotic metals or new plastics—they mirrored the materials chemists were developing for the first time. The 'positronic brain' got its name from a subatomic particle discovered in 1932. And so on and so on. Modern-day robot stories might feature 3D printing, cyborgidization, and graphene films. (Indeed, in 2015 Chinese scientists developed a sheet of graphene that behaves as a tiny 'origami robot' able to fold up and walk away on command![2]) Whatever level of technology the authors used, however, the basic notion of robots was of a device built to help people achieve their goals. Typically those goals would involve work tasks, in which case the robots were glorified tools, but they might involve emotional or even spiritual goals. And since humans tend to adapt their environment to make it more comfortable for themselves, most robots were quite naturally depicted as taking a more or less human form.

As Asimov's robot stories developed, however, it became increasingly clear that a focus on a robot's external form was missing the point. The robot *mind* was important, not the particular construction of its body. The interesting question then becomes: can a machine think? If we create an artificial intelligence (AI), if we develop a mechanism that possesses self-consciousness and self-awareness, then we will have done something truly amazing—irrespective of where that intelligence and consciousness is housed.

If AI *is* developed it seems likely to come from advances in computing, but this is a field in which real-world progress has outstripped the science fictional imagination. The SF authors who marvelled in 1946 at ENIAC, the first electronic general-purpose computer, were taken in by the sheer size of this 'giant brain': the thousands of vacuum tubes and diodes, relays and resistors, filled a large room. In general, they didn't foresee the situation in which we now find ourselves—a world in which toddlers have more computing power in their electronic toys than the scientists using ENIAC had at their disposal. One notable exception was Murray Leinster's story "A Logic Named Joe" (1946), which appeared just a few weeks after ENIAC was switched on and described a society in which personal computers are in widespread use and are networked together—much as we have today. In the story, a 'logic'—essentially a computer—develops a sense of self-awareness and ambition, but in a networked environment this leads to problems. With "A Logic Named Joe", Leinster was astonishingly prophetic.

Many other SF stories of that era now seem, at best, naive. In Dickson's "The Monkey Wrench" (1951), for example, the operative of a computer called the Brain claims that his machine is impregnable. The Brain contains twenty banks—what we would now call processing cores, I guess—and most of the time a single bank is all that's needed to solve any problem. If it encounters a more difficult problem then it simply assigns a second bank to the task; for a *really* hard problem the Brain would throw all twenty banks at it. Eventually, by devoting all its resource, the Brain would solve the problem. A smart-aleck visitor gives the following command to the Brain: 'You must reject the statement I am now making to you, because all the statements I make are incorrect.' The Brain immediately devotes the rest of its existence in an attempt to resolve the paradox. Captain Kirk in *Star Trek* was particularly fond of 'talking computers to death' in this way—he saved the day at least four times by making computers or robots face up to some form of paradox. Well, all I can say is if a computer shuts itself down when faced with self-contradiction then it doesn't possess much in the way of artificial intelligence.

Some science fictional computers clearly *are* vastly intelligent. In "Answer" (1954), a 250-word short-short story by Fredric Brown, all the computers in the universe are linked together and humanity asks the question: 'Is there a God?' The computer gives a shocking answer. It's the same answer given in Asimov's "The Last Question" (1956), which begins with the Earth-spanning computer Multivac being asked the question 'can the net amount of entropy of the universe be massively decreased?' and ends with ... well, in case you haven't read this or the Brown story I can't spoil it for you. The point is, these authors suggested a computer—or rather a vast network of computers—could possess unimaginably more intelligence than people.

What would an artificial intelligence of great power actually *do*? Arthur Clarke, in "Dial F for Frankenstein" (1965), imagined how in 1975 the telephone system with its billions of switches and interconnections would match the intricacy of the human brain and somehow 'wake up' and become conscious. The newborn intelligence causes chaos as it examines its surroundings, but when humans realize what is happening and try to switch it off—well, they find it's too late for that. The intelligence has taken care to ensure its continuing existence. (The basic premise of the story would seem to be flawed: the present day internet is much more complex than the telephone system

Clarke described, but it hasn't yet 'woken up'. If consciousness does arrive automatically with a sufficiently complex network then we clearly haven't yet reached the necessary level of complexity. What's interesting, however, is that Sir Tim Berners-Lee cites the story as one of the inspirations that made him think about networks.) Clarke of course went on to create HAL 9000, the famous computer in Kubrick's movie *2001: A Space Odyssey* (1968). HAL is a sentient artificial intelligence that controls the *Discovery One* spacecraft and makes the craft safe for the human crew. Unfortunately HAL develops a malfunction and when it realizes the astronauts plan to shut it down it tries to kill them (much to Asimov's disquiet when he watched the movie; HAL was breaking First Law).

Perhaps the most disturbing representation of what an artificial intelligence might do was given by Harlan Ellison in the story "I Have No Mouth, And I Must Scream" (1967). Three superpowers each develop a supercomputer to help them fight the three-way war in which they are engaged. One of the supercomputers becomes self-aware, absorbs the other two supercomputers, and becomes a vastly powerful—yet deeply damaged—consciousness. This single supercomputer destroys all of humanity except for four men and one woman, and then spends all its time torturing these five people: the computer hates—*really* hates—humanity and all that's left of it. One of the men makes the only possible attempt at freedom. He kills the other four but before he can kill himself the supercomputer prevents the suicide. It then ensures the man will never be able to harm himself—the computer wants to keep that ability for itself. It intends to torture the last survivor forever.

Somehow, stories of disembodied AI tend to be more sombre than robot stories. And yet a useful robot would surely have some element of AI built in. The two concepts, robot and AI, are essentially the same thing. So will the robots of the future have Asimovian Laws built in, preventing them from harming humanity? Or do we really have to fear the coming of our robot overlords?

8.2 Real-Life Robots

There's an American company called iRobot®, which was founded in 1990 by roboticists from MIT. In Asimov's *I, Robot* (and I presume the real company's name is a nod to Asimov) we learn that US Robots and Mechanical Men, Inc., the main manufacturer of robots in his fictional universe, was incorporated just 8 years earlier, in 1982. Fiction and reality are following similar timescales. And just as the founding roboticists at US Robots and Mechanical Men want to widen the appeal of robots to the public, the scientists who founded iRobot® did so with the vision of manufacturing robots for use in everyday life. They are beginning to succeed, and are perhaps making a better job of it than the fictional company. A quarter of a century after its founding, iRobot® has sold more than 10 million home robots—not, it must be admitted, humaniform robots such as R. Daneel Olivaw, but small robotic devices that clean people's homes. These vacuuming robots can be put in a room and then just left while they clean the floor, using a clever algorithm to avoid obstacles and walls. However, these vacbots have been sold to an unsuspecting public without the safeguards provided by Asimov's Three Laws. In January 2015 a South Korean woman took a nap on the floor and woke in agony to find her vacbot was trying to suck up her hair. It took a call to the fire department before she was freed from this robot rage attack.[3] (I should point out that I don't know whether it was one of the iRobot® range of devices that attacked, or a device from one of its competitors.)

That a household device should try to eat a woman's hair sounds vaguely comical (of course, I'm sure it wasn't funny for the victim). It does, however, highlight the point made by the early generation of SF writers, and increasingly by present-day scientists and engineers: perhaps robots, or more generally AI, pose a threat. Before considering this, it's worth taking a brief look at the history of robotics in the real world. Because iRobot® wasn't the first company to build industrial robots—there's evidence for saying the first industrial robot was constructed even before Asimov wrote his first robot story.

8.2.1 Industrial Robotics

The definition of 'industrial robot' is rather dry. The International Organization for Standardization defines it to be 'an automatically controlled, reprogrammable, multipurpose manipulator programmable in three or more axes'. The first machine to meet this definition was constructed in Toronto in 1937 by Griffith P. Taylor, an Australian-born student who was the son of the geologist on Scott's ill-fated Antarctic expedition.[4] Taylor even called his construction a robot—'The Robot Gargantua', to be precise. Taylor's handiwork was quite fantastic: an automatic crane built from Meccano, driven by a single motor, and capable of being programmed through punched tape. His robot could pick up wooden blocks, rotate them, and place them to build structures. The plans for and photographs of The Robot Gargantua were lost for about half a century, but came to light when the Meccano factory in Liverpool was demolished. Only after Taylor's death did people recognize he had developed the first industrial robot—a couple of years before Asimov started writing about them.

The first patent for an industrial robot was granted to the American inventor George Devol, who filed the patent for Unimate in 1954 and had it granted in 1961. At a cocktail party in 1956, Devol met the physicist and engineer Joseph Engelberger and the two of them started to discuss Asimov's robot stories. Engelberger realized that Devol's invention, Unimate, had potential as a robot in the manufacturing sector and the two men formed a business partnership that led to the world's first robot manufacturing company: Unimation. The Unimate robot worked on an assembly line in one of the General Motors factories, taking die castings from one place and transferring them to another. This seems straightforward, but these objects were welded onto car shells, which made it a potentially dangerous work environment for people—the process generated toxic fumes, and there was a risk of accidents to hands and arms. Unimate could perform its task accurately and safely, time after time. In 1966, Engelberger gave presentations about robotics to Japanese manufacturers and this sparked the development of the hugely successful robotics industry in Asia.[5] Robots were soon to be found in automobile production plants around the world; now, as we all know, our cars are put together in large part by robots.

Robots have many advantages over humans in assembly line procedures. They are more powerful than humans; they can perform the same repetitive task without tiring, without losing concentration, without requiring breaks; and they perform tasks accurately. A survey conducted in 2010 showed the wide array of activities that robots were able to undertake. The commonest task for robots was in handling operations: moving pallets, machining metal, and moulding plastic. The second most common was welding: robots can spot weld and arc weld, and do so mainly in the automotive industry, but increasingly as prices drop they can be found doing small shop work. A third category of robotic use was in assembly: fixing, inserting, dis-assembling, and so on. Other applications include dispensing (painting, gluing, sealing, spraying) and processing (mechanical and laser cutting, for example). In the past these tasks were performed by humans. Indeed, children of my age were advised when leaving school how mastery of these tasks could lead to a decent job. Now, such tasks are performed mainly by robots (Fig. 8.4).

Assembly line production is an ideal environment for a robot. If a line is producing steel widgets, say, each one of which must be identical within given tolerances, then a robot worker will outperform a human worker. But robots are moving out from the assembly line environment. In the forthcoming years, as robots become increasingly intelligent and mobile, many tasks currently undertaken by humans will be done instead by our metal friends. For example, farmbots have already found a role inspecting crops and picking weeds; in 2016 a trial started in Australia for a farmbot to herd livestock and check the

Fig. 8.4 Robots, real-world industrial robots rather than the clunking human-like robots of our imagination exemplified by the illustration in Fig. 8.3, are a common sight in many manufacturing fields. They are used in aerospace applications, the automotive industry, palletizing operations, ... pretty much everywhere (Credit: KUKA Systems GmbH)

quality of grazing pasture.[6] Robots are being used in surgery, in military missions (particularly in bomb disposal), in duct cleaning, in underwater exploration, in Martian exploration... robots already perform many activities better or more quickly than humans, and with each passing year the range and reach of robot activity increases. The future foretold by Asimov, where people must fear for their jobs because of competition from robots, is certain to come to pass for a large section of the work force. For some, it already has.

Asimov was a technophile. Yes, he understood how some people would resent the march of robots; but he believed if robots did the boring work they'd free up humanity to engage in pleasurable activities, to undertake research, to pursue artistic endeavors. Asimov pictured a future in which robots brought about a new golden age. It's a picture I'd like to agree with, but recent thinkers strike a more cautious note[7] and warn of a forthcoming class of people who might be called 'useless'. This is not a moral definition. It's not that those people will be less worthy of respect than anyone else; they'll be fully deserving of value and consideration. It's just that the 'useless' class won't fit in to our current economic and political systems. They'll have no productive employment, no job requiring them to get up in the morning and do something. And the 'useless' people won't only be those who currently have manual jobs: many jobs that rely on cognitive skills, the sort of jobs currently being filled by university graduates, will soon be done better, faster, and more economically by robot. So will the rise of the robots lead to a miserable future in which most of humanity lacks focus and a sense of purpose? Or will we, as Asimov hoped, join forces—with robots doing what they are good at while humans continue to do the stuff we excel at? We'll soon find out. I suspect the answer might well depend more upon political and economic factors than scientific ones. While we still have the chance, society needs to think deeply and discuss widely how robot-produced wealth should be distributed amongst the population. Science fiction can provide valuable input into those thoughts and discussions. Society needs to take SF seriously here.

Industrial robots started on industrial assembly lines, branched out into moving and stacking pallets, and are now taking over hitherto graduate-level jobs. But they are still *industrial* robots. They follow pre-programmed instructions, typically act in well-defined environments, and are generally kept away from humans. They aren't exactly R. Daneel Olivaw. But even that is changing—the latest generation of robots are beginning to operate in the messy environment of the real world.

8.2.2 Run, Robot, Run

ASIMO—Advanced Step In Innovative Mobility—is a robot to rival Robby in terms of cuteness (Fig. 8.5). It's a humaniform robot, stands a non-threatening 1.3 m high, and can perform a number of sophisticated actions: walk (at a respectable pace) both on the flat and up stairs; kick a ball (better than a number of English soccer players I could mention); and respond to voice commands. It was developed by Honda in 2000, and with each iteration its abilities have improved. ASIMO is a glimpse of how robots could move away from specific tasks and environments out into the human realm. It's easy to imagine how such a robot could *cooperate* with humans on a task rather than *replace* humans who currently do a task—they'd be cobots, I guess.

I'd love to have ASIMO trotting round the house, but Honda doesn't sell the devices—and even if it did I doubt that I could afford one. At least, I couldn't afford one *now*. In a few years time, however, ASIMO and his cousins will become cleverer and thus useful (at present, despite the gimmicks,

Fig. 8.5 The Honda-built robot ASIMO shakes hands with the Serbian TV presenter Zoran Stanojevic at the 2012 'Days of the Future: Robotics' festival held in Belgrade (Credit: Honda)

ASIMO is merely adorable). The more useful such robots become, the more it will make sense for people to own them. The more people want or need to own them, the more a legal framework will be put in place to facilitate such ownership (at present, I suspect the reason Honda doesn't sell ASIMO has more to do with legal worries than anything else). And once there's a large enough market the cost of production will plummet. This technology will scale well. Soon robots won't be stuck in factories; they'll be out there in the wide world.

Boston Dynamics, an American robotics company, has developed a number of quadruped robots. BigDog (a DARPA-funded machine designed to carry packs for soldiers in rough terrain) and Cheetah (a robot that can run on the flat at 45 km/h) aren't as cute as ASIMO but they seem to be more practical. As it turned out, the BigDog project was shelved: it was too loud for use in military work. But BigDog proved a robot can operate in challenging conditions: it can walk across rubble, trudge through snow and water, climb slopes of up to 35 degrees—all while carrying a 150 kg load (Fig. 8.6).

With earlier robots, even with ASIMO, if you push them then they fall over. The impressive thing about BigDog, even more impressive than its capacity to carry heavy packs over uneven ground, is its ability to maintain its balance after suffering a knock. In order to demonstrate this ability Boston Dynamics engineers filmed themselves kicking a BigDog called Spot. They posted the video on YouTube in 2015, and the footage caused uproar. When Spot was

Fig. 8.6 Two BigDog robots, developed by Boston Dynamics (Credit: DARPA)

kicked it briefly stopped, its machinery whirred a little, and instead of tipping over it regained its balance. But for a brief instant, when it stopped, Spot seemed to cower; the whirr sounded like a whimper. The footage upset me, as it upset many people who posted comments on the video. My head tells me that feeling sympathy for a robot dog being kicked is as silly as feeling sympathy for the laminate flooring I walk on every day: in neither case is any suffering involved. It's particularly silly in this case, since the whole point of BigDog was to send it into theatres of war—places where it could suffer much greater damage than a gentle nudge with a foot. My heart, though, responded with sympathy. The kick looked like an act of cruelty.

As robots become more advanced, able to do more things, perhaps we'll increasingly have to ask whether it is ethical to order an object such as ASIMO to engage in dangerous acts. As long as robots are tools, as long as they have no capacity for suffering, then I suppose the hard-headed response must be to put our sympathy on hold. However there is, of course, the more pressing question when it comes to ethics and robots: how can we ensure *they* don't harm *us*?

8.2.3 The Robot's Dilemma

As mentioned earlier, Asimov argued that whenever people develop a new tool they also develop safeguards to ensure the tool can be used efficiently and without danger. He believed the same approach would be adopted with robots, and his Three Laws were an attempt to explore how such safeguards might work with tools that possess intelligence. Three quarters of a century after Asimov started writing about robots, we are now having to grapple with the moral and ethical issues surrounding robot use.

As we shall see in Chap. 9, and as many readers will be aware, governments and private companies are scrutinizing the issues surrounding driverless cars. Already, numerous companies have some form of driverless vehicle on our roads. Driverless vehicles are essentially robots—not humaniform robots, but robots nonetheless—and if they operate in the messy world of humanity then they'll have to make difficult decisions. What Laws of Robotics will be programmed into these vehicles? If a driverless car is forced to decide between swerving left (and killing a child), swerving right (and killing a pensioner) or going straight on (and killing its passenger)—what will it do? What *should* it do? It's easy to come up with any number of such scenarios, situations that put the Three Laws to the test; it's far harder to decide on what should be the correct course of action, and make explicit *why* that is the correct ethical choice.

The American military are grappling with similar issues. They already deploy drones to direct missiles against terrorist groups. At present, as far as I'm aware, the drones are controlled remotely by soldiers. Soon, though, those drones will be able to work autonomously; they'll be told to find a target, use its judgement as to whether the target has indeed been found, and then eliminate the target. That killer drone will be a robot, albeit a robot without First Law. The military are painfully aware of the difficulties surrounding this technology, because they are already facing them. What happens if the target, perhaps a terrorist responsible for many deaths, hides in the midst of a group of terrorists—is it acceptable to eliminate the target if the rest of the terrorist group will also be killed? What if the target hides with his family, who are all quite innocent—is it still acceptable to attack? What if the target hides in a nursery full of infants—still acceptable?

The American military have asked Ronald Arkin, a roboticist at Georgia Tech, to develop rules for robots to follow in combat situations. Arkin's lab has developed an 'ethical adapter'—a program that monitors the degree of destruction a weapon is causing against how much was expected. If the difference is too large the ethical adapter—an emulation of guilt—causes the robot to stop using the weapon. It's not First Law exactly, but it's an attempt to build some sort of moral compass into a robot. As Arkin himself admits, many situations don't reduce to a shoot/don't shoot decision, but he argues that an ethical adapter will make fewer errors than human soldiers because it won't be affected by panic, confusion, or fear. Perhaps a robot might make more consistent moral choices than humans.

It's encouraging that scientists are attempting to achieve something similar to Asimov's Three Laws. It shows we are trying to reach a situation in which robots can mingle safely with humans. Nevertheless it's worth remembering that Arkin, whose lab worked on the ethical adapter, was funded in his research by the military; it's worth bearing in mind that the most advanced robots are military robots. And, ultimately, the whole point of military robots is to kill people. Not much First Law going on there. We need to remember, too, that clever people can also be evil people. Seemingly impregnable computer systems have in the past been hacked by criminals, terrorists, and rogue governments. A robot is essentially just a computer with extra peripherals, so even a Three-Laws-abiding robot might be hacked by bad guys and turned into a weapon.

Robots are coming. Perhaps, just as with people, the majority of them will be honorable and upstanding. But, just as with people, there might be some bad ones in the mix.

8.2.4 All Hail Our Robot Overlords?

The robots we have at present are restricted to circumscribed environments where they perform well defined tasks: they are glorified widget-shifters. The robots of tomorrow will interact with us in the human sphere. Whether that will be a good thing (with robots offering companionship and care for the elderly, and freeing young people from the drudgery of repetitive labor to concentrate on high-value tasks) or a bad thing (with robots rendering billions of people jobless) remains to be seen. But what about the day after tomorrow—what will robots look like then? Some futurists believe computational power will continue to increase and result in machines that think faster, better, and more deeply than people. Humans will no more be able to understand the motivations and achievements of these robots than monkeys can appreciate the reasons for constructing the Large Hadron Collider. Robots will be our intellectual masters. In which case—what? The best-case scenario seems to be they ignore us. The worst-case scenario is they get rid of us.

A large number of extremely bright people consider the development of advanced robot intelligence to be inevitable, and that artificial intelligence will eventually surpass human intelligence. After that point AI will have the capacity to develop even more powerful AIs, and then some sort of runaway effect—an intelligence explosion—will take place. A common prediction is for this event to occur some time later this century.

At first glance the prediction of a so-called Singularity, an upcoming event involving ever-increasing levels of machine-based intelligence, might seem outrageous. But that's because our minds tend to imagine the future by making linear rather than exponential extrapolations. If the world's computational capacity continues to increase exponentially then fantastic things will, presumably, become possible. The prediction of a Singularity is rooted in science rather than fiction. But is the coming of the Singularity as inevitable as the futurists believe?

One problem with the whole notion of AI is the lack of a full understanding of what it is that constitutes intelligence. Humans, it seems, have different types of intelligence: a general-purpose intelligence, certainly, but also specific modes of intelligence that we apply at different times and in different situations. Intelligence is much more than the number gained on an IQ test. Computers have long been able to beat the best human at chess, for example, and in 2016 a computer beat the best human at the boardgame Go. Are these game-playing computers intelligent? Presumably not—although the humans who undertook the research to make these computers possible are certainly intelligent. And the current level of commercial AI, the sort of 'smart' assistant that makes purchasing recommendations, are so stupid as to be irritating. Present-day computers are undeniably far more advanced than earlier versions, but they certainly don't possess anything approaching a general-purpose intelligence. Tomorrow's computers will be even more advanced, but there's no roadmap for how they might arrive at machine-based general-purpose intelligence, much less superintelligence.

Even if some sort of machine with advanced intelligence *is* developed, it doesn't necessarily follow that the same machine would also develop volition. For example, the artificial neural networks in current labs can learn to recognize dogs in photographs. The learning process involves exposing the networks to hundreds of thousands of photographs, and the result is far from impressive—typically, a neural network is worse at the dog-recognizing task than a young child—but to recognize a dog at all is an achievement when one considers how just a few years ago no computer would have possessed the power to attempt the task. In the future, as processors become even quicker and algorithms improve, computers *will* get better at recognizing dogs in photographs. But even if they get so good at the task they can outperform humans, they still won't be recognizing 'dog-ness'. They won't understand what else is going on in the photograph. They'll bring none of the underlying associations a human would bring to the interpretation of the photo. They won't have a favorite photo. The dog-recognizing robot might outperform a human at that one particular task, just as chess-playing robots currently outperform humans at chess, but it wouldn't because of that success necessarily have a sense of volition. It wouldn't be sentient. It wouldn't be conscious.

I don't believe I'm being a luddite when I write that intelligent machines will not possess volition. It seems to me there's a widespread assumption that piling more and more processing power into a smaller and smaller volume will inevitably lead to consciousness. But intelligence and consciousness are two quite different things. Scientists might eventually agree on an approach that could lead to intelligent robots, but at present there isn't even an agreed definition of what consciousness *is* let alone a well defined research program to deliver the phenomenon. The scientists who work in the field of AI are themselves all extremely intelligent, and perhaps it's natural for them to consider intelligence to be the defining characteristic of humanity. But humans are conscious beings, and the characteristics which go alongside—compassion, love, hope; and darker compulsions, too—are just as important as intelligence to our success as a species. An intelligent robot without those qualities is just an intelligent robot: its actions will continue to be guided by human orders. Tell a robot to clean your house and that's what it will do. It won't, as a human might, trash the place, steal your money, then go on a drunken bender.

Personally, then, I don't think we need fear a robot uprising. Even if the threat is greater than I'm making out, we still have time to understand the dangers and put safeguards in place. Science fiction has long warned us of the perils of AI; I'm with Asimov in believing we can reap the benefits of intelligent robots while mitigating the risks.

8.3 Conclusion

The robots are coming. There can be little doubt of that. Indeed, they are already here. The question is, how long will it be before they start to move widely among the human population? And, when they are ubiquitous, how will our society respond to them? Robots only need a *little* bit more intelligence to be better at many of the tasks currently performed by humans—will robots then replace human workers, or will their presence among us lead to jobs and activities that are as yet undreamed of? We'll soon find out which of those scenarios, already well sketched by science fiction writers, will come to pass.

As for the possibility of robots somehow *replacing* humans—well, as I've explained, personally I find that unlikely. I don't see where the drive for a robot rebellion would come from. Intelligence is not the same thing as consciousness. What I do find plausible is that, rather than AI posing a threat, IA will instead prove to be our species' salvation. IA (where the acronym stands for intelligence amplification rather than Isaac Asimov, though it's a topic Isaac discussed in his stories) will take the best of human and artificial intelligence to create beings surpassing the capabilities of robots and people. By joining forces—human *and* machine rather than human *or* machine—we will possess the tools required to solve the problems that currently seem so impervious to solution. Together, we might survive into the distant future.

Notes

1. For a complete set of Asimov's positronic robot short stories you need to seek out a number of anthologies: *I, Robot* (1950) and *The Rest of the Robots* (1964) are the best known and probably the easiest to access, but later stories can be found in *The Complete Robot* (1982). The novels are *The Caves of Steel* (1954), *The Naked Sun* (1957), *The Robots of Dawn* (1983), and *Robots and Empire* (1985).
2. For details of the origami robot, see Mu et al. (2015).
3. See McMurray (2015) for details of the vacbot attack.
4. Taylor's achievement in constructing 'The Robot Gargantua' was described in Meccano (1938).
5. For the early history of industrial robots see the book by Engelberger (1983), which contains a foreword by Asimov.
6. For details of the Australian farmbot trial, see Klein (2016).
7. See for example Harare (2016) for a discussion of what the rise of robots might do to the world of work.

Bibliography

Non-fiction

Engelberger, J.: Robots in Practice: Management and Applications of Industrial Robots. Springer, Berlin (1983)

Harare, Y.N.: Home Deus: A Brief History of Tomorrow. Harvill Secker, London (2016)

Klein, A.: Robot ranchers monitor animals on giant Australian farms. New Scientist (May 20, 2016)

McMurray, J.: South Korean woman's hair 'eaten' by robot vacuum cleaner as she slept. The Guardian (February 9, 2015)

Meccano: An automatic block-setting crane. Meccano Mag. **23**, 172 (1938)

Mu, J., Hou, C., Wang, H., Li, Y., Zhang, Q., Zhu, M.: Origami-inspired active graphene-based paper for programmable instant self-folding walking devices. Sci. Adv. **1**(10), e1500533 (2015)

Fiction

Anonymous: Stranger Than People. Young World, London (1968)

Asimov, I.: Strange Playfellow. Super Science Stories (September 1940)

Asimov, I.: Liar. Astounding (May 1941)

Asimov, I.: Runaround. Astounding (March 1942)

Asimov, I.: I, Robot. Doubleday, New York (1950)

Asimov, I.: The Caves of Steel. Doubleday, New York (1954)

Asimov, I.: The Last Question. Science Fiction Quarterly (November 1956)

Asimov, I.: The Naked Sun. Doubleday, New York (1957)

Asimov, I.: The Rest of the Robots. Doubleday, New York (1964)

Asimov, I.: The bicentennial man. In: del Rey, J.-L. (ed.) Stellar-2. Ballantine, New York (1976)

Asimov, I.: The Complete Robot. Doubleday, New York (1982)

Asimov, I.: The Robots of Dawn. Doubleday, New York (1983)

Asimov, I.: Robots and Empire. Doubleday, New York (1985)

Asimov, I., Greenberg, M.H.: Isaac Asimov Presents the Great SF Stories 1 (1939). DAW, New York (1979)

Bates, H.: Farewell to the Master. Astounding (October 1940)

Binder, E.: I, Robot. Amazing (January 1939)

Brown, F.: Answer. In: Angels and Spaceships. EP Dutton, New York (1954)

Čapek, K.: R.U.R. (translated into English 1923) (1920)

del Rey, L.: Helen O'Loy. Astounding (December 1938)

Dick, P.K.: Second Variety. Space Science Fiction (May 1953a)

Dick, P.K.: Impostor. Astounding (June 1953b)

Dick, P.K.: Do Androids Dream of Electric Sheep? Doubleday, New York (1968)

Dickson, G.R.: The Monkey Wrench. Astounding (August 1951)

Ellison, H.: I Have No Mouth, and I must Scream. IF: Worlds of Science Fiction (March 1967)

Kelley, J.E.: Rust. Astounding (October 1939)

Leinster, M.: A Logic Named Joe. Astounding (March 1946)

Phillips, P.: Lost Memory. Galaxy (May 1952)

Sheckley, R.: The cruel equations. In: Can You Feel Anything When I Do This? Doubleday, New York (1971)

Vincent, H.: Rex. Astounding (June 1934)

Williams, R.M.: Robots Return. Astounding (September 1938)

Williamson, J.: With Folded Hands. . .. Astounding (July 1947)

Visual Media

2001: A Space Odyssey: Directed by Stanley Kubrick. [Film] USA, MGM (1968)
Blade Runner: Directed by Ridley Scott. [Film] USA, Warner Bros (1982)
Forbidden Planet: Directed by Fred M. Wilcox. [Film] USA, MGM (1956)
I, Robot: Directed by Alex Proyas. [Film] USA, 20th Century Fox (2004)
Lost in Space: [TV Series] USA, 20th Century Fox (1965–68)
The Day the Earth Stood Still: Directed by Robert Wise. [Film] USA, 20th Century
    Fox (1951)
The Terminator: Directed by James Cameron. [Film] USA, Orion (1984)
Westworld: Directed by Michael Crichton. [Film] USA, MGM (1973)

9

Transportation

The trouble with most forms of transport, he thought, is basically one of them not being worth all the bother.

The Restaurant at the End of the Universe
Douglas Adams

The building of nations and empires depends upon the capacity to move goods and people. The nineteenth and twentieth centuries gave us many examples of how transportation affects economics, politics, and the workings of democracy. It's hardly surprising, then, that SF writers have a long history of imagining new forms of transport: they were hardly going to ignore one of the main drivers of societal change. Authors interested in telling tales of galaxy-spanning civilizations had an absolute requirement to discuss transportation: how else could the adventures unfold without some new form of transport? Authors who wanted to tell more local tales of the future still needed technology to get their protagonists from A to B as quickly as possible, just as earlier generations of storytellers used flying carpets and capricious genies to move their heroes, and stories, along.

I discussed space travel in an earlier chapter, so here I'd like to look at more mundane forms of transport. A number of authors of hard SF were interested in exploring the forms that new transport technologies might take and in how those new technologies would affect individuals and society. Heinlein, for example, in "The Roads Must Roll" (1940), postulated a future in which moving roadways were common—and the chaos that would be caused if roadway maintenance engineers went on strike. Asimov, in *The Caves of Steel* (1954), imagined a day when people might move about their cavernous

© Springer International Publishing AG 2017
S. Webb, *All the Wonder that Would Be*, Science and Fiction,
DOI 10.1007/978-3-319-51759-9_9

cities using rolling strips—and the different uses to which people would put those strips. The technology sketched in such stories was sufficiently futuristic to promote that all-important sense of wonder, but imagined so vividly it seemed we'd live to see. A small number of authors made a variety of exciting predictions about how we'd travel—and it's the spectacular failure of some of those predictions that the mainstream holds against SF, and even against science itself. Journalists like to point out, as if we hadn't noticed, how we lack the flying cars and hoverboards of the *Back to the Future* films. The title of one recent book encapsulates the disappointment: *Where's My Jetpack?* (2007). We were promised shiny metallic suits for use with our personal flying packs, but all we got was cars with better paint jobs and incrementally improving petrol consumption figures. Where did it all go wrong?

Actually, science fiction's record of transport-based predictions is slightly more nuanced than all the 'jetpacks and flying cars' criticism would imply. And the current situation regarding transport is more fluid than ever—the perfect time for science fictional input.

9.1   Around the World in 80 Ways

In 1763 an anonymous English author published a book called *The Reign of George VI, 1900–1925*. The author, looking a century and a half ahead, foresaw a rosy future for the canal barge as an all-round transport medium. From our vantage point it's easy to laugh at the naivety of that author's prediction, but we should remember that he (I'm assuming the author was male) lived in a world in which most people moved about this planet using the same methods that were available in the time of Jesus. People walked, or they made use of animals, or they sailed.

The opening of the Stockton and Darlington railway in 1825 saw the world begin to open up. As I mentioned in Chap. 1, it was the railways that saw my home town spring into being and it was the railways that allowed my ancestors to move to that town. But much more importantly than that, the railways opened up Europe and America and allowed humankind to vastly increase its reach. The steam engines that powered trains could also be used to power ships; in 1840, Isambard Kingdom Brunel developed the screw propeller for his ship the *Great Eastern*, and the oceans themselves seemed to shrink in size. It's hardly surprising, then, that the revolutions taking place in transport and the possibilities opening up for exploration, tourism, and trade fired the imagination of authors such as Jules Verne.

Verne was fascinated by devices that made travel possible. His first novel was *Five Weeks in a Balloon* (1863). A few years later, at about the time he was publishing an expanded version of *Journey to the Centre of the Earth*, he wrote an account of his voyage on Brunel's *Great Eastern*. In 1870 he published the classic *Twenty Thousand Leagues Under the Sea*, and for many readers the most interesting character in this book was not the megalomaniac Captain Nemo but the futuristic vehicle—the submarine *Nautilus*. Verne followed this 3 years later with another classic, *Around the World in 80 Days*.

Verne's tales of air, submarine, and subterranean exploration demonstrated how the world had changed. When it came to transportation, the canal barge no longer represented the outer limit of human imagination. What, then, did the SF imagination promise us when it came to travel?

9.1.1 Jetpacks

At first glance the jetpack—a device strapped to your back that lets you fly off like a bird—is a quintessentially science fictional notion. Indeed, articles and cartoons routinely use the term 'jetpack' as a shorthand for science fiction.

The jetpack dates back at least to E.E. Smith's description of the flying harness in his 1928 novel *The Skylark of Space*. (Figure 2.1 depicts the harness.) At around the same time, Buck Rogers was making his first appearance. His intertron-based antigravity belt, as discussed in Chap. 2, was a form of jetpack. The 1949 production *King of the Rocket Men*, the sort of Saturday-morning cinema serial my father's generation enjoyed and that I lapped up as a child when it was shown later on television, featured a scientist-hero packing all the necessary accoutrements with which to combat the evil Dr Vulcan: ray gun, bullet-shaped helmet, and rocket backpack. And of course there was the famous example of James Bond using a jetpack to escape the baddies in *Thunderball* (1965).

When sitting down to write this chapter, I assumed I must have read hundreds of stories involving jetpacks. But, you know, I really can't think of too many decent SF stories featuring a jetpack. There's a jump harness in Heinlein's novel *Stranger in a Strange Land* (1961), but Heinlein didn't really provide much elaboration—and it wasn't even a proper flying jetpack. Science fiction comes in for far too much jetpack-related criticism: despite what the critics say, the jetpack has never been a big deal in SF. Let's move on.

9.1.2 Planes, Trains, and Automobiles

Some of my favorite science fictional vehicles are ships: the side paddle wheeler *Not for Hire* built by Samuel Clemens in Philip Farmer's wonderful *The Fabulous Riverboat* (1971); the steamboat *Fevre Dream* in George Martin's novel of the same name (1982); the 'cruiser' *Selene* in Arthur Clarke's *A Fall of Moondust* (1961). A ship is a ship, though, even if in the case of *Selene* it floats on seas of lunar dust. A ship somehow doesn't *feel* particularly science fictional. Here, therefore, I'd like to look in a little more detail at something slightly more modern—the airplanes, trains, and cars of SF.

9.1.2.1   Air Travel

SF writers had been contemplating the future of flying vehicles long before the Wright brothers pioneered heavier-than-air flight in 1903. In 1886, for example, Jules Verne wrote about an electrically-powered aircraft in his novel *Robur the Conqueror; Or, A Trip Around the World in a Flying Machine*. And in 1899, H.G. Wells published his novel *When the Sleeper Wakes*. Where Verne was interested in the mechanics of flight, and how it might be achieved, Wells was more interested in exploring the use to which mankind would put the ability to fly. Wells reached the conclusion that aircraft would be used for warfare, and *When the Sleeper Wakes* describes spectacular aerial battles. Wells returned to the subject of aerial warfare several times, including the novels *The War in the Air* (1908) and *The Shape of Things to Come* (1933). The latter novel describes a world destroyed by wars and the rise of a benevolent 'Air Dictatorship'. For once, though, Wells was not the originator of an SF idea: Rudyard Kipling, an author not best known for his science fiction, wrote about the Aerial Board of Control and the establishment of a 'Pax Aeronautica' in his stories "With the Night Mail" (1905) and "As Easy As A.B.C." (1912). So it's clear that, as a field, SF was addressing the possibilities inherent in aviation from the very beginning. The field's interest in aviation did not last long, however.

In April 1926 Hugo Gernsback began the modern era of science fiction by establishing *Amazing Stories*. It was the first magazine devoted to the genre. Three years later, in July 1929, he founded the magazine *Air Wonder Stories*. In its first editorial Gernsback promised his new magazine would publish 'solely flying stories of the future, strictly along scientific-mechanical-technical lines' with the intention of preventing 'gross scientific-aviation misinformation from reaching our readers'. The magazine's covers were wonderful creations by

Frank R Paul; many of the original stories were from established *Amazing* authors, including the great Jack Williamson; and Gernsback also used the opportunity to reprint some well known tales of aviation-based action–adventure. (Of the latter, perhaps the best known was George Allan England's novel *The Flying Legion* (1920).) And yet the magazine lasted only 11 issues. Gernsback quickly realized the market could only take a limited number of stories about airplanes. The rate of technological progress was already removing the elements of heroism and romance from aviation and, of course, within a few decades of the Wright Brothers' pioneering flight at Kitty Hawk civil aviation was a reality. The readers of *Air Wonder Stories* wanted stories about *interplanetary* travel. The imagination of science fiction writers quickly moved from air travel to space travel. Gernsback understood this. He merged his aviation magazine with *Science Wonder Stories* and called the hybrid *Wonder Stories*.

Perhaps the speed with which aviation became a part of everyday life explains why aircraft play a significant role in so few SF stories. At least, I struggle to remember reading much science fiction about air travel. Arthur Clarke's novel *Glide Path* (1963) immediately comes to mind—but then I remind myself that this wasn't SF. *Glide Path*, Clarke's only non-SF novel, was based on his wartime service with the RAF and tells the story of a radar-based airplane landing system. There's Fritz Leiber's multi-award winning story "Catch that Zeppelin!" (1975)—but this belongs not in the aviation but in the alternative-history subgenre. In the story, the protagonist's sighting of a blimp moored to the Empire State Building triggers memories suggesting he has lived a quite different life in some alternate universe. And I can't forget John Varley's superb short story "Air Raid" (1977), which he later expanded into a novel *Millennium* (1983). (As so often happens, a good novel formed the basis of a mediocre film of the same name.) In "Air Raid" airplanes play a key role, but this is a time-travel story rather than a tale of aviation. Could it be that science fiction stories about airplanes pretty much ended with the demise of *Air Wonder Stories*?

9.1.2.2   The Locomotive

Richard Trevithick ran the first railway steam locomotive in 1804 along a track at the Penydarren Ironworks in Wales. When, a few years later, George Stephenson and others improved the technology, humanity was propelled, as I've mentioned before in this book, towards an industrial world. The achievements of Trevithick and Stephenson to the pre-Victorian age must have seemed like those of Gagarin and Armstrong to us. Perhaps their influence was greater: the Victorians had the imagination to see how railways could change the world whereas our crewed exploration of the solar system appears to have halted. So it would have been surprising if early SF writers had failed to explore railway-related themes or chosen not to use the locomotive as a leitmotif in their work.

In *The Mummy!, or a Tale of the Twenty-Second Century* (1836), written by a pioneer of science fiction, Jane Louden, and published just 11 years after the opening of the Stockton–Darlington railway, imagines a future in which houses move across the land on railways. An 1846 French novel by Emile Souvestre, translated into English as *The World as it Will Be*, imagines a steam locomotive that can travel through time as well as space—a sort of steam-powered Tardis. And Verne wrote about locomotives (albeit with the addition of his own speculative technological improvements) in the 1863 novel *Paris in the Twentieth Century*. In the case of railways, however, the imagination of science fiction writers outstripped the rapid pace of technological innovation. Even more quickly than with planes, stories involving trains and railways began to drop out of fashion (though they continued to provide a backdrop for mainstream writers).

There were occasional speculations about the future of trains. Hugo Gernsback, for example, wrote a novel *Ralph 124C 41+* (1911). You can read most of the novel for free, online, but to be honest it's about as exciting its title suggests. Nevertheless, if you're willing to put aside the clunky writing and the absence of a decent plot, you'll delight in the cornucopia of futuristic gadgetry that Gernsback foretells—including a trans-Atlantic subterranean magnetic train running from New York to Brest. But in later years, if an SF writer had to choose between setting a futuristic story on a train or on a space ship...well, it was no contest. Science fiction pretty much abandoned stories with a railway theme.

There were some exceptions. My all-time favorite SF tale involving trains is a short story by A.J. Deutsch.[1] Armin Deutsch was an astronomer who, when commuting to work at Harvard one day, started to ponder the complexity of the Boston subway system. His daydream led to him writing "A Subway Named Mobius" (1950), a tale in which the addition of a new railway line makes the topology of the Boston subway system so complex that a train vanishes into a fourth spatial dimension. It's a fun story, even if the mathematics is inaccurate in some places. Although the events described in the story of course couldn't happen in reality, I'm always reminded of "Mobius" whenever I have to navigate the London tube system. (Figure 9.1 shows a

Fig. 9.1 The famous designers Charles and Ray Eames developed their 'Mathematica' exhibition in the early 1960s. One of their exhibits was a train that ran forever on a track containing a half-twist. This is a replica of the original, which is on show in the New York Hall of Science. The exhibit brings to mind Deutsch's story "A Subway Named Mobius" (Credit: Ryan Somma)

museum exhibit, a train that loops forever along a Möbius track.) For the most part, however, Golden Age SF paid little attention to trains and railways.

It was six decades after the appearance of *Ralph 124C 41+* that trains made something of a comeback. Harry Harrison revived Gernsback's notion of a transatlantic railway tunnel, although his *A Transatlantic Tunnel, Hurrah!* (1973) was hardly a piece of SF in the Gernsback tradition; rather, it was a satire taking place in an alternate history in which Washington lost the Battle of Lexington and America remained a British colony. A year later the novel *The Inverted World* (1974), by Christopher Priest, described a post-apocalyptic city that moved on a railway track. Priest's novel has one of the most intriguing first sentences in all of science fiction: 'I had reached the age of six hundred and fifty miles'. The reader is immediately drawn in, and learns how the city's inhabitants must take up track behind the crawling metropolis and lay it down in front, in order to keep their home moving—the city is doomed if it stops. Larry Niven (in *A World Out of Time* (1976)) and John Varley (in *The Ophiuchi Hotline* (1977)), in novels filled with descriptions of technological marvels, took the chance to update Gernsback's concept of a subterranean gravity train. And Greg Bear, in his novel *Eon* (1985), produced in passing an imaginative description of a future train: a carriage carried by a giant, trackless, aluminum millipede!

Railways were perhaps never at the cutting edge of science fiction. But neither did they ever entirely disappear from the field.

9.1.2.3   The Internal Combustion Engine

Vehicles powered by the internal combustion engine have transformed societies—for both good and ill—across the planet. Own a car and you have a freedom of travel your forebears would have thought impossible, while the vast fleets of commercial vehicles have revolutionized logistics and given you almost instant access to a plethora of goods. On the other hand, the cost of that freedom and that ease of access to goods might be catastrophic: engine emissions could cause irreversible environmental damage.

The issues thrown up by near-universal levels of car ownership—big issues, such as personal freedom of movement versus harm to the global environment—are precisely the sorts of question SF tries to address. So although I could recall reading few stories involving cars, I thought a little research would quickly uncover dozens of prescient stories: comic pieces about rush-hour gridlock; dystopian stories bemoaning the creation of continent-spanning motorways; satires ridiculing our devotion to automobiles. But I failed to find many good Golden Age examples. It's as if SF writers failed to appreciate what was happening under their noses: they were living through a revolution and they either didn't notice or didn't much care. Perhaps they failed to appreciate that this was a technology capable of scaling, even though it was car manufacturing that framed the very definition of mass production.

There were, of course, exceptions to this science fictional silence relating to cars. "Sally" (1953), one of Asimov's positronic robot tales, depicted a future in which cars have built-in 'brains'. In the story it is these artificial brains, rather than humans, who drive the cars. Given the current state of technology

Asimov here seems quite prescient. He was less prescient when describing car-to-car communication: rather than use the sophisticated wireless communications we have come to expect, the positronic brains honk horns and cause engine knocking in order to pass messages. Rather than dwell on technical issues, however, "Sally" raises the question of what would happen if our intelligent automobiles decide they no longer want to chauffeur us humans around. Perhaps it's a question we'll soon have to address.

And there was "Killdozer!" (1944), the only story Theodore Sturgeon wrote while the US was at war. I remember coming across the tale in numerous anthologies. It was an interesting story, complete with technical descriptions of the workings of a construction vehicles, but it was hardly Sturgeon's best work. Furthermore, the vehicle of the title—a bulldozer possessed by a sentient killing machine developed by a long-dead civilization—was hardly going to be the automotive vehicle of the future.

I fondly remembered reading Ray Bradbury's story "The Pedestrian" (1951), which is set in a future where people don't do much except stay indoors and watch TV. The protagonist has an unusual hobby: he's the only person who likes to walk, in a city whose roads have fallen into disrepair. A robotic police car, the only car in the city, comes across him one night; it can't understand why this man would be outside, walking. It's a typically enjoyable Bradbury story—but it was hardly an accurate forecast of our vehicular future.

Those few stories were the only ones on the theme that made any impression on me. After some searching I came across a 1928 story in *Amazing*, called "The Revolt of the Pedestrians". It was the first published story by David H. Keller, a psychiatrist who went on to write for Gernsback's magazines. The story rather ignores the lessons of Darwinian evolution by positing a future in which a ruling elite lose the use of their legs because of their reliance on cars, but it was arguably a more realistic forecast of our vehicular future than Bradbury's "The Pedestrian". In a later story, "The Living Machine" (1935), Keller tells of an inventor who is inspired to develop a self-driving car when a moron driving a traditional automobile fails to spot him and knocks him into a ditch. As far as I'm aware, this story is the first appearance in SF of the self-driving car.

And that's pretty much it. Written SF must contain more car-related stories than mentioned here, but nowhere near the number the concept deserves.

Films are more interesting, if only because they often feature a 'proper' SF concept: flying cars. *Star Wars* (1977) showed us the landspeeder—a car-like vehicle that hovers just above ground. (The limitations of special effects departments back when *Star Wars* was produced meant Luke Skywalker's landspeeder really *was* a car-like vehicle: it was based on the chassis of a three-wheeled British car called a Bond Bug, with angled mirrors hiding the wheels.) Ridley Scott's *Blade Runner* (1982) featured 'spinners', a police vehicle that could be driven like a normal car but could take off and land vertically as well as hover and cruise above ground. The detail in Scott's film gave it an air of realism; this was a 'lived-in' future. The film's timescales, however, were hopelessly wrong: the action was set in 2019 and spinners won't be available by then (if ever). There's *Back to the Future* (1985), of course, which features the iconic flying DeLorean. The film's sequel starts with Marty saying 'Hey, Doc, we better back up. We don't have enough road to get up to 88 [mph].' Doc replies 'Roads? Where we're going, we don't need roads!' Doc has already traveled from 1985 into the glorious road-free future that will be 2015. The low-budget *Death Race 2000* (1975) had a quite different vision of the future of automobiles. It caused controversy when it was distributed, but it had a sharp satirical edge: a car race across America in 2000 has as its winner the driver who kills most pedestrians.

Thoughtful explorations of the interaction between individuals and their cars, between society and its motor vehicles, appear more often in mainstream literature than in SF. J.G. Ballard, for example, who in the early 1960s wrote several excellent SF stories and novels, examined the human fascination with automobiles in his mainstream novel *Crash* (1973). His novel was not only more outrageous than anything found in the science fiction of the time (one publisher's reader delivered the verdict 'This author is beyond psychiatric help') it was also more sensitive to one of the great transformations of the twentieth century: the rise of car ownership. For the most part, the SF I read as a youth simply didn't address the transformations that might be wrought by universal automobile ownership. This is surely one of the biggest failures of prediction in SF (rivaling the failure to foresee that, having reached the Moon, humankind would rapidly lose interest in crewed space exploration).

9.1.3 Teleportation

Perhaps the failure of the collective science fictional imagination when it came to mundane matters of transport wasn't a failure at all. Perhaps SF writers simply weren't overly interested in how one day we might travel from Woking to Dorking, or from Littlefield to Lubbock, and make the journey in under one hour with favorable traffic. After all, who wants to read about how hero reaches heroine by navigating a contraflow on the A247? Authors in any genre face the same problem: in getting their protagonist from A to B they risk halting the action and boring their readers. But SF writers had a solution. They could effectively ignore the issue and have their protagonists 'move' by means of teleportation or matter transmission. *That's* the science fictional approach to transport!

The word 'teleportation' started out as a synonym for 'telekinesis'—the ability to move a physical object by the power of the mind. The word 'teleport' as a verb in this context first appeared in the novel *Lo!* (1931), written by that collector of all things weird, Charles Fort. Eventually, though, the word came to mean the ability to move oneself by the power of one's mind: you would start at point A and then shift instantaneously to point B. Edgar Rice Burroughs was one of the first to use teleportation in his stories, but it was merely a facilitating device and he gave no serious thought to its implications. By the 1950s, however, John Campbell's infatuation with psi-related matters was in full swing and his influence in the field was such that many authors began to explore seriously the consequences of teleportation.

One of the best such explorations was Alfred Bester's novel *Tiger! Tiger!* (1956), set in a world where the emerging use of personal teleportation disrupts the existing social and economic order. (In Bester's novel, teleportation is called 'jaunting'—it's the psionic ability to think yourself somewhere else. I first came across the term 'jaunte' as a child, in the TV series *The Tomorrow People* (1973–1979). The series told the story of youngsters with a variety of special mental abilities, including the ability to teleport or 'jaunte'. I didn't realize until later that the scriptwriters must have borrowed the term from Bester's novel.) Bester described a particularly vivid consequence of teleportation, the 'blue jaunte', which occurred if someone failed to visualize their destination properly: an arrival inside solid material resulted in an explosion and a messy demise. Daniel Galouye's story "The Last Leap" (1960) played with this idea: Galouye imagined a world in which teleportation was so easy that thought *automatically* translated into travel—so one had to be very careful to avoid certain mental pictures! This deadly concomitant of teleportation was used to trick one of the beasts in James Schmitz's *The Lion Game* (1973), a novel in which predators use telepathy to track their prey and teleportation to attack.

Many other well known SF authors have employed teleportation in their stories. Gordon Dickson, for example, wrote "No More Barriers" (1955), in which he imagined that mastery of teleportation would lead to the removal of national borders and their replacement by organizations devoted to services. Theodore Sturgeon, in one of his greatest works, *More Than Human* (1953), wrote about twins who develop the ability to teleport. Anne McCaffrey, in *Dragonflight* (1968) and the rest of her Pern series, wrote about dragons with the ability to teleport: the creatures can enter hyperspace and, if they have a suitably strong mental image of their destination, can reappear anywhere on the planet Pern. Vernor Vinge in *The Witling* (1976) describes an alien race whose members have differing teleportation abilities; those who can't teleport at all are called witlings, and form an underclass in their society. Walter Jon Williams in his novel *Knight Moves* (1985) describes an advanced race of aliens who teleport unwitting creatures as part of some strange game.

Teleportation, then, is a common theme in science fiction. Nevertheless, even though it features in some terrific stories, teleportation is not a theme I personally much cared for. Although some authors attempted to provide a scientific justification for the phenomenon, such as having protagonists learn how to manipulate a fourth spatial dimension, teleportation is one of those 'psi' activities for which there is no evidence. Teleportation is wish fulfillment rather than science. Having your character jaunte somewhere is little different from having a genie pop out from a magic bottle and take them there.

What about matter transmission? At first glance the transmission of people and goods through space, in much the same way that we currently use radio waves to transmit information, is no more plausible than teleportation. But it *sounds* more plausible. Perhaps that's why I preferred stories about matter transmission to those about teleportation; authors who used the phrase 'matter transmitter' at least seemed to be making an effort. Or perhaps I simply preferred the concept because of the transporter in *Star Trek*. It's no surprise that the phrase 'Beam me up, Scottie' has entered the language: *Star Trek*'s transporter, which enabled characters to get from ship to shore in an instant, was such a terrific device.

The notion of matter transmission has a surprisingly long history. Victorian writers were exploring the consequences of matter transmission going wrong decades before Bester wrote of 'blue jauntes'. Edward Page Mitchell's "The Man Without a Body" (1877) postulates an invention capable of transmitting cats, people, and other objects through wires in much the same way as words were being transmitted through the recently invented telegraph. The inventor meets a gruesome end when, midway through a trans-Atlantic transmission, the device loses power; only his head arrives. (Unfortunately the author gave his matter transmitter such a terrible name—the telepomp—he pretty much ensured his story would remain obscure.) Robert Duncan Milne's "Professor Vehr's Electrical Experiment" (1885) told how a scientist hooked up a device to a telegraph line and transmitted a young man in order to be with his betrothed. Needless to say, when the scientist tried to bring the couple back, things went wrong—the young couple are lost in a telegraphic otherworld. This particular idea, of matter transmission going wrong, was rich enough for George Langelaan to produce a fresh twist in his famous story "The Fly"

(1957), which has given rise to two films of the same name. George O. Smith's story "Special Delivery" (1945) tells of a device that scans an item at an atomic level and stores the atoms in a 'matter bank'. Information is beamed to a distant receiver, where the item is recreated from a second 'matter bank'. If what is transmitted is not a physical object but information about how to *reconstitute* that object then immediately one can think of a host of possible plot complications. For example, matter might be accidentally duplicated rather than transmitted, an idea explored by Clifford Simak in his novel *The Goblin Reservation* (1969). Smith himself thought of other pitfalls. Small wonder *Star Trek*'s Dr McCoy hated going in the Enterprise's transporter.

Many stories address the question of how society would react if routine transportation could be undertaken by a working system of matter transmitters. For example, in 1973 Robert Silverberg commissioned three original novellas—by Larry Niven, John Brunner, and Jack Vance—on the theme. Niven wrote "Flash Crowd", which built on ideas that first appeared in his famous novel *Ringworld* (1970). He came up with the idea of a 'displacement booth'. For the end-user, the experience of using a displacement booth would be much like our experience of using present-day ATMs: simply insert a card and type in some numbers. The difference is, instead of cash appearing the user disappears—and in an eye blink reappears at a point of disembarkation. Niven went on to write four more stories containing the displacement booth as transport technology, all of which appear in his collection *A Hole in Space* (1974). John Brunner's contribution to Silverberg's anthology was the story "You'll Take the High Road", and Brunner went on to write two novels (*Web of Everywhere* (1974) and *The Infinitive of Go* (1980)) about matter transmission as transport mechanism. Jack Vance's contribution to the anthology was "Rumfuddle", which offered yet another take on the theme of instantaneous transportation—in this case, portals can convey a person to any place or time, in this world or in parallel universes. All three novellas in the anthology were written to address Silverberg's question: 'What sort of society would develop where Arabia is an eye blink away from Brooklyn, where one can step from Calcutta to the Grand Canyon between two heartbeats?' That's the sort of question that SF is interested in: how would such a radical change impact on the quality and texture of life?

One of the most interesting studies of matter transmission is Lloyd Biggle Jr's *All the Colors of Darkness* (1963). Biggle begins by imagining what a working system might look like. There'd be a transmitter and a receiver, along the lines of the broadcasting technology of the day. It would be rather like the system envisaged earlier by George O. Smith. As technology improved, however, the system would work without the need for a receiver station: the system would function much like *Star Trek*'s transporter. Finally, Biggle imagined the technology improving to the stage where a transmitter could transmit itself without the need for a receiver—in which case a device for local transportation would have evolved into a means for reaching the stars. Several other authors have proposed matter transmitters as a method of space travel, including Robert Heinlein in *Tunnel in the Sky* (1955) and Clifford Simak in *Way Station* (1963), but this fictional transportation technology permits more surprising applications. For example, it could be used to construct a variety of interesting buildings. Bob Shaw's story "Aspect" (1954) and Roger Zelazny's novel *Today We Choose Faces* (1973), both feature structures in which doorways are matter transmitters leading to rooms on different worlds. James Schmitz's *The Lion Game*, mentioned above, features a hotel in which matter transmitting doors lead to various tourist attractions. And Philip José Farmer's novel *A Private Cosmos* (1968) imagines cells manufactured in rock deep underground that can be accessed only by matter transmitter—the perfect prison.

Science fiction's treatment of current transport technologies—planes, trains, and automobiles—might have lacked imagination. Even clearly science fictional technologies, devices such as jetpacks, flying cars, and rolling roads, are in essence merely slight extrapolations of existing methods of transport. But matter transmission—that's a true science fictional concept.

9.2   Transport Today

Writing about transportation now, in the second decade of the twenty-first century, is an awkward task. On the one hand, it seems as if all is changing: automobiles, mass transit systems, and planes all seem to be evolving rapidly and are perhaps on the cusp of a radical transformation. On the other hand, it has to be admitted that current methods of transport would be readily recognisable to authors from the Golden Age of SF. Railways were rolled out in the 1830s; the petrol engine dates back to 1876; the first practical diesel engine was made in 1892; and the patent for the jet engine was granted in 1932. Transport technology is better now than it was in the Golden Age of the 1950s—engines are more efficient, fuels are of higher quality, and improved designs for almost all vehicle components can rely upon lighter and stronger materials—but the technology isn't fundamentally *different*. Indeed, the Victorians developed a transport innovation whose basic design we use unchanged to this day: the bicycle.

The English industrialist John Kemp Starley produced his 'safety' bicycle in 1885 and we still use Starley's idea of equal-sized front and back wheels, a steerable front wheel, chain-and-cog drive to the rear wheel, and a triangle-shaped body. From the outset the machine was simple, efficient, reliable, and cheap, which perhaps explains why the bicycle often comes top of polls that ask for the greatest invention of recent times.[2] Furthermore, the invention bestows a number of benefits on the user. I bicycle ten miles in my commute to and from work and thereby get my daily quota of exercise without having to pay for gym membership; I know my journey time to within a minute or so, irrespective of how road traffic is flowing; and I get the smug feeling that comes with knowing I'm not polluting the planet. H.G. Wells, perhaps the greatest of all SF writers, put it best: 'When I see an adult on a bicycle, I no longer despair for the human race.'

So bicycles and cars, planes and trains—they've been around for more than a century and we're using them still. Is it likely they'll ever be replaced by the jetpacks, flying cars, or moving sidewalks? Or with something even more science fictional?

9.2.1 Jetpacks—Really?

Jetpacks have been around for decades. In 1958, two engineers at Thiokol Co. developed the Jump Belt, a device to pump nitrogen through two downward-directed nozzles, thus enabling the wearer to jump higher than normal. (Heinlein's jump harness in *Stranger in a Strange Land* was probably based on this invention. Conversely, it wouldn't surprise me if the Thiokol engineers had been influenced by the propulsion devices of Buck Rogers and King of the Rocket Men.) And by the early 1960s there was a real 'King of the Rocket Men'-type jetpack. The Bell Rocket Belt, developed by Bell Aerosystems on behalf of the US Army, could fly a distance of 120 m (which, admittedly, was probably not far enough to be useful in thwarting the evil Dr Vulcan). The jetpack used by James Bond in *Thunderball* was a Bell Rocket Belt. The Bell Belt underwent gradual evolution, culminating in a jetpack consisting of three cylinders (one containing high-pressure nitrogen gas, the other two containing concentrated hydrogen peroxide); the nitrogen pushes the H_2O_2 onto a catalyst, which facilitates a chemical reaction resulting in superheated steam and oxygen. The gas blasts out of two nozzles and—hey presto—you have a jetpack. There are other designs, too. Yves Rossy, a Swiss adventurer, is known as Jetman.[3] He flies with a jet-powered carbon fiber wing strapped to his back (although it isn't quite the jetpack of the type featured in

Fig. 2.1: Jetman must first gain altitude by means of a conventional airplane or helicopter).

So jetpacks exist, which raises an obvious question for me: why wasn't I able to strap one to my back this morning and fly like an eagle into work? The answer comes almost instantaneously: jetpacks are a terrible idea. Let's face it—personal jetpacks were never going to fly.

The first drawback with jetpacks hardly needs to be spelled out: having one's nether regions in close proximity to twin columns of extremely hot gas is not to be recommended. The pain caused by a badly adjusted bicycle saddle would be as nothing compared to the agony caused by a mis-adjusted jetpack nozzle. Then there's the issue of power. We've already covered this in earlier chapters: Earth's gravitational attraction means it is much, *much* easier to travel horizontally than it is to move vertically. It's that issue, once again, of how things scale. If you want your jetpack to get you from the ground up into the clouds, and then take you from London to Paris, say, then you are going to need a *lot* of fuel. Convenience is a consideration, too: using a jetpack, you'd struggle to bring your weekly shop back with you from the supermarket. Then there's the safety issue. If your car runs out of fuel then you call your breakdown agency and suffer nothing except perhaps the inconvenience of a missed meeting. If your jetpack malfunctions at an altitude of 500 m then a missed meeting will be the least of your worries. It's the same with collisions: these days people often walk away unhurt from a car crash, but a collision between two jetpack wearers would almost certainly kill the jetmen and quite possibly kill any people on the ground unfortunate enough to be directly underneath the collision.

It's not that jetpacks don't exist. They just don't exist in a form that makes them remotely usable as a means of transport for the likes of you and me. They are great for James Bond or for putting on a show. But for everyday transport? Nah.

That doesn't stop people dreaming. As I write this, a New Zealand based company called Martin Aircraft has announced it will soon begin production of a jetpack.[4] The company claims its jetpack could allow a person of average weight to take off and land vertically, reach an altitude of 1 km, and fly for up to half an hour at a speed of up to 74 km/h. It's the sort of toy a multimillionaire might buy, and Martin Aircraft believe it could be of real use by first responder emergency crews. Perhaps. But I believe SF's treatment of the jetpack was spot on: it's a fun idea, but not one we need to take too seriously.

9.2.2 Getting from A to B

We've looked at what SF predicted for some of the commonest forms of transport—planes, trains, and automobiles. There isn't much to add when it comes to aircraft. We're no longer seeing revolutionary advances in flight speed or in the materials used to build craft, but we *are* seeing a steady increase in the use of digital technology in the aviation industry: computers are increasingly used to design, manufacture, and fly planes. That trend is likely to continue, but in the next few years we are unlikely to see any profound changes in aviation. The situation is more interesting when it comes those other methods of transport we considered: trains and cars. Let's look first at mass transport.

9.2.2.1 Mass Transport

The railway was the most transformative development in transportation since Appius Claudius Caecus had the idea of building a proper road between Rome and Brindisi. Is there a future for the transport system that plunged us fully into the Industrial Age? Or, more generally, is there a future for *any* mass transit system?

Combine people's increasing aspirations with an increasing world population, the majority of which will soon be living in cities, and it's hard to avoid the conclusion that some sort of cheap and efficient system of mass transit, whether by bus or train, is a necessity. However, public transport systems suffer from the so-called 'first and last mile' problem: although the systems are extremely effective in shifting groups of people from fixed point to fixed point there are many individuals—the physically disabled, those needing to carry lots of luggage, those for whom the weather is too cold, too hot, or too inclement—who find it difficult to reach those fixed point stations. That's why an efficient mass transit system of the future is also likely to feature a 'mobility on demand' network: a variety of vehicles capable of making short-distance, low-speed trips around the city and to the pick-up/put-down points. Those vehicles might include some sort of programmable, autonomously driving 'pods', electric scooters—and also bicycles. SF pointedly ignored the bicycle as an element of the future, but if we do develop new types of train and bus system then the humble bike will surely continue to play a role!

That visionary writer William Gibson once wrote that 'the future is already here—it's just not very evenly distributed'. Well, perhaps some of the elements of our transport future are already here—and to be seen in China. As I write this, China is in the middle of a program of building 90 new mass transit rail lines; it's experimenting with on-demand buses that replace fixed routes with pick-ups/drop-offs based on user requests; it's even looking at a so-called 'straddling bus', an electric and solar-powered vehicle capable of transporting its 1400 passengers on the same roads as normal traffic but designed so bikes, cars, and normal buses simply drive beneath it. The irony here is that the driving force for many of these innovations is the move by Chinese people away from the bicycle and to the car: their thousands of kilometers of freshly laid motorways are already gridlocked, leading to devastating levels of air pollution and long journey times. A megacity such as Beijing urgently requires the outline sketched above—the combination of mass transit system with a mobility-on-demand network—before the automobile chokes it.

China is also at the heart of rumors of vast railway tunnels, connecting it to Russia, Europe, and the US via high-speed train links. Personally, I find these rumors far less plausible than on-demand pods, say, or straddling buses. Although tunnels have a fascination for the SF community—Harry Harrison's *A Transatlantic Tunnel, Hurrah!* wasn't the only novel to feature subterranean railway lines—they are hugely expensive. The world's longest and deepest traffic tunnel, the Gotthard Base Tunnel, has a route length of only 57 km and yet it cost $12 billion to construct. The cost of making a transatlantic tunnel, or any of the other rumored tunnels, would be crippling even if the immense engineering challenges they pose could be overcome. Furthermore, by their very nature they have a limited carrying capacity. And if the system breaks, whether through fire, mechanical failure, or deliberate human action, then it really does break: I regularly use the Channel Tunnel, which is just over 50 km long, and have suffered delays because of repairs due to fire damage, locomotive breakdowns, and the attempts of migrants to reach England from France. Tunnels of course can be a godsend but, given their drawbacks, it's unlikely anyone would be willing to finance a Sino-American train tunnel.

Other rail-based systems are being researched, although some of the proposals stretch the meaning of the word 'rail'. For example, Tubular Rail, a Texas-based company founded by Robert Pulliam, dreams of reconfiguring the traditional elements of a railway.[5] A Tubular Rail system would consist of a series of rings, each fitted with an electric motor, and the train would have wheels powered by the rings. In essence, the train would be passed from ring to ring. In another example, in 2014 the British engineering firm Arup tried to imagine what the railway network would look like in 2050. It arrived at a vision of robots driving the trains and of swarms of small robots maintaining old rail infrastructure and building new. Predictably, if frustratingly, the head of the transport workers union responded by saying 'this report and its proposals are just dangerous nonsense straight out of some barmy work of science fiction'.[6] Sigh. And then there's Elon Musk's Hyperloop.

Musk made a mint by founding PayPal and he used some of his fortune to create science-fictional sounding companies such as SpaceX and Tesla Motors. In 2013, Musk outlined a new transport concept—Hyperloop—which he believes will drastically shorten journey times between major cities. For example, consider the journey between Los Angeles and San Francisco. By car, on a good day, the trip takes about six hours; by plane it takes about 80 minutes; by Hyperloop—a system of pods traveling at near-supersonic speeds through a low-pressure tube suspended above the existing California highway—the journey time would be just 30 minutes: an impressive saving.

The basic idea behind Hyperloop is that a compressor on the front of each pod passes air to the rear and also creates a cushion of air underneath, upon which the pod rides; induction motors in the tube accelerate and decelerate the pods. Now Musk, with his interest in space travel and electric cars, is a pioneer in transport technology. His Hyperloop proposal—which he has made open source so that others can develop the technology—is perhaps his most interesting contribution in this field. At this point, however, it's difficult to judge what sort of future Hyperloop will have. On the one hand, certain elements of the technology have already been shown to work: in 2016, a small test track built in the Nevada desert hosted a successful journey (Fig. 9.2). Furthermore, several countries have expressed an interest in Hyperloop-type systems. The attraction of reducing the journey time between Helsinki and Stockholm, for example, has convinced Finland and Sweden to study the options; Hyperloop One plan to build a system capable of getting from Dubai to Abu Dhabi in 12 minutes. On the other hand, there are obvious difficulties with the proposal. For example, the original idea of linking Los Angeles and San Francisco requires the authorities to be persuaded that building an untried transport system above a car-laden California highway would be a Good Thing. That seems to me to be a task for Sisyphus. Besides, if it really is important to shift masses of people between Los Angeles and San Francisco quickly then engineers could use proven technology to develop a mass transport system with a journey time of less than two hours. A 'bullet train' of the type currently in everyday use in Japan could (assuming a straight track and a direct service) make the journey in 105 minutes—longer than the Hyperloop journey time, certainly, but not hugely so. The cost of a Hyperloop system is also likely to be an issue. The development of an entirely new transport infrastructure is unlikely to be cheap. And then there are human factors: at the speeds envisaged for Hyperloop, even small curves in the path will generate lateral forces of a magnitude that tends to induce nausea. Even if you don't suffer motion sickness yourself, you

Fig. 9.2 In 2016, on a specially prepared test track in the Nevada desert, the Hyperloop One project succeeded in accelerating an electrically propelled sled to a speed of over 160 km/h in 2 seconds. Many technological and regulatory hurdles must be overcome before Musk's original Hyperloop vision can become a reality; this sled ride constitutes only the first step on a long journey towards a high-speed rapid transport system (Credit: Hyperloop One)

wouldn't want to be stuck in a pod with other people who do get sick. Technical, financial, and ergonomic factors might well mean Hyperloop remains in the category of 'cool but impractical' ideas.

The futuristic systems mentioned above—the straddling bus, Tubular Rail, Hyperloop—are all systems of mass transit. They need R&D and investment to put the required infrastructure in place, but once working they'd have the capacity to move large numbers of people from one point to another. The trouble is, many of us are wedded to our *personal* transport system. Despite the many disadvantages of the internal combustion engine, we love our cars. It's not just the 'last and first mile' problem that keeps many people from using mass transit systems; cars give people control over their transport plans. So is there anything on the horizon that might take the advantages of the two different approaches to transportation, the public and the private? Well, perhaps some sort of personal rapid transit (PRT) system might get results.

Several different types of PRT have been proposed. Essentially they consist of lightweight pods, big enough to hold a few people, which move along tracks but not along prescribed routes: passengers arrive directly at their destination without having to stop. Described in this way a PRT seems similar to the 'mobility on demand' system described at the beginning of this section; indeed, it could form part of such a system. But it might also do more.

A company called skyTran, spun off from the NASA Ames Research Center, has an interesting proposal along these lines (Fig. 9.3). Imagine two-passenger pods traveling at speeds of up to 150 km/h using solar-powered maglev. (A magnetic levitation, or maglev, transport system uses magnets to both lift a vehicle above a guideway and propel it. The Transrapid maglev in Shanghai is currently the world's fastest commercial train, with a top speed of 430 km/h.) Because maglev means the pod is not in contact with the rail there is no friction; this not only enables the train to reach high speeds, it means there is less wear and tear—skyTran would require relatively little maintenance. The rails would be lightweight, modular and snap together like pieces of Lego; the tracks they form could insinuate themselves throughout a city, running alongside existing roads and even going through buildings. A future city with skyTran would look something like the opening credits of *Futurama*. A passenger would order one of the computer-controlled pods using an app; the pod would move to a side rail while its passengers boarded and would then rejoin the traffic flow without causing any other pods in the network to slow.

Fig. 9.3 An artist's conception of the skyTran personal rapid transit system. The two-person pods would travel on lightweight modular rails that could insinuate themselves into existing cityscapes (Credit: skyTran)

As I write this, skyTran are constructing a 500 m demonstration track on an industrial campus in Tel Aviv. If the demonstration succeeds then perhaps we might live to see our city skylines one day transformed into something resembling those gorgeous *Astounding* and *Amazing* covers of the pulp era. On the other hand, it's difficult to imagine this technology working in the densely packed, twisting lanes that characterize, for example, so many villages and small towns in Europe.

9.2.2.2   Automobiles

Earlier in this chapter I argued that pre-1985 SF, my focus in this book, had failed to foresee the coming dominance of the automobile. While wearing his non-fiction hat, however, Asimov seems to have predicted with clear-eyed accuracy our automotive future. In 1964, Asimov visited the World's Fair in New York, following which he wrote an essay for *The New York Times* imagining a visit to the year 2014. Concerning cars he wrote: 'Much effort will be put into the designing of vehicles with 'robot-brains'—vehicles that can be set for particular destinations and that will then proceed there without interference by the slow reflexes of a human driver.'[7]

As if on cue, Google were putting huge effort into the design of autonomous self-driving vehicles. The company's work in this area had started some years before. In 2004, DARPA (a US Defense department agency, whose earlier projects played a key role in the development of the internet) launched its Grand Challenge series: a competition between American companies and university departments to develop autonomous vehicles capable of navigating a variety of circuits and different types of terrain. The goal of the first competition was to complete a 240-km route; of the 15 competitors, the farthest any vehicle travelled was 11.7 km. Needless to say, the competition garnered much cynical and humorous reporting on TV. In the second competition, in 2005, 5 of the 23 competitors finished the course. The challenge was won by a team from Stanford University, led by Sebastian Thrun. Two years later, Thrun began to work with Google on their Street View project and then went on to found the autonomous vehicle project mentioned above. Perhaps what Asimov missed in his 1964 piece of fortune telling was the way in which a company such as Google would attempt to develop a self-driving car: Google simply hired the best people from universities doing research in this area—not just research in robotics, but in computer science, big data, and automotive engineering—and then gave them access to what counts: computational and data resources (of which the company has lots).

The Google car that evolved from the project is packed with technology enabling it to propel itself while doing all the things you'd expect of a moving vehicle—avoiding stationary obstacles, taking account of other moving objects, obeying relevant traffic laws, and so on. The most important instrument on the car is a roof-mounted, 64-beam lidar (a kind of laser-based radar), which creates a three-dimensional map of its immediate surroundings. This map is continuously compared with existing maps, allowing an onboard computer to generate data models that permit safe driving. On fast roads the Google car maintains a safe distance from other traffic through the action of four smaller, traditional radars on the front and rear bumpers. A camera, mounted by the rear-view mirror, detects traffic lights. The car's location, and its overall journey, is tracked by GPS and monitored by other onboard sensors. The original Google car came with steering wheel and pedals, allowing a human passenger the option to take control. More recently, Google have experimented on private roads with an entirely autonomous vehicle: there's no steering wheel, no accelerator pedal, no brake pedal. It simply can't be driven by humans. (A car that's incapable of being operated by a human driver under any circumstances possesses 'Level 5' autonomy on the scale developed by the Society of Automotive Engineers. See the box for further details. Google, and companies such as Ford, are focusing on 'Level 4' autonomy: the car drives itself under specific circumstances. Other companies have developed 'Level 2' autonomous vehicles that have already driven many millions of kilometers.)

Currently the Google car is expensive. The roof-mounted laser system alone costs more than any car I've ever been able to afford. The raw computational power required to navigate the real world is also expensive. But this is a technology capable of scaling well. When the big automobile manufacturers begin to mass produce radar systems then the cost of those systems will plummet. The price of computational power will plummet anyway; it always does. Google has demonstrated how autonomous vehicles are possible: its self-driving cars have covered a million kilometers in those jurisdictions that permit them and, at the time of writing, they have been the cause of just a single accident. (A Google car travelling at 3 km/h pulled out in front of a bus travelling at 24 km/h and scraped the side of the bus. No one was hurt.)[8] Soon, self-driving cars will be affordable.

So, will we all soon be chauffeured around in our autonomous vehicles? There are many reasons why we *should*. Off the top of my head, here are a few advantages to self-driving cars. They don't drink and drive. They don't succumb to road rage. They don't get tired. They won't crash because they were paying more attention to their latest text message than they were to the road ahead. They'll allow blind people more freedom to travel. They'll allow more freedom for those who, through age or illness, find it difficult to drive. Their greater driving efficiency will produce environmental benefits. They'll permit passengers to use drive time more productively or more enjoyably. They'll allow passengers to drive to the pub and enjoy something stronger than orange juice. Their more efficient use of roads will save public money by reducing the costs associated with maintenance. And it's not just human transportation they could transform: they can also bring benefits to the transport of goods. What's not to like about self-driving cars? Indeed, given the benefits outlined above (and I'm sure with some thought it would be possible to come up with many more benefits) surely their introduction is inevitable?

Although there are indeed benefits to self-driving cars, their introduction would also pose questions—problems that should have formed the basis for stories in SF's Golden Age, but somehow didn't.

Let's start with the question of what happens when things go wrong. Every revolution in transport technology has led to innocent people being killed. In 1827, William Huskisson MP got run over by Stephenson's *Rocket* and became the first to die in a train-related accident. In 1896, Bridget Driscoll became the first person to die in a petrol-engined car accident. In 1908, Thomas Selfridge became the first to die in a plane crash. And in 2016, Joshua Brown was the first to die in a self-driving car. Brown's was a Tesla Model S, with an autonomous cruise control feature called Autopilot—a 'Level 2' grade of autonomy which human drivers unfortunately, it seems, treat as if it were a 'Level 3' technology. Brown was watching a movie when the car drove itself at 97 km/h under the trailer of an articulated lorry that had turned across the road in front of it. The car's cameras failed to distinguish the white side of the trailer from the bright Florida sky. What might be the outcome of Brown's death? Presumably, tragic though it was, this first fatal accident involving an autonomous car will have the same impact as Huskisson's death had on trains, Driscoll's death on automobiles, and Selfridge's death on planes. The company Tesla, although of course saddened by the death, noted they strongly advise drivers to keep control of the vehicle at all times and not rely upon Autopilot; these are *partially*-autonomous vehicles. (In development, Tesla have taken a different approach to Google. Whereas Google cars trundle rather slowly around city streets, being exposed to lots of risk and uncertainty but seldom travelling fast enough to cause a spectacular crash, Tesla have opted for the much simpler environment of the motorway. A motorway is more straightforward situation for an autonomous car to navigate, but if things go wrong then the higher speeds involved mean the consequences are likely to be severe.) Furthermore, Tesla pointed out that in the USA there is one fatality for every 150 million kilometers driven in cars; semi-autonomous vehicles using Autopilot covered 210 million kilometers before the first fatality. Statistically, therefore, semi-autonomous cars would appear to be safer than normal cars—and their safety record will improve as the industry learns from each mile driven. Even though things will sometimes go wrong with the technology, enthusiasts claim autonomous cars have the potential to save a million road deaths a year, worldwide.

Six Levels of Driving Automation

The Society of Automotive Engineers has developed a standard that defines various terms relating to automated driving systems and describes various levels of driving automation (SAE 2014).

Level 0—no automation—the driver is in complete control

Level 1—driver assistance—for example, the car can be set to maintain a constant distance from the vehicle in front; ultimate control is with the driver

Level 2—partial automation—the car has control in a specific use case, but the driver must monitor the system at all times; the driver is still in charge

Level 3—conditional automation—the car has control in a specific use case, and the driver is not required to monitor the system; but if the system detects a situation it can't manage, it passes responsibility back to the driver

Level 4—high automation—during a defined use case, a driver is not required; the car is in charge

Level 5—full automation—a human driver is not needed

Personally, I'd be happy at either extreme: Level 0 or Level 5 autonomy. It's the middle ground where I'd be anxious. What happens if I'm in a Level 3 car, I've fallen asleep because my input has not been needed, and then the vehicle hands back control to me because it can't cope with a changed road condition?

But can fully autonomous cars ever be truly safe? The current generation of lidars struggle at distances greater than about 300 m; if the vehicle is moving fast and something happens in front of it, the driving system might not have time to react. Furthermore, the lidars are less effective in rain, fog, and snow, and rumors suggest they are less than perfect at identifying potholes. Investigators armed with cheap low-power lasers have already shown how the lidar can be fooled. Then there's the issue of standardization: vehicles will need to communicate constantly with other vehicles, transmitting information about location and velocity. However, the automobile industry is notoriously competitive. Getting companies to collaborate on an open source software platform for their cars will be difficult. And once a software platform has been agreed, or the industry comes up with some other solution to permit driverless cars on the road, it's only a matter of time before some antisocial misfit hacks into a car's system and causes havoc. Security analysts have already shown how it is possible to take control of some vehicles by hacking into internet-connected navigation or entertainment systems; once any car is connected it can, in principle, be hacked. And what happens when pedestrians realize that cars are programmed not to hit them? If they know cars will stop, what's to prevent people—to the great frustration of drivers—just wandering across the road whenever they wish?

Of course, technology improves all the time. What appears to be an insurmountable problem now probably won't be a problem a decade from now. Nevertheless, if there are a billion driverless cars on the road there will inevitably be 'incidents', as the English police euphemistically refer to crashes, collisions, and pile-ups. And that raises the question: who will be to blame when a driverless car causes an incident? With Joshua Brown's accident the situation was clear: Brown's vehicle possessed a 'Level 2' grade of autonomy and therefore he should have been driving. Who's to blame if there is no driver?

If you are a passenger in your driverless car, and the car crashes and kills someone, are you—the owner—to blame? That hardly seems fair. If the car contains no steering wheel and no brake pedal then nothing you could have done would have prevented the crash. So is the car manufacturer liable? Does liability rest with the manufacturer of a specific part that failed? Or with the authors of the car software, who perhaps failed to spot a minor glitch in one line of a multi-million line piece of code? Just as airplanes have 'black box' data recorders, perhaps driverless cars will need to be fitted with devices that monitor all aspects of a journey in order to assign liability if things go wrong. But that then raises issues of privacy: would you want a hacker, or even a legitimate government agency, to get access to all of your car journeys? These sorts of questions can keep lawyers, as well as science fiction writers, in a job.

And then there are ethical issues. The software controlling driverless cars will have as a priority the need to avoid a collision with pedestrians. Imagine, then, a scenario in which a child chases a football and runs out into the road. There's no time for the vehicle to stop. If the car continues straight ahead the child dies. If it swerves right, into oncoming traffic, it risks killing its own passengers and those in other vehicles. If it swerves left, it will mow down a dozen geriatric gentlemen enjoying an afternoon stroll. In that situation, what should the algorithm tell the car to do? Who would be responsible for developing the algorithm? The automobile industry already have philosophers and ethicists debating these issues, and are quite rightly looking for guidance from society in this area—but I can only wish them good luck with this. A dozen people are going to give a dozen different responses. Even if everyone agrees on a default setting for an ethical algorithm what if I, as the car's owner, disagree with it? Should I be permitted to override those settings? Asimov addressed these sorts of questions in his robot stories, not specifically within the context of cars (with the exception of "Sally", mentioned earlier) but the underlying issues are the same. What are we to do when our technology reaches the stage where ethical decisions must be made?

These are challenging questions, and we need to face them right now because driverless cars are soon going to be on our streets. The arguments in favor of using them are too strong for the technology not to be employed.

And as for those other science fictional cars? The amphibious car driven by Captain Zeppos, which so fascinated me as a young child? The glorious landspeeder in *Star Wars*, which I longed for as a teenager? The superbly realized spinner in *Blade Runner*, which seemed to be such an accurate forecast of the future? Well, the Clarkson, Hammond, May team on BBC's *Top Gear*—the world's most popular factual television program, which I guess in itself is testament to our love of cars—built a variety of motor vehicles over the years. Their motto, 'ambitious, but rubbish', sums up for me what happens to the car once it leaves terra firma. Amphibious vehicles exist, for example, but few of us would use their water-going capabilities to make it worthwhile owning one. And several flying-car-like vehicles have been built over the years, but none have succeeded. As an idea, the flying car is about as sensible as the jetpack. (Drones might fare better: they already deliver goods, and it's now possible to buy a passenger-carrying drone.) Nope—cars are made for the ground, not for trips over water or flights in the air. But autonomous, self-driving cars—*that* is a future we're going to see, and roughly on the timescale given by Asimov's prediction following his trip to the World's Fair.

9.2.3 Quantum Teleportation

Star Trek's 'Beam me up, Scottie' approach, as I argued earlier, is the most science fictional method of transport. What can science tell us about the possibility of such an approach?

Let's consider a particular scenario. Imagine Alice has in her possession a box. Let's call the box A, and let's suppose it contains some object. Alice's friend Bob also has a box, which we may as well call B, and we want to move the object from A to B. No problem. Simply take the object out of A and use some form of transportation—segway, jetpack, whatever—to move it to

B. But think of all the hassle involved when wheeled vehicles or rockets are involved. As SF tells us, it would be much simpler if we had a technology such as *Star Trek*'s transporter. In that case we'd just 'scan' the object in A, measuring the state of all the particles that make it up. (Presumably, the object would be destroyed in the scanning process; we wouldn't want two copies of someone or something.) Then we'd transmit all the state information via phone, radio link, or whatever, and use the information to reconstruct the object at B. The transporter would thus be rather like a fax machine: it wouldn't transmit the object itself, but rather sufficient information about the object for it to be reconstructed. (For the benefit of younger readers I should perhaps explain that fax machines were once a popular technology for scanning a paper document and transmitting the written information on the paper over the telephone network. A fax machine at the receiving end produced a copy, or facsimile, of the original. I haven't seen a fax machine for years, and I assume they've gone the way of typewriters. (A typewriter was a keyboard-based printing device, lacking a screen and with 0 Gb of memory.))

According to classical physics, it's possible to construct a transporter. To be sure, it would be extremely difficulty to build such a device—but, in principle, a sufficiently gifted engineer could make one. However, deep down, the universe doesn't work according to classical physics. Instead, it obeys the laws of quantum mechanics. And for a long time physicists believed quantum mechanics ruled out the possibility of a *Star Trek*-type transporter. The reason for this belief was one of the characteristic features of quantum mechanics: the Heisenberg uncertainty principle.

The uncertainty principle tells us we could never scan an object and measure accurately all the information we'd to measure in order to transport the object successfully from A to B. Consider the simplest case: the object to be transported is a single electron. According to the uncertainty principle, we cannot simultaneously measure with perfect accuracy the position and momentum of the electron. The teleported electron would thus not be a perfect copy of the original. That's the simplest case. A macroscopic object contains trillions upon trillions of atoms, and the process of scanning the object would generate uncertainty in the information attached to all those atoms. After reconstruction, the result of teleporting all this imprecise information would be... well, *mush*, I guess. Quantum mechanics would seem to rule out the possibility of teleportation.

At least that's the way things stood up until 1993, when Charles Bennett and Gilles Brassard, along with their colleagues Claude Crepeau, Richard Jozsa, Asher Peres, and William Wootters, showed how 'quantum teleportation' is possible after all. In certain circumstances it's possible to take an unknown quantum state in Alice's box A and teleport the state to Bob's box B. This is teleporting information rather than matter, but this should satisfy us: two particles with the same set of quantum numbers are completely indistinguishable, so if we give the particle in B the same quantum numbers as the particle in A then in essence we've teleported the particle from A to B.

So how does quantum teleportation work? Well, it makes use of two other characteristic features of quantum mechanics: *superposition* and *entanglement*. Without going into great detail, let's look at these two concepts.

As outlined in Chap. 6, everything you can know about a quantum particle such as an electron, say, is contained in its wavefunction. And one of the perplexing features of the quantum world is that the wavefunction of an unobserved particle seems in some way to 'try out' all possible options. For example, an electron possesses an attribute called spin. As long as you don't take it too literally, you can picture an electron as being a tiny rotating billiard ball, and when we measure its spin along some particular axis we always find the spin can point along the axis (in which case we might say the electron is spin-up) or the opposite way (spin-down). In the macroscopic world, a billiard ball has its spin axis either pointing up *or* pointing down. In the quantum world, until you make an observation to determine its spin along a particular axis, an electron is spinning up *and* down *at the same time*. Both realities co-exist. If there's an electron in box A then, until we make a measurement, it is in a superposition of states: along any particular axis it is spinning both up *and* down. The wavefunction gives the *probability* of observing a particular outcome when we make a measurement. Once we make a measurement, though, the wavefunction 'collapses' and the electron is in the state we observe.

Superposition applies to more than just one particle. Suppose we create a pair of electrons in such a way that the total spin of the two particles is zero: if one electron is spin-up then the other must be spin-down. Until we make a measurement, both electrons are in the superposition state—both of them are spinning up *and* spinning down, but overall the spins cancel. If we observe one of the electrons, and its wavefunction collapses so we measure spin up, say, then the other electron is immediately spin down. This phenomenon is known as entanglement, and Einstein himself called it 'spooky'. It's spooky because it doesn't matter how much distance separates the two electrons—there's an invisible but unbreakable connection between them. It's as if they were a single entity. The two electrons could be in neighboring rooms or neighboring galaxies and their entanglement would still apply: if the electron is measured to be spin-up *here* then the other electron is immediately spin-down *there*. This might seem to be an example of faster-than-light communication, and depending upon your favored interpretation of quantum mechanics it could indeed be seen as ftl communication. Nevertheless, entanglement doesn't violate special relativity because there's no possible way of using the effect to transmit useful information. The point is, before you make the measurement of the electron *here* you have no way of knowing what you'll find: 50% of the time you'll measure spin-up and 50% of the time you'll measure spin-down, and the nature of quantum mechanics means the observed result is entirely random. When it comes to spin, or any other quantum property, before you make a measurement you aren't in possession of any formation you could usefully transmit.

Entanglement doesn't allow us to communicate at ftl speeds—but Bennett and his colleagues showed how it permits teleportation.

We begin the process of teleportation by preparing an entangled pair of electrons. We give one electron to Alice, to put in box A, and the other to Bob, to put in box B. Alice and Bob can be separated by any distance: Alice might be in London and Bob in New York, or perhaps Alice is on Earth and Bob is somewhere in orbit—it really doesn't matter, as long as the entanglement holds.

Suppose we want to teleport a particle X from A to B. Well, Alice puts the particle in the box A so that it enters into a quantum superposition with the entangled electron. This imprints the entangled system with information about the particle X. Bob, meanwhile, puts a 'blank' particle Y into his box. Alice then measures the superposition and thereby extracts certain (classical) information about the combined state of the particle X and the entangled electrons. This scanning process completely disorders the quantum numbers of particle X (since quantum mechanics doesn't allow us to have duplicate copies of an object) but it also causes the entangled electron in B to assume the properties of its entangled partner in A. Alice sends the result of the classical measurement to Bob by radio, telephone, or even snail-mail. The 'blank' particle Y is then made to interact with the entangled electron and Bob combines this with the measurement information from Alice. This combination of classical measurement information and entangled quantum information is sufficient to transfer the complete set of quantum numbers from particle X (in box A) to particle Y (in box B). And since particles with the same set of quantum numbers are indistinguishable, Alice and Bob have succeeded in teleporting the particle from A to B.

This recipe for teleportation might sound too arcane for it to possibly succeed, but the recipe works just fine. Once experimental physicists had got their head around Bennett's proposal they were able to teleport a quantum state over a distance of a few meters; as I write, the record distance for quantum teleportation is 143 km^9; as techniques improve, the distance over which teleportation can take place will undoubtedly increase. Experimenters first demonstrated quantum teleportation with single photons; with current technology they can quantum teleport materials such as atoms; as techniques improve, it will surely be possible to teleport larger systems. Quantum teleportation is already a reality, and it's an active field of research.

And a *Star Trek* transporter? Well, unfortunately, the transporter will probably remain an impossibility. For quantum teleportation to work, the transmitter and receiver must have once been in contact: it requires particles to be entangled. Captain Kirk can't simply beam down to Alpha Aradoni II when he feels like it; the link between *USS Enterprise* and the landing point must have been prepared in advance. Furthermore, it's one thing to quantum teleport an atom, but quite a different matter to quantum teleport an object as large as a virus, say. The difficulties in quantum teleporting a system as large and as complex as a human being are overwhelming.

Quantum teleportation is set to become a key technology—but its importance will reside in applications involving computing and cryptography, rather than transport. And for me that's a shame, because the *Star Trek* transporter was such a wonderful device.

9.3   Conclusion

When it comes to transportation, humankind is entering a science fictional future that science fiction did not foresee. The way in which digital technology scales means computers can be built into cars and those computers will drive more safely, more efficiently, and more economically than human drivers. The accidents they'll prevent and the money they'll save make the introduction of autonomous vehicles an inevitability. On the other hand, as we've seen, the introduction of autonomous vehicles raises hundreds of problems—technical, moral, and ethical—that have yet to be resolved. Our automotive future is going to be interesting!

It is hard to foresee other existing transport technologies scaling in a way that will transform our lives. Consider railways, for example. In my own country, a political battle has raged for years over whether the so-called HS2 should be constructed. HS2 would be a high-speed railway line linking London with Birmingham and then the northern cities of Manchester and Sheffield. If the proposal overcomes the significant opposition to it, the plan is for the work to be completed in 2033. One respected economics institute estimates the cost of HS2 will be over £80 billion. This huge sum of money will provide a service that shaves 32 minutes off the journey time between London and Birmingham. That's good news if you are one of the people who needs to get from London to Birmingham, but HS2 will do nothing to improve the journey time between, say, London and Portsmouth or London and Bristol. (Indeed, the cost of HS2 is so great that it's quite possible that other services will have to be reduced in order to pay for it.) HS2 is an example of a technology that doesn't scale and thus cannot produce a fundamental shift in most people's lives. A system such as skyTran, however, *does* have the capacity to scale. Could the appearance of our cities in future be dominated by the poles and narrow guideways of a personal rapid transport system?

Notes

1. Armin Joseph Deutsch (1918–1969) was a professional astronomer, known for his work on Doppler tomography. "A Subway Named Mobius" was his only SF story. For readers who don't know, a Möbius strip is a surface with only one side and one boundary. It's easy to construct a Möbius strip: take a long, thin piece of paper, give it a half twist, and then glue the ends together to form a loop. To demonstrate the strip has only one side, draw a line along the length of the loop: the pen returns to its starting point without having left the paper. Deutsch's story formed the basis of the film *Moebius* (1996), which tells the story of a train vanishing into the Buenos Aires underground.

2. In 2005, the bicycle was chosen by 59% of listeners who voted in BBC Radio 4's online poll to find the most significant invention since 1800 (BBC News 2005). The bicycle won a similar poll carried out a year earlier by The Times Online, and 3 years earlier by the UK Patent Office. Perhaps bicycle riders ran orchestrated campaigns to increase the bike's vote—but even if that's the case it's telling that car drivers and train passengers didn't feel sufficiently enthused to run a similar campaigns.

3. Yves Rossy (1959–), or 'Jetman', is a Swiss ex-fighter pilot who has developed a number of personal jetpacks. In 2012, on BBC's *Top Gear* motoring program, Rossy raced against a rally car driven by a professional rally driver. The car followed an 8-mile rally dirt track; Rossy followed the track at an altitude of 8000 feet. Rossy won.

4. According to the latest reports at the time of writing, the Martin Jetpack will be constructed from carbon fibre and aluminium, and be propelled by fans driven by a 2-litre petrol engine. There'll be a parachute for when things go wrong. See, for example, Guardian (2015).

5. For details of Tubular Rail, visit the website tubularrail.com
6. For the Arup report on the future of railways, see Arup (2014). For the details of the union response quoted in the text, see Institution of Mechanical Engineers (2014).
7. For Asimov's prophecy of what the world would look like 50 years after his visit to the World's Fair, see Asimov (1964).
8. For details of the first crash caused by a Google car, see BBC (2016).
9. Details of the experiment in which quantum teleportation was demonstrated over a distance of 143 km can be found in Ma et al. (2012).

Bibliography

Non-fiction

Arup: Future of Rail 2050. Arup, London (2014)
Asimov, I.: Visit to the World's Fair of 2014. New York Times (16 Aug 1964)
BBC News: Bicycle Chosen as Best Invention. http://news.bbc.co.uk/1/hi/technology/4513929.stm (2005)
BBC: Google Self-Driving Car Hits a Bus. www.bbc.co.uk/news/technology-35692845 (2016)
Guardian: The World's First Commercial Jetpack Will Cost $150,000 Next Year. www.theguardian.com/technology/2015/jun/26/worlds-first-commercial-jetpack-next-year (26 June 2015)
Institution of Mechanical Engineers: Arup Predicts Robotic Future for Railways. imeche.org (2014)
Ma, X.S., et al.: Quantum teleportation over 143 kilometres using active feed-forward. Nature. **489**(7415), 269–273 (2012)
SAE: Standard J3016_201401. http://standards.sae.org/j3016_201401 (2014)
Wilson, D.H.: Where's My Jetpack?: A Guide to the Amazing Science Fiction Future that Never Arrived. Bloomsbury, New York (2007)

Fiction

Anon: The Reign of George VI, 1900–1925: A Forecast Written in the Year 1763. Niccoll, London (1763)
Asimov, I.: Sally. Fantastic (May/June 1953)
Asimov, I.: The Caves of Steel. Doubleday, New York (1954)
Ballard, J.G.: Crash. Jonathan Cape, London (1973)
Bear, G.: Eon. Tor, New York (1985)
Bester, A.: Tiger! Tiger! Sidgwick and Jackson, London (1956)

Biggle Jr., L.: All the Colors of Darkness. Doubleday, New York (1963)

Bradbury, R.: The Pedestrian. The Reporter (August 1951)

Brunner, J.: You'll take the high road. In: Silverberg, R. (ed.) Three Trips in Time and Space. Hawthorn, New York (1973)

Brunner, J.: Web of Everywhere. Bantam, New York (1974)

Brunner, J.: The Infinitive of Go. Ballantine, New York (1980)

Clarke, A.C.: A Fall of Moondust. Gollancz, London (1961)

Clarke, A.C.: Glide Path. Harcourt Brace, London (1963)

Deutsch, A.J.: A Subway Named Mobius. Astounding (December 1950)

Dickson, G.R.: No more Barriers. Science Fiction Stories (September 1955)

England, G.A.: The Flying Legion. McClurg, Chicago (1920)

Farmer, P.J.: A Private Cosmos. Ace, New York (1968)

Farmer, P.J.: The Fabulous Riverboat. Putnam, New York (1971)

Fort, C.: Lo! Cosimo, New York (1931)

Galouye, D.F.: The Last Leap. If (January 1960)

Gernsback, H.: Ralph 124C 41+. In: Modern Electrics (serialized April 1911–March 1912) (1911)

Harrison, H.: A Transatlantic Tunnel, Hurrah! Tor, New York (1973)

Heinlein, R.A.: The Roads must Roll. Astounding (June 1940)

Heinlein, R.A.: Stranger in a Strange Land. Putnam's, New York (1961)

Heinlein, R.A.: Tunnel in the Sky. Scribner's, New York (1955)

Keller, D.H.: The Revolt of the Pedestrians. Amazing (February 1928)

Keller, D.H.: The Living Machine. Wonder Stories (May 1935)

Kipling, R.: With the Night Mail. The Windsor Magazine (December 1905)

Kipling, R.: As Easy as A.B.C. The London Magazine (March–April 1912)

Langelaan, G.: The Fly. Playboy (June 1957)

Leiber, F.: Catch that Zeppelin! Fantasy & Science Fiction (March 1975)

Louden, J.C.: The Mummy!, or a Tale of the Twenty-Second Century. Colburn, London (1836)

Martin, G.R.R.: Fevre Dream. Posiedon, New York (1982)

McCaffrey, A.: Dragonflight. Ballantine, New York (1968)

Milne, R.D.: Professor Vehr's Electrical Experiment. The Argonaut (January 1885)

Mitchell, E.P.: The Man Without a Body. The Sun (March 1877)

Niven, L.: Ringworld. Ballantine, NewYork (1970)

Niven, L.: Flash crowd. In: Silverberg, R. (ed.) Three Trips in Time and Space. Hawthorn, New York (1973)

Niven, L.: A Hole in Space. Ballantine, New York (1974)

Niven, L.: A World Out of Time. Holt, Rinehart and Winston, New York (1976)

Priest, C.: The Inverted World. Faber and Faber, London (1974)

Schmitz, J.H.: The Lion Game. DAW, New York (1973)

Shaw, B.: Aspect. Nebula (August 1954)

Silverberg, R. (ed.): Three Trips in Time and Space. Hawthorn, New York (1973)

Simak, C.D.: Way Station. Doubleday, New York (1963)

Simak, C.D.: The Goblin Reservation. Berkeley, New York (1969)

Smith, E.E.: The Skylark of Space. Amazing (August/October 1928)
Smith, G.O.: Special Delivery. Astounding (March 1945)
Souvestre, E.: The World as It Will Be. Coquebert, Paris (1846)
Sturgeon, T.: Killdozer! Astounding (November 1944)
Sturgeon, T.: More than Human. Farrar, Straus and Giroux, New York (1953)
Vance, J.: Rumfuddle. In: Silverberg, R. (ed.) Three Trips in Time and Space. Hawthorn, New York (1973)
Varley, J.: The Ophiuchi Hotline. Dial, New York (1977a)
Varley, J.: Air Raid. Asimov's (Spring 1977b)
Varley, J.: Millennium. New York, Berkeley (1983)
Verne, J.: Five Weeks in a Balloon. Hetzel, Paris (1863a)
Verne, J.: Paris au XXe Siècle. Hetzel, Paris (1863b)
Verne, J.: Twenty Thousand Leagues Under the Sea. Hetzel, Paris (1870)
Verne, J.: Around the World in 80 Days. Hetzel, Paris (1873)
Verne, J.: Robur the Conqueror; Or, A Trip Around the World in a Flying Machine. Hetzel, Paris (1886)
Vinge, V.: The Witling. DAW, New York (1976)
Wells, H.G.: When the Sleeper Wakes. George Bell and Sons, London (1899)
Wells, H.G.: The War in the Air. George Bell and Sons, London (1908)
Wells, H.G.: The Shape of Things to Come. Hutchinson, London (1933)
Williams, W.J.: Knight Moves. Tor, New York (1985)
Zelazny, R.: Today We Choose Faces. Signet, New York (1973)

Visual Media

Back to the Future: Directed by Robert Zemeckis. [Film] USA, Universal (1985)
Back to the Future, Part II: Directed by Robert Zemeckis. [Film] USA, Universal (1989)
Blade Runner: Directed by Ridley Scott. [Film] USA, Warner Bros (1982)
Captain Zeppos: Written by Lode de Groof. [TV series] Belgian Radio and Television (BRT) (1964–1968)
Death Race 2000: Directed by Paul Bartel. [Film] USA, New World Pictures (1975)
King of the Rocket Men: Directed by Fred C. Brannon. [Movie Series] USA, Republic Pictures (1949)
Millennium: Directed by Michael Anderson. [Film] USA, 20th Century Fox (1989)
Moebius: Directed by Gustavo Mosquero. [Film] Argentina, Universidad del Cine (1996)
Star Wars: Directed by George Lucas. [Film] USA, 20th Century Fox (1977)
The Fly: Directed by Kurt Neumann. [Film] USA, 20th Century Fox (1958)
The Tomorrow People: Created by Roger Price. [TV series] UK, Thames Television (1973–1979)
Thunderball: Directed by Terence Young. [Film] USA, United Artists (1965)

10

Immortality

Immortality isn't living forever. That isn't what it feels like.
Immortality is everybody else dying.

Dr Who, "The Girl Who Died"

I'd guess the dream of immortality appears in all human cultures. The Greeks had their list of immortal gods, demons, and minor deities; mortals such as Sisyphus who dreamed of immortality were made to suffer. Early followers of Taoism believed in an elixir of life, and thought the ingestion of cinnabar or gold would lead to immortality; in an attempt to stave off death Qin Shi Huang, the first emperor of China, ate so much cinnabar he died of mercury poisoning. In ancient Sumer, the immortal Utnapishtim tells Gilgamesh that longevity can be obtained by eating a special plant; Gilgamesh finds the plant but promptly loses it to a snake and so doesn't find out whether it would have worked. In Hindu belief, eating a nectar called amrita confers immortality; in Norse mythology it's the eating of golden apples; in Japanese mythology it's the eating of a fish called ningyo. Christianity, of course, is replete with characters who gain immortality, including Sir Galahad (who touches the Holy Grail), the Wandering Jew (who taunted Jesus while he was *en route* to his crucifixion), and the Three Nephites (who were granted immortality so they could continue their ministry until the second coming of Christ).

The dream of immortality, as the partial list above demonstrates, holds a firm grip on the human imagination. And immortality is indeed a great idea, spoiled only by the fact it lasts forever. The myth of Tithonus highlights one problem with the idea. Tithonus (in Tennyson's version) asks for immortality, a gift the gods grant, but forgets to ask in addition for the gift of eternal youth.

© Springer International Publishing AG 2017
S. Webb, *All the Wonder that Would Be*, Science and Fiction,
DOI 10.1007/978-3-319-51759-9_10

Tithonus lives forever, but wastes and withers with each passing day. Even if you stay healthy, there'd be the problem of boredom: there's never anything decent on TV, so how would you fill a never-ending succession of days?

It's not surprising that SF writers have played with the idea of immortality. Sometimes they play with the classic myths of immortality (for example, the Wandering Jew appears in Walter Miller's classic *A Canticle for Leibowitz* (1960), but they have also devised a number of interesting variants of the theme based on science rather than wish fulfillment.

10.1   How to Live Forever

As I mentioned in the Introduction to this book, one of my earliest encounters with SF was through the Anderson's marionette puppet TV show *Captain Scarlet and the Mysterons*. I can still recall the end production credits, which consisted of ten paintings by the artist Ron Embleton each showing Captain Scarlet in a moment of peril—tied up and weighed down by blocks as sharks circled him, squeezed between closing spike-filled walls, face-to-face with a spitting cobra—all while the terrific theme music played. Embleton's paintings beautifully captured the essence of the series: the adventures of a special agent, Scarlet, who was killed by the evil Mysterons and then brought back to life with the wonderful power that any physical injuries healed within hours. For me and my young playmates, Captain Scarlet was terrific: he wasn't just immortal, he was *indestructible*.

For a child, Scarlet's adventures were just fine—there was something comforting in the knowledge that Scarlet couldn't be killed and therefore was always going to defeat the Mysterons. It was only when I began reading SF, and encountered so many wonderful stories in which immortality played a role, that I began to understand the many drawbacks associated with immortality: how lonely life would be if your friends and family died while you lived on; how boring it would be if instead everyone was immortal and you kept on seeing the same old faces; how terrible an immortal life would be if health deteriorated while the body persisted. Nevertheless, despite the downsides to extreme longevity, people still dream of cheating death. Woody Allen speaks for many of us when he says: 'I don't want to achieve immortality through my work; I want to achieve immortality through not dying. I don't want to live on in the hearts of my countrymen; I want to live on in my apartment.'

SF writers have considered numerous ways in which immortality might be achieved, and we'll look at some of them—such as genetic engineering, cryonics, and mind uploading—later. (I shan't consider an approach to immortality based on one's achievements at work, the approach derided by Woody Allen, except to say Frederick Pohl explored this in his hilarious "Shaffery Among the Immortals" (1972).) First, though, let's look at what SF writers have to say about the simplest method of achieving immortality: living forever through the expedient of not dying.

10.1.1 Life Eternal (or at Least Life Long-Lived)

For most practical purposes, we can equate immortality with longevity. Although advances in nutrition, environmental health, and medical science have extended the average life expectancy to around 80 years in the West, there seems to be a hard biological limit when it comes to human lifespan. Only one person in history is known with certainty to have reached the age of 120. Jeanne Calment, a French woman who lived all her life in Arles and who once sold colored pencils to Vincent Van Gogh, died in 1997 at the age of 122 years 164 days. So as far as I'm concerned, if there really had lived someone such as Methuselah, a character for whom the Bible reports a lifespan of 969 years, then we could consider him to have been effectively immortal.

Breaking through that 'twelve decades' barrier would seem to require some new ideas. Later we'll look at how SF writers have investigated some physical approaches for lengthening lifespan. First, though, let's look at a few ideas grounded in biology.

Heinlein, in *Methuselah's Children* (1958), suggested longevity would come by breeding people from those with long-lived ancestors. In the novel, a character who made a fortune in the 1850s California gold rush dies childless in his fourth decade. He leaves his fortune to a trust that provides a financial incentive to young people with long-lived grandparents to marry each other and have offspring. In this way, after three centuries have passed, members of certain families have a life expectancy of 150 years or more. The first version of the story appeared in Campbell's *Astounding* in 1941, when the average life expectancy of an American male was about 60 years. An increase in life expectancy from 60 to 150 must have seemed science fictional to Heinlein, even though the advance took place over three centuries. But in the time since Heinlein wrote the story—a duration equivalent to a typical Western lifespan—the average life expectancy has already increased from about 60 years to about 80 years. In the UK, the Office of National Statistics estimates that one in three babies born in the UK in 2013 will live to be 100. We can reasonably expect that by the time of the twenty-second century, which is when the action in *Methuselah's Children* takes place, a lifespan of 122 years—the age reached by Calment—would not be particularly unusual. Perhaps by then a few people will reach the age of 150, an achievement gained without any of the funny business described by Heinlein.

Methuselah's Children is noteworthy for the number of ingenious ideas Heinlein tosses nonchalantly to his readers, but the method it proposes for achieving longevity is clumsy and far from convincing. Indeed, although one of Heinlein's most successful characters—Lazarus Long—makes his first appearance in this story, Long's very existence highlights the problem with the proposed approach. Long is supposed to have been born in 1912, the result of the third generation of the selective breeding program described above, and he turns out to have a natural lifespan of 250 years or more. But Heinlein himself acknowledges that a mere three generations of selective breeding is nowhere near enough to produce such a long-lived individual, so he has a character remark that Long is 'a mutation'. Perhaps *that* is how immortality—or at least longevity—might be achieved? By random mutation?

In *Fury* (1950), Henry Kuttner imagined a society in which the ruling class possess exceptionally long life spans, the consequence of a heritable genetic mutation caused by atomic war. The picture painted by Kuttner is not pretty: the same atomic war that produced genetic mutations also rendered Earth uninhabitable, and the survivors live in domed cities at bottom of the oceans of Venus (Kuttner wrote the book before astronomers learned about conditions on Venus). Not only that, society is stagnating: normal people cannot explore beyond the confines of their submerged cities while the ruling class, who view matters from the perspective of those who have centuries to think things through, prefer procrastination to action.

James Gunn, in *The Immortals* (1962), also postulates that a random genetic mutation can lead to effective immortality. The novel's protagonist is a lucky person who never gets ill, never catches cold, never ages. But Gunn asks whether such a person could be considered 'lucky' if he lived in a world where everyone else is mortal, where even the richest and most powerful inevitably die? Once the rich and powerful learn of the possibility of immortality, they'd stop at nothing to learn the secret. The blood of the lone immortal would be the most precious substance on Earth. There'd be those willing to bleed him dry to learn the secret of long life.

Heinlein, Kuttner, Gunn (and many other SF writers) considered immortality in terms of an individual's genetic make-up. Other writers have taken a more science fictional approach to the problem, and wondered whether symbiosis—usually with an alien creature—might lead to immortality. In one of my favorite stories, George Martin's evocative "A Song for Lya"

(1974), humans willingly wander into caves in order to join with a parasite lifeform called the Greeshka. To onlookers this appears to be a particularly horrible way to die, but the Greeshka absorb not only the body but also the consciousness of a victim. The Greeshka offer a form of everlasting life. For some, this is enough to enter the caves and be absorbed. F. Paul Wilson's novel *Healer* (1976) has a not dissimilar setting—a human enters a cave, joins with an alien, and becomes immortal—but this setup is the springboard for a quite different tale. The now-immortal human, along with his symbiotic partner, travels across the galaxy and learns enough to become a renowned doctor. Anne McCaffrey's novel *Crystal Singer* (1982) tells the tale of humans who develop a symbiosis with a spore native to the planet Ballybran, a relationship that makes humans extremely long-lived. Again, the symbiosis isn't without its drawbacks.

Another approach to extreme longevity is a variant of the elixir of life—a potion, such as the Hindu amrita or the cinnabar-based concoctions of the Chinese, which confers upon the drinker eternal youth. SF is replete with drugs offering similar effects. For example, James Blish's *They Shall Have Stars* (1956) postulates how the development of an anti-agathic ('anti-ageing') drug could be used to help humans cope with the centuries-long journey times involved with interstellar travel.[1] Frank Herbert's epic *Dune* (1965) is set on the planet Arrakis, a desert world notable because it's the sole source of the most valuable substance in the universe: melange, better known as 'spice'. Spice is made from substances secreted by giant worms that live under the sands of Arrakis; when ingested, the drug improves health, bestows improved mental faculties on the user, and extends life span.[2] Larry Niven's "Known Space" is the setting for many novels and short stories, and one of the key technologies in that setting is 'boosterspice'—a drug for reversing the effects of ageing. With a large enough supply of boosterspice, and a capacity to endure the boredom of a long life, a human could in principle live forever. In *World Without Stars* (1966), Poul Anderson postulates the existence of antithanatic— 'against death'—drugs: take them when you reach adulthood and you then live forever. (As with most such elixirs, taking an antithanatic doesn't mean you can't die. A bullet through the heart, an axe through the head, or a ton weight dropping on you from a height are all going to kill you. Taking an antithanatic won't turn you into Captain Scarlet. But with melange, boosterspice, or antithanatics you can live indefinitely so long as you avoid accidents or acts of violence.) The characters in Anderson's novel face a problem that perhaps all long-lived humans would face: as one ages one accumulates memories. Anderson supposes that a person's memories must be edited and the trivia removed, in order for new and potentially more important memories to be laid down.

10.1.2 Pantropy

As we have seen, James Blish suggested 'anti-agathic' drugs might lead to longevity. In a number of stories, collected together in *The Seedling Stars* (1957), Blish also contemplated another technology: pantropy. The idea behind pantropy—literally, 'turning everything"—was simple: to engineer the human body so it can survive in alien environments. (Blish did offer an alternative to pantropy: terraforming. This would change the environment to make it suitable for human life. But altering a human body is probably easier than altering a celestial body.) A combination of pantropy and anti-agathics, Blish suggested, might be necessary if humanity were to colonize planets beyond the solar system: a changed body would be necessary to live on planets with unfamiliar atmospheres, different climates, or excessive surface gravities, while longevity would be required to reach those planets in the first place. Pantropy by itself, however, might provide a route to immortality! The methods for achieving longevity outlined in the previous section—selective breeding, random mutations, drugs, and so on—seem rather careless. They either take too long, are hit-and-miss, or possess deleterious side effects. Why not *engineer* the human body for longevity? Pantropy for immortality.

10.1.2.1 Cyborgs

One way of engineering the body is to take a 'hardware-based' approach: replace fallible human tissue with something rather more resilient. One of the earliest SF tales dealing with this idea is Clifford Simak's "Desertion" (1944), in which colonists on Jupiter go through some sort of biological converter that gives the human body the same form as indigenous Jovian life. Once transformed, their bodies are strong enough to let them leave the protection of their domed cities and live freely in this alien environment. A later, much grittier, version of this idea is Fred Pohl's satire *Man Plus* (1976). Pohl describes the transformation a man undergoes in order to live on the Martian surface. The astronaut is gelded; his lungs are scooped out and replaced with something more suitable for the thin Martian atmosphere; he is given a cloaca; his various sensory organs are improved; his super-strengthened body is powered by huge solar-power-collecting wings attached to his shoulders... the only parts of him left untouched are certain areas of his brain.

The transformation described in the short story "Camouflage" (1945), one of the Kuttner/Moore collaborations attributed to Lewis Padgett, is less graphic than that in *Man Plus* but is perhaps even more extreme. The central character in "Camouflage" suffers an accident that destroys his body but leaves his brain intact. His brain is wired into a spaceship, which he then controls. This combination of human brain and spaceship is sometimes called a 'brainship', an idea developed by SF luminaries such as Anne McCaffrey (see for example "The Ship Who Sang" (1961)) and Larry Niven (see for example "Becalmed in Hell" (1965)).

These examples of pantropy are what Blish envisaged: alterations in the human body that permit humanity to live and explore hostile environments (in the examples above, respectively Jupiter, Mars, and space itself). Catherine Moore's "No Woman Born" (1944), one of the first cyborg ('cybernetic organism') stories, was a more sensitive investigation of the topic. The story's multi-talented protagonist, a beautiful actress, singer, and dancer, dies in a theatre fire. A scientist, prompted by the actress's manager, saves her brain and implants it in a magnificent metallic body. The reactions of the scientist and the manager—both men—to her new body are thought provoking and, despite the flowery language typical of 1940s pulp fiction, still ring true.

Asimov considered the concept of the cyborg in a couple of stories. "Segregationist" (1967) describes a future in which humans who have been granted the right to longevity can opt for metallic replacements when their organs give out; similarly, robots can choose to have artificially-created organic body parts. Humans and robots are thus becoming increasingly similar. A more famous Asimov story is "The Bicentennial Man" (1976), which won that year's Hugo and Nebula awards. This is the story of a robot who undergoes the sort of transformation described in *Man Plus*, but in reverse: he submits to a series of procedures to make him increasingly human, culminating in an operation that causes his positron brain to decay and eventually cease function. By dying, he becomes fully human.

Television made full use of the cyborg idea, of course. The TV movie *The Six Million Dollar Man* (1973) spawned a series that ran for five seasons. The series was based on the novel *Cyborg* (1972) by Martin Caidin, the story of a test pilot whose crash leaves him blind in one eye and with only one functioning limb. Scientists replace the limbs, eye, and other body parts with prosthetics—'bionics'—which are better and more efficient than the originals. The series was successful enough to afford a spin-off, called *The Bionic Woman* (1976). And of course *Dr Who* battled the Cybermen (cyborgs in which only the brain is of biological origin) and the Daleks (organic creatures that live within a metallic shell).[3]

Few of the stories mentioned above explicitly reference immortality or even longevity. Asimov in "The Bicentennial Man" addresses immortality by implication: the robot protagonist deliberately undergoes procedures that will lead to his early death, and without those procedures it would presumably live on indefinitely. But it's natural to assume a character who starts out in life as a robot will have a much longer life expectancy than someone who begins life with an organic body. What of characters who gradually transition from human to cyborg? Can we assume they will be immortal?

One might expect the individual replacement parts of a cyborg to have a longer shelf-life than their organic equivalents. But cyborgs in fiction almost always are left with one original organic part: the brain. Presumably, the authors are arguing that if you have your arm replaced by a prosthetic device then 'you' are still 'you'. Similarly for other body parts. But for 'you' to remain 'you', according to these authors, the brain itself must remain intact. And the brain, as with other organs, does age. So a cyborg body might be immortal—it would presumably at least permit longevity—but the brain it houses would age and die. That's what happens to the beautiful actress in Moore's "No Woman Born".

10.1.2.2   Genetic Engineering

A hardware-based approach to increasing lifespan might not be the best way forward. It's little use having bionic arms and legs if your organic brain is fading away. Perhaps genetic engineering offers a better approach? Perhaps we can alter our genes so our body and brain can better withstand the degradations caused by age?

In *Last and First Men* (1930), Olaf Stapledon told the stories not of individual characters but of entire species of humanity. In Stapledon's novel, our own species and civilization goes by the name of First Men. In the far distant future we evolve into Second Men, which in turn give way to the Third Men. This long-lived species of humanity is interested in the creation and design of living organisms, and they succeed in developing the Fourth Men. The Fourth Men are cold, calculating, rather nightmarish beings (and probably the most interesting of the Men conjured by Stapleton's imagination) and are essentially immortal: immortality has been achieved through genetic engineering. Although immortal, the Fourth Men die out because they genetically engineer a new human species: the Fifth Men. These are super-intelligent beings and, upon clashing with the Fourth Men, they use their more advanced technology to wipe out their creators. The Fifth Men aren't quite immortal, but they do live for between 3000 and 50,000 years—quite creditable, I'd say. And so it continues. Varieties of humanity come and go; some are reach incredible heights, others descend into savagery. The use of genetic engineering is not a constant through Stapleton's future history, but it does play an important role in those first few stages up to the Fifth Men. And later, the Eighth Men genetically engineer the Ninth Men so they are adapted for living on Neptune (an example of pantropy long before Blish coined the term). The final and most advanced species of humanity is Eighteenth Men. They are essentially immortal, and die only through accident, murder, or suicide. When the Sun dies, however, even the era of Eighteenth Men must end. As a swan song, they create a basic life form that they send out to seed new species throughout the Galaxy—life itself, then, becomes immortal.

Stapledon wrote of events happening over billions of years. Robert Heinlein was usually more interested in what might happen in the near future. In *Beyond This Horizon* (1948), Heinlein builds on the approach described in *Methuselah's Children* and imagines a society in which generations of humans are genetically engineered to produce healthy, intelligent, and long-lived humans. The novel was originally a two-part serial published in *Astounding* in 1942. In the first part, the protagonist is revealed to be an almost perfect genetic specimen. The second part involves the birth of his genetically engineered—and genetically perfect—son. The son is effectively a superman, with perfect health to go with advanced mental abilities, who can expect to live for many centuries. At the end, we learn the protagonist will also have a genetically engineered daughter—and someone in government arranged all this because souls, it seems, are reincarnated: the government official will be reincarnated in the daughter's body. (Reincarnation, indeed, is another means of achieving immortality. I'll discuss that option later in the chapter.)

Jack Vance's novel *To Live Forever* (1956) contains a neat twist on the idea of genetic engineering and immortality. Vance imagines a near future in which scientists have discovered how to halt the ageing process. There's a problem with this, however, a problem affecting all the stories mentioned above but acknowledged and addressed by only a few authors: if fewer people die then overpopulation will result, and lead to catastrophe. In Vance's novel, society has responded to this problem by rationing the anti-ageing treatments based on a form of meritocracy. An individual, through 'striving', can reach higher and higher levels of social standing, and at each level is awarded extra years of life. After a certain number of years a person will be killed—unless they reach the highest level of society, Amaranth, at which point they are granted the right to live forever. Not only do Amaranths receive longevity treatments, genetic engineers create clones of them. That way if an Amaranth dies through accident or murder, their life can continue essentially uninterrupted by transferring their consciousness into their clone.

The idea of immortality via clone is, to my mind, the neatest of the ideas in *To Live Forever*. If it is indeed possible to transfer one's mind and consciousness from one substrate to another (and we'll look at this in more detail later) then there's really no need to mess around with longevity treatments. When your decrepit body is about to shuffle off this mortal coil, simply transfer your consciousness to a fresh clone and you're good to go for another few decades. And this surely does qualify as immortality: your clone, after all, would be genetically identical to you while the transfer of your memories and your consciousness would presumably mean you'd recognize yourself as still being *you*. (Various philosophical and technical objections could be raised at this point, but let's park them for the moment.)

John Varley played around with this idea in his "Eight Worlds" series of stories. The wonders of advanced genetic engineering mean it is trivial to undergo body modification: characters tweak their bodies with the same ease you tweak your smartphone by adding apps. Immortality, or at least its illusion, is obtained through means of mind recordings and clones. Varley has great fun with the possibilities, and in his novel *The Ophiuchi Hotline* (1977) he poses the interesting question of what 'murder' means in a society in which the victim can be revived and live again in a body genetically identical to the original.

10.1.3 Corpsicles

The stories mentioned so far are set in the far future—a few centuries ahead in the case of Varley's *The Ophiuchi Hotline*, many aeons ahead in the case of Stapledon's *Last and First Men*. That timescale for immortality might seem realistic, but it doesn't help those of us alive right now. Is there any hope of immortality for us? Well, SF has long offered one possible way of cheating death: if a person can't be granted immortality *now*, perhaps they can be kept in a state—some sort of suspended animation—that will allow them to be revived when a cure for their particular ailment is found, and when immortality treatments have been developed.

The main character in Pohl's *The Age of the Pussyfoot* (1969) is a man who was born in the latter part of the twentieth century and is fatally injured in a fire. His body is placed in a cryopreservation chamber—in SF, freezing is the usual method for suspending animation—and becomes, to use Pohl's phrase, a 'corpsicle'.[4] Five hundred years later, his body is revived and the man is able to continue his life. Cryonic preservation also plays a key role in Norman Spinrad's *Bug Jack Barron* (1969). Spinrad's novel describes how the Foundation for Human Immortality will, for a large sum of money, cryogenically preserve a dying customer for revival whenever a cure is found for whatever it was that killed them. One of the best treatments of the theme is Clifford Simak's *Why Call Them Back From Heaven?* (1967). Simak imagines a near-future world in which a corporation, the Forever Center, has persuaded people of the inevitability of immortality within a decade or two. The recently deceased are cryogenically preserved, to be reanimated when immortality can be conferred upon them. What makes the novel so thought-provoking is the way in which Simak examines the impact of this development upon society. It's not just the dead and dying who are frozen; vast numbers of individuals *choose* to freeze in order to have their chance at immortality. Except, of course, that after a decade or two immortality is *still* not available. What would happen to the economy in this scenario? How would people pay for the cost of preservation? How would they ensure they'd have income after reanimation? Simak provides some downbeat, but plausible, answers to these questions.

Larry Niven, in "The Defenceless Dead" (1973), points out a horribly plausible problem with cryogenic preservation. The point is: the perished can't protect themselves. In the story, a law is passed stating paupers who have been frozen are legally dead—and therefore their organs can be harvested for use by the living. (Niven wrote other stories exploring the relationship between immortality and freezing. For example, his story "Wait It Out" (1968) tells the tale of an astronaut who, marooned on the surface of Pluto, removes his helmet and quickly freezes to death—except the extreme cold of Pluto's night turns his brain into a superconductor. During the day the temperature rises and his consciousness switches off; at night, though, thoughts flow. He hopes to stay this way until he is rescued and taken back to Earth to be thawed, however long it takes.) For one last take on the idea, consider Greg Benford's wry story "Doing Lennon" (1975): a millionaire pays to have himself frozen and, after a century or so has passed, revived into a world of greater life expectancy—at which point he plans to impersonate John Lennon. Not all, however, is as it seems.

Numerous stories, then, use cryopreservation as a form of time travel—as a shortcut to a future in which (it is hoped) science has cracked the problem of immortality. It's a means of putting death on hold.

10.1.4 Mind Uploading

The wish for immortality is not so much the wish for an everlasting body but for an unbroken sense of self. After all, putting one's body in deep freeze will ensure one's flesh is protected from decay but that's of little consolation if the consciousness once residing in the body has died. Conversely, a body might change over the years—a transplant here, a limb lost there—but if the person who identifies with that body maintains a sense of self then the person survives. It's immortality of the mind that's important. In this view the body is merely a receptacle, so if you can pour your mind into some other receptacle—your clone, someone else's body, a computer—then you can achieve immortality.

Roger Zelazny's *Lord of Light* (1967), which was awarded the 1968 Hugo award, describes precisely this approach to achieving immortality. The crew of a starship develop a technology that permits them to transfer the soul of a person from one body to another. Mind transfer here offers a form of reincarnation. The crew become effectively immortal and, over the millennia, start to behave as gods. Robert Silverberg's novel *To Live Again* (1969) has a slightly different take on the idea. Silverberg describes a society in which the great and the good, if they can afford it, regularly make a backup of their personality. When a person dies they can, if they've been backed up, transfer their personality into another person's body. The wrinkle here, though, is that the host body already has its own personality. And since it becomes a matter of prestige to have more than one personality, a body can thus become host to multiple personalities. Silverberg movingly explores the complications this technology would bring.

We have already touched upon the idea of mind uploads into clones or corpsicles. Mind upload into computers seems to offer another route to immortality. Arthur Clarke, in *The City and the Stars* (1956), offers one of the more interesting variations on this theme. The city of the novel's title is Diaspar, which as far as its inhabitants know is the only city left on Earth. Diaspar is a completely enclosed environment, which has kept its people safe from the outside world for a billion years or more. An artificial intelligence called the Central Computer runs Diaspar, and one of its functions is to store the minds of people in its vast memory banks. At any point in time, only a fraction of the total number of people are alive and living in Diaspar; the rest are in storage, so to speak. When a body dies, the person's mind returns to the memory bank; and when this happens, to even up the numbers, the Central Computer creates a new body for someone else, and downloads a mind into the body. This has gone on for countless ages, and thus the people of Diaspar are effectively immortal—they live an endless succession of bounded lifespans. And then a coincidence occurs: two important individuals find themselves living in the city at the same time. Khedron, a so-called Jester, has the job of disrupting normal behavior in the city—the Central Computer uses Jesters as a means of preventing social stagnation. Alvin, a Unique, is even more interesting: he is the first person for millions of years never to have lived before. This haunting novel describes the adventures of Alvin as he strives to learn what is outside of Diaspar.

Philip Farmer's award-winning novels *To Your Scattered Bodies Go* (1971) and *The Fabulous Riverboat* (1971) have yet another take on mind uploading and immortality. In these novels, everyone who has ever lived is reincarnated on the banks of a river that appears to stretch on forever. If you get killed in this afterlife you find yourself reincarnated yet again, somewhere else on the banks of the river. Once dead, it seems you can't die again. Farmer has a great cast of characters from which to draw upon—every human being who ever lived—and he follows a few key characters as they try to learn the secret behind this mysterious world.

The Clarke and Farmer novels are on my longlist of favorite SF novels, but they aren't particularly realistic. As the use of computers became more pervasive, and people began to think more about the possibilities they provide, SF writers developed a grittier and more down-to-earth approach to the issue of mind uploading. For example, in *Software* (1982) Rudy Rucker considers the possibility of immortality via the transfer of mind into computer code—software, in other words. One of the themes of William Gibson's *Neuromancer* (1984), an influential novel discussed elsewhere in this book, concerns immortality. Part of action in the novel springs ultimately from the efforts of two people to achieve a form of immortality. One person pursues the cryogenic route—immortality of the body—and fails. The other uploads her personality to an artificial intelligence—immortality of the mind—and has more success, although whether the attempt truly succeeds is open to debate. The transfer of human consciousness into machine is explored by Pohl in *Heechee Rendezvous* (1984)—and he raises questions future lawyers will perhaps need to debate. If your consciousness is transferred into machine at the time of your body's death, are you still alive from a legal perspective? Do you retain control of physical assets or do they pass to your next of kin?

In *Time Enough for Love* (1973), Heinlein took a typically thought-provoking approach to the question of immortality and mind transfer. The novel features Lazarus Long, the character from *Methuselah's Children*, who by this time is more than a thousand years old. But the Heinlein twist occurs when he describes how an artificial intelligence transfers its mind (actually, it's a 'female' mind) into a human body engineered to have the best possible genes. So there are two directions for mind transfer: human mind into computer or artificial mind into human body.

The story that takes the possibility of mind transfer and immortality to its ultimate conclusion is Asimov's "The Last Question" (1956). It was Asimov's favorite of his own stories, and it's easy to see why. In the story, over vast reaches of time, every human mind is transferred to and amalgamates with a galaxy-spanning computer. The combination of computer with humanity develops unimaginable power, and eventually... well, I won't spoil the strength of the story's closing line.

10.1.5 The Dream of Immortality

I've barely touched on the variety of stories relating in some way to the theme of immortality. I haven't mentioned Jonathan Swift's *Gulliver's Travels* (1726) and its depiction of the struldbruggs—humans who are immortal but who nevertheless age at the usual rate and who are thus condemned to suffer infirmity and ill-health. Then there's the Borges story "The Immortal" (1949), in which life's endless repetition makes the protagonist long for death. And there's Wowbagger, the Infinitely Prolonged, a character in Douglas Adams's *Life, the Universe and Everything* (1982) who accidentally becomes immortal and doesn't much care for it. Wowbagger resolves to keep himself occupied by delivering an insult to every living creature in the universe, in alphabetical order.

Authors typically depict only the negative aspects of long life—the inevitable frailty accompanying longevity unless the natural ageing process is also halted, the inevitable ennui after you've experienced everything there is to experience. Nevertheless, despite the many and obvious drawbacks, the dream of immortality—or at least of greatly extended lifespans—retains a powerful hold on our imagination. Is science any closer to making those dreams a reality?

10.2   The Time of Our Lives

Over the past century the average life expectancy, in developed countries at least, has seen a marked increase. When my paternal grandfather was born in 1896 he came into a world in which, at birth, the average male life expectancy in the UK was 45.4 years; for a boy born in 2014 in the well off borough of

Kensington and Chelsea that figure had risen to 83.3 years. In some ways, this improvement in life expectancy represents a better rate of progress than Heinlein predicted in his novels. However, life expectancy is defined as the median age at death—it's a number referring to populations rather than to individuals. It says nothing about the maximum life span that can be achieved. At present, 125 years seems a hard upper limit for the human lifespan. Can we increase the limit, so some of us might live to be 150 years or more? Can we perhaps even achieve the ancient dream of immortality? Society is throwing resources at the problem: biologists are engaged in longevity research, engineers are developing ever-improving devices to augment our bodies, and computer scientists are producing evermore powerful computers that might augment our brains. Scientists are still a long way from realizing the dreams (or the nightmares) of SF writers, but with each passing year they are learning more about how—or rather if—human longevity might be increased. Let's start with the problem of ageing.

10.2.1 The Dying of the Light

Benjamin Franklin, in a letter written in 1789, observed that 'In this world nothing can be said to be certain, except death and taxes'. We've learned in recent years how the super-rich possess any number of clever ways of making taxes optional—but no matter how rich they might be they can't avoid death. Even if they evade injury, even if they have 'good' genes, even if they live a life devoted to minimizing their risk of suffering the shocks that flesh is heir to, they are still going to die. No matter what they do, they *will* age and they *will* then eventually depart. Why, though? Why do humans—rich and poor, male and female, tall and short—age?

In one sense, ageing is entirely to be expected: the second law of thermodynamics tells us *everything* decays. Mountains erode; iron rusts; rocks crumble. Disorder rules, simply because there are so many more ways for things to be disordered than ordered. Life, though, seems to be a wonderful exception to this rule. Biological cells, which in essence are fantastically complex nanoscale factories, are capable of repair and renewal. For a brief but glorious period a collection of cells is capable of bringing order from disorder: Michelangelo can create a statue, Shakespeare a play, Einstein a theory. Eventually, however, individual cells themselves age. Human cells can divide, and thus produce new cells for growth and tissue repair, roughly 50–70 times—after which they enter a senescence phase, and either stay in the body as malfunctioning cells or else die. But why? If the second law of thermodynamics can be held at bay for the Biblical three-score-years-and-ten why can't it be kept at bay indefinitely?

Ageing turns out to be a complex process, but fundamentally there are two different types of explanation: a physics-based explanation (which in essence says ageing is the inevitable outcome of the second law of thermodynamics) and a biology-based explanation (which in essence says evolution has ensured that organisms age and die in order to make way for new generations). Let's look at the biology-based explanation first.

10.2.1.1   It's All in the Genes

In *Methuselah's Children*, Heinlein suggested human life expectancy might be increased through a form of genetic engineering. There certainly seems to be a genetic component to life expectancy—if your parents are long-lived then you yourself are more likely to be long-lived—and, indeed, biologists estimate that our genes account for about 30% of our age at death. These genetic factors don't kick in until relatively late in life (when you're young, factors such as bad nutrition, bacterial infection, and lifestyle choice all play a dominant role in life span), but once you are in your sixth decade then you can count your blessings if your genes give you some extra protection against stroke, cancer, and diseases of the heart and blood vessels. From an evolutionary point of view, however, why should individuals (of any species, not just humans) age and die? What evolutionary forces led to a situation where organisms age?

This is not an easy question to answer.[5] One's first thought is likely to be that death is somehow 'programmed' into us—perhaps so that population size is limited or so that the time between generations is reduced and thus organisms can better adapt to changing environments. In other words, this argument suggests there might be a 'death gene' that kicks in after an organism has reproduced and raised its offspring. This idea doesn't work, however, because very few animals in the wild live long enough for ageing to be a cause of death: they typically die young, and they die because they starve, get eaten, succumb to infection or cold, or whatever. Indeed, the reason human life expectancy has increased so rapidly in developed countries over the past century is that advances in medicine and public sanitation have almost eliminated the risk of death from infectious disease—a risk factor that in previous generations was constant and essentially independent of age. The reduction in death caused by constant environmental factors meant the median age at death increased, but the maximum human lifespan was unchanged. Some other factor must be at work to account for ageing.

A more plausible biology-based explanation is that the many external causes of mortality—starvation, predation, infection—mean natural selection has little material with which to work on when it comes to long-lived animals in the wild. In turn, there's nothing to halt the accumulation of heritable mutations with deleterious effects in the organism's old age. A competing explanation is that some genes can influence more than one trait: a particular gene might offer beneficial effects for a young organism (for example, the gene might produce oestrogen and thus enhance fertility) but have deleterious effects when the same organism is old (hormone production might increase the risk of developing breast cancer, for example). From an evolutionary point of view those early beneficial effects can outweigh the late detrimental effects, even if death in old age is the result. Yet a third explanation is that there's a trade-off between allocating resources for reproduction and resources for maintaining the body. For many organisms, the chances are high they'll die early. It makes no sense for them to have developed complex systems to guard against the degradations of old age if most of them get eaten while they are young; it's more important for their metabolism to be geared for early and successful reproduction. Or perhaps some combination of these various factors is in play.

If the forces of evolution really have shaped the mechanisms of human ageing, then perhaps we can tackle ageing via genetic engineering, as Heinlein and others suggested? The specific Heinleinian approach to extending the human lifespan, which involves 'improving' a person's genetic heritage, would be impractical because of the length of time before results could be seen. A more efficient approach would be to try to identify the genes involved in ageing and then promote 'good' genes (or remove 'bad' genes, or both).

In recent years, geneticists have been using model organisms to investigate ageing. One organism often used in this regard is the roundworm *C. elegans*, a tiny creature which was the first multicellular organism to have its entire genome sequenced. The worm has a typical lifespan of just 20 days, and so many generations of the worm can be studied over a relatively short period. Scientists have investigated a number of genetic variants in *C. elegans* that might play a role in longevity, and there are a couple of interesting candidate genes. For example, the *daf-2* gene encodes for a protein that plays a role in metabolism and in *C. elegans* allows the worm to survive in situations where food is scarce. If mutations in *daf-2* inactivate the gene then the worm's lifespan doubles (which is perhaps why some geneticists call it the 'grim reaper gene'). If *daf-2* mutations are combined with mutations in certain other genes then the worm's lifespan can be made to increase by a factor of five.

Increasing an organism's lifespan by a factor of five sounds impressive, even if the organism involved is only a 1 mm-long worm. And it's not just worms in which specific genes have been linked to longevity: scientists have extended the average lifespan of fruit fly and mouse by altering particular gene activity. If geneticists could do the same in humans as they can in *C. elegans*, and extend lifespan by a factor of five, then the maximum human lifespan would be extended to about 625 years. Not exactly immortality, but not bad.

Needless to say, things are not so simple. For a start, even if genes are responsible for ageing it's not clear whether meddling with them can produce the same lifespan increase in a naturally long-lived organism (humans are among the longest-lived animals on the planet) as it can in a naturally short-lived organism. Even identifying the genes involved in human ageing would be a daunting task. Scientist can't conduct controlled laboratory experiments on humans over long periods so it's almost impossible to control for various factors that might affect longevity. And even if scientists *do* find an association between the presence (or indeed absence) of a particular gene amongst extremely long-lived people, the association would not imply causation. Future advances in genetics will undoubtedly help many people lead longer and healthier lives: the human 'healthspan' will increase. But it isn't clear how or if these advances will add vastly to the maximum human lifespan.

10.2.1.2   Ticking Clocks and Telomeres

It's possible there's a particular biological cause of ageing—a ticking clock within each cell that determines how long the cell lives. This clock is related to cell division, the process whereby, through division, new cells are produced so body tissue can be repaired. The genetic materials controlling cell division are to be found in the cell nucleus, in strands of chromosomes. At both ends of a chromosome strand are caps, known as telomeres—think of a shoelace and you'll have a good mental picture: in the same way plastic end caps prevent a shoelace from fraying so the telomeres, which are essentially just repetitive sequences of DNA, shelter the chromosome from damage. Without telomere protection, the chromosomes would unravel and stick to each other, and thus scramble an organism's genetic information. The problem is, each time a cell divides its telomeres become slightly shorter. A cell begins life with long telomeres but, as it gets older and undergoes repeated divisions, its telomeres shorten. Eventually, the telomeres are gone and the chromosomes have no protection against damage. All the usual cellular processes break down. Our cells get old, so we get old.

It's not universally accepted that telomere shortening causes ageing. The shortening might be a symptom of age rather than a cause. But if you believe telomeres really are the culprits involved in ageing then you have a possible means of attacking the problem of ageing: develop a way to maintain those telomeres. It would be like halting the ticking of a biological clock.

It's clear that reproductive cells must have some way of maintaining the length of telomeres: so much cell division takes place during gestation that, if telomere shortening took place in the womb, children would be born with body cells incapable of further division. The next generation of children would not be conceived, and humanity would quickly become extinct. Well, it turns out that reproductive cells don't undergo telomere shortening because they produce an enzyme called telomerase: if the telomeres lose a section of DNA during cell division, the telomerase immediately replaces it. Reproductive cells don't age. Essentially, they are immortal. (The so-called germ line, after all, has been dividing since life itself got going on Earth.) So here is another suggestion for how to increase the maximum lifespan of humans: insert the telomerase gene into all our cells.

Inserting telomerase directly is not a good idea. A problem occurs if the gene enters at the wrong site. You see, there's another type of immortal cell, another type of cell in which telomerase maintains telomere length: the cancer cell. Inserting the telomerase gene could turn a healthy cell into a cancerous cell that might grow to become a killer. Since the human body contains trillions of cells, it's likely that a telomerase-insertion treatment designed to lengthen a person's lifespan would instead end up killing them. Fortunately, there's no need to go down this route because all our cells already contain the code for telomerase—it's just that in every cell except the reproductive cells (and cancer cells) the telomerase gene is repressed. If a non-toxic drug could be found that turns on telomerase and doesn't cause cancer—well, who knows? Perhaps the pharmaceutical companies would have a cream not for 'the visible signs of ageing' but for ageing itself!

10.2.1.3   Wear and Tear

A physics-based approach says we don't need to tie ourselves in knots working out how natural selection could have brought about senescence. Ageing is due to the cumulative effects of 'wear and tear' to the genome. Ageing is the inevitable outcome of the second law of thermodynamics applied to the intricate machinery within our cells.

It has long been known that DNA can suffer damage in a number of ways. For example, when DNA replicates mistakes can creep in; usually the body's DNA repair mechanisms fix the mistake, but mistakes sometimes slip through and the errors accumulate over time. Exposure to ultraviolet radiation can also damage DNA. So too can free radicals. All these effects might play a role, but let's look just at the free radical effect in a little more detail.

A free radical is an atom, molecule, or ion with an unpaired electron in the valence shell. That unpaired electron makes a free radical extremely reactive, and when it comes close to another molecule the free radical tries to become more stable by pulling an electron away from the molecule. This causes the molecule to become a free radical itself, and that molecule in turn will interact with another molecule and turn *that* into a free radical. Within the close confines of a cell, a free radical will eventually cause a biologically important molecule to lose an electron—and, if that molecule can then no longer perform its biological function, the cell is damaged. Unfortunately, our bodies produce free radicals in a variety of metabolic processes. The simple act of breathing is a problem; our bodies require oxygen, of course, but molecules containing oxygen can be highly reactive free radicals. Breathing, a process that keeps us alive, at the same time creates free radicals capable of sweeping through a cell and causing devastation. It's possible, then, that the damage caused by free radicals is one of the reasons our bodies age. Perhaps young cells have a biochemical defence system to mop up the damage caused by free radicals, and the system doesn't work so well in older cells. If that's the case then cellular damage would mount up. Once a cell suffers too much damage, it dies. When too many cells die, we die.

The free radical theory of ageing is not universally accepted (no theory of ageing is universally accepted), and its difficult to determine whether it's the free radicals or the ageing that comes first. Perhaps the ageing process permits free radicals to start to damage cells? But if you believe in the free radical theory of ageing then an obvious route to greater longevity suggests itself: get rid of free radicals in your body. You can't stop free radicals from being produced (as mentioned, the simple act of breathing generates them) but some molecules—antioxidants—can neutralize free radicals. Some common antioxidants are vitamins C and E, and the molecule beta-carotene which is found in many plants and fruit. So if you stuff your face with fruit and veg will the antioxidants you consume ward off the effects of free radical damage? It's a nice thought, but medical trials have failed to show the anticipated effects. Indeed, if you take antioxidant supplements then you might make matters worse because those supplements can impede the body's natural mechanisms for coping with free radicals. Perhaps one day medicine will develop a simple pill to stop free radical damage, but for the moment the traditional advice for healthy living is the same as it has been for years: don't smoke, take exercise, and eat a varied and sensible diet. The advice won't guarantee a long life, but it's as good a way as any of stacking the odds in your favor.

However, even if free radical damage could be prevented—and if damage from ultraviolet radiation and the imperfect mechanism of DNA repair could be stopped—it *still* might not be enough. In addition to all this, damage is caused within cells simply because of the temperature at which we all live our lives. Water molecules within cells constantly bustle around due to random thermal motion, and so they constantly bang into the precision protein-based machines in charge of operating our cells. The higher the temperature, the greater the thermal motion. If thermal motion really is a cause of ageing then, on this basis, we might predict that people living at different internal temperatures would tend to enjoy different lifespans: those with higher internal temperatures would suffer greater levels of thermal-induced damage and thus have shorter lives. Clearly, this is not something we can confirm through experiments on humans! The equivalent experiments have been done with simple creatures such as *C. elegans*, however, and the results reflect the prediction: worms raised at low temperatures have a greater maximum lifespan than worms raised at high temperatures.[6]

Perhaps the biological explanations for ageing, then, miss the point. The physics-based, wear-and-tear model of ageing suggests the biological effects we observe as cells age are responses to an underlying source: the accumulation of damage caused by thermal motion and other effects. This accumulating damage will manifest itself in different ways in different people—as we age we'll all develop different diseases and die of different things—but the underlying source will be the same: wear and tear; or, slightly more formally, the second law of thermodynamics. If this argument is correct then there will be a limit to what can be achieved to increase the human lifespan: as *Star Trek*'s Commander Scott was fond of pointing out, we canna change the laws of physics. Research into specific aspects ageing might give us a few extra years or even decades but, if a physics-based explanation of ageing is correct, it won't lead to immortality. (Such research would be useful anyway, because even if it doesn't lead to an increased human lifespan it might well lead to an increased human healthspan—and surely this is more important.)

I should mention that there is a technique that *has* been shown to extend maximum life span, at least in short-lived species such as mice: caloric restriction. Studies in rodents have regularly shown that caloric restriction can increase the maximum lifespan by up to 40%. It's unlikely the technique works to the same extent in humans, but as far as I'm aware caloric restriction is the only widely available technique for those interested in longevity. The idea is simple: drastically reduce one's calorie intake while ensuring the body is not malnourished and has all the essential nutrients it needs. It's not clear why caloric restriction should extend lifespan, but one possible mechanism is that with less energy going into the body there is less free radical generation. Although the technique has not yet been shown to increase human longevity (the naturally long lifespan of humans makes these studies difficult) I'm prepared to believe there are health benefits to a near-starvation diet. Personally, though, I can't see the attraction. Call me deathist, but I'd rather trade the extra couple of years a caloric restricted lifestyle might offer me for the pleasure given by food and drink.

None of the arguments mentioned so far provide much hope for those interested in immortality. Humans are ingenious, however. Even if we can't prevent ageing there might be ways we can dodge it. For example, we might improve on our troublesome, flesh-based bodies.

10.2.2 Human Enhancement

Prosthetics—the fitting of artificial body parts—is an ancient activity. In the Rigveda, a Hindu sacred text written somewhere between 1500 and 1200 BC, mention is made of a woman called Vishpala who lost her leg in battle and was given a new one made of iron. The story is impossible to confirm but we do know that in Egypt, just a few centuries after the Rigveda episode, a woman was fitted with a replacement big toe on her right foot: the prosthetic was found on a mummy from Thebes. Those early prosthetics were probably desperate attempts to improve a person's ability to function; modern prosthetics, of course, can be highly refined pieces of engineering. Vishpala's iron leg might have allowed some limited motion; Paralympic runners wearing J-shaped carbon-fibre prosthetics are able to perform at the same level as Olympic athletes.

In recent years, the possibilities for prosthetics have increased at an astonishing rate. Recently, advances in neuroscience and prosthetic technology allowed a patient to wiggle individual fingers on an artificial arm—using the power of his mind.[7] The technology is in its infancy, but it might not be too long (in rich Western countries, at least) before those missing a limb can be fitted with prosthetics they control by thought. The technology has the potential to transform beyond recognition those who, through accident or disease, lose function in hands and feet, arms and legs. And it's not just limbs that can be replaced. In 2011, for example, surgeons in Sweden implanted an artificial trachea into a patient. They took stem cells from the patient, incubated them on a plastic model of the patient's trachea, then replaced the diseased original with the synthetic substitute. For decades, cochlear implants have been inserted into those who are profoundly deaf or severely hard of hearing. And research is ongoing into the development of artificial organs to replace, or at least support the natural functioning of, heart, liver, lungs and pancreas.

Clearly, although these various developments will grant extra years and quality of life to increasing numbers of patients, we are a long way from seeing a 'bionic man'. And suppose biotechnologists do eventually manage to develop artificial organs and limbs, muscles and bones; suppose surgeons can successfully implant those replacements; and suppose neuroscientists learn how all of these developments can be controlled by a patient's nervous system... would any of that increase the maximum lifespan of humans? There would *still* remain the problem faced by the actress in Catherine Moore's "No Woman Born": the brain ages and dies, just as other organs in the body age and die. For immortality, or even longevity, we need to preserve the brain as well as the body.

10.2.3 Cryonics

The technical and medical advances discussed so far all have the potential to prolong life. But it's far from clear when we will see the benefit of all this in terms of enabling humans to live for centuries rather than decades. A different approach is to take a short cut through time and try to reach a future in which the problems surrounding immortality have already been solved. It's the approach discussed in those stories by Pohl, Simak, Niven, and others. And it's an approach being taken by a number of people in the real world.

Cryonics—the freezing of the recently dead at ultra-cold temperatures, in the hope future technology will be able to resuscitate them and restore them to health—is an idea first popularized by Robert Ettinger, a decorated World War II veteran who later became a physics teacher. Ettinger came up with the idea of cryonics after reading Neil R. Jones's story "The Jameson Satellite" in the July 1931 issue of *Amazing Stories*. In the story, Professor Jameson wants to preserve his body after death and so has it launched into orbit in a capsule. Forty million years later his body is discovered by alien cyborgs called the Zoromes; they repair Jameson's brain, put it in a cyborg body, and reboot him. Ettinger, after three decades of thinking about the concept, self-published a book in which he outlined the principles of cryonics, his basic idea being that in many cases death is a process that happens gradually. Furthermore, we know if we suffer a massive heart attack in the middle of the Sahara our prospects are poor, but suffer the same attack while in a well equipped hospital and the chances of survival might be good. So the yardstick for death is certainly a function of place; Ettinger argued it is also a function of time. An illness that's fatal *now* might be trivial for future doctors; and the damage caused to the body by freezing might be easily repaired by future medics. The hope was, by deep-freezing a patient's body immediately after death, the patient at some future date would wake up as if after a night's sleep with memories, thought patterns, and feelings all intact. Ettinger sent his book to a number of establishment figures, which caused something of a sensation and attracted the attention of the publishers Doubleday, who sent a copy of Ettinger's book to Asimov for comment. Asimov seems to have found the ethics problematic, but he told Doubleday the basic idea was not outlandish. Doubleday then published *The Prospect of Immortality* (1964), and Ettinger went on to publicize cryonics and became President of the Cryonics Institute.

The first person to be cryopreserved after death (and, for obvious reasons, the process must take place after a person is declared legally dead) was James Hiram Bedford, a professor of psychology who died from cancer in 1967. His frozen cadaver is currently still kept in liquid nitrogen, under the care of Alcor—one of the leading institutions in the field of cryopreservation. Some other patients were less fortunate than Bedford: the same person who organized Bedford's cryopreservation allowed several other bodies to thaw. Several other clients have been thawed and reburied after payments ceased—and payments are necessary, of course, since there's an ongoing cost associated with this procedure. The state of cryopreservation must be constantly maintained.

Ettinger himself died of respiratory failure in 2011 at the age of 92, and became the 106th person to be preserved by the Cryonics Institute. He joined his wife, who was cryopreserved in 2000, and his mother, whom the Cryonics Institute treated in 1977. And the technology he popularized, while not exactly mainstream, is currently being applied: several hundred people are in a state of cryopreservation and about two thousand people have made arrangements with the various service providers.[8]

So, does the practice of cryonics make any sense? Well, with current technology it's impossible to revive a human who has been cryopreserved. With future technology there's probably very little chance a cryopreserved human can be revived. On the other hand, that 'very little chance' has to be compared with the zero chance of revival of a person who has been buried or cremated. From that viewpoint, someone who opts for cryopreservation is making a rational choice. There's another question: if people *are* revived by some far future technology, would their personalities and memories be intact? In other words, if you were revived, would you still recognize yourself as being 'you'? This is an undiscovered country—the only answer to this question must be: we have no idea.

10.2.4 The Singularity

We've looked at various paths to immortality, or at least increased longevity, and the problem with all of them is they have to cope with the fallibilities of a biological body. Even if cryopreservation turns out to be practicable, and terminally ill patients who are frozen now are revived in some future world in which medicine can treat them, they will presumably once again continue the ageing process. And even if future medicine has developed some 'elixir of life' capable of halting the ageing process, the fallibility of the human body still threatens the dream of immortality. Millions of people die each year in accidents, disasters, acts of god (and no amount of technological advance will prevent all such fatalities); millions more die from war and violence (and in a future where death from old age has been abolished, I'd guess struggles over finite resources would be common); further millions die from famine (and with more mouths to feed on a warming planet the threat of famine would be real indeed). In other words, death would remain a concern for anyone whose home is flesh—which is all of us.

Perhaps the real route to immortality, therefore, is to upload a person's mind—memories, consciousness, sense of 'I'—into some sort of computational device. The mind might be placed into a robot, or perhaps into a virtual world, or maybe into bioengineered tissue of some kind. Of course, a device such as a computer is also subject to shocks—physical destruction, power outages, wear and tear—but one could keep multiple copies of a person's mind in different locations to minimize the risks. Perhaps then, if you can upload your mind to a computer, immortality could be yours. (If you did this, and the upload resembled a copy-and-paste rather than cut-and-paste operation, would there be two of you? While both copies existed, who would have priority?) But is it likely you'll ever be able to upload your mind?

The challenges involved in modelling and uploading a mind are staggering.

The human brain is the most complex object science has encountered. A typical brain contains just under 90 billion neurons (brain cells, in other words), and each neuron can be connected to as many as 10,000 other neurons through a complex web of axons and dendrites. A neuron can communicate with another neuron—typically by sending out an electrical signal along an axon where it is received by a dendrite of the other neuron, although other signal pathways are possible. The connecting points between neurons are called synapses, and it's the patterns of synaptic connections between neurons that act as the storehouse for our memories. A whole slew of chemical and electrical activity takes place in this complex network and somehow, in a process that is still quite mysterious, consciousness emerges.

To reconstruct a mind, then, we'd presumably need access to a variety of different types of information about the brain that plays host to the mind. First, we'd need a wiring diagram of all the synaptic connections between neurons. Second, we'd need a detailed account of the electrical activity taking place in the brain. Third, we'd need to understand in detail the structure and function of each neuron, axon, and synapse. Fourth, we'd need to understand how all these elements respond to changing environments. None of this is inherently impossible, but neuroscience is far from having that information to hand.

Let's take the first item on the list: the wiring diagram. In 2015, neuroscientists published[9] the wiring diagram of a small piece of mouse brain tissue that contained 1700 synapses. This achievement is the current state of the art, but the human brain has tens of billions more synapses than that sliver of mouse neocortex. As for the second item, it requires not only that we have the wiring diagram but that we track the strength of signals travelling long those wires and monitor them over time: signal strength and patterns change as we learn and they change on a variety of time scales. Those other items require much more research in order to reach the point where we can predict in real time how neurons, axons, and synapses would respond to a particular input. There's a long way to go before we have the necessary level of understanding of the brain. So is the prospect of mind uploading so distant that we might as well consider it to be fantasy? Some people argue not.

Advocates of mind uploading point to the increasing power of computers. It's a commonplace that the smartphone in your pocket is more powerful than the computer that helped land Apollo 11 on the Moon, but it's difficult for many of us to develop an intuitive feel for this development. We tend to make straight-line extrapolations from the past into the future, but the history of computing is one of exponential increase. Back in 1965 the co-founder of Intel, Gordon Moore, pointed out how the number of components in an integrated circuit was doubling every year or so. Many numbers relevant to digital electronics—the quality-adjusted price of IT equipment, memory capacity, pixels per dollar, and so on—follow a form of Moore's law. Such exponential increases are not something most people have a natural feel for.

Consider the following conundrum.[10] A pond has water lilies floating on its surface. The surface coverage doubles every day and, if the lilies are not cut back, the water will be covered in 30 days time; at that point all other living things in the pond will die. The gardener keeps watch on the pond, notes that for the first few days the number of lilies is small, and so decides to wait until the lilies cover half of the pond's surface before cutting them back. The question is: when will the gardener have to cut back the lily pads? Most people when asked this question don't really pay attention, and thus give an answer of 15 days. The correct answer, of course, is 29 days. When exponentials are involved, things start off slowly then take off with a bang. And this is what's happening with computers. The amount of computing power required to model a mind is so vast that at first glance the prospect seems to be centuries or millennia away; but if exponential increases continue, the necessary power could be with us much earlier than we expect.

Indeed, there is a hypothesis that some day a computer or a network of computers will become intelligent (whatever that word actually means), with the capacity to develop a computer more intelligent than itself. This next generation computer would be capable of building an even more intelligent computer—and the stage would then be set for an exponential, runaway event leading to the construction of an intelligence of such puissance that unaugmented humans such as us would be unable to fathom it. We would hit the Singularity (as discussed in a different context in Chap. 8).

The mathematician and SF writer Vernor Vinge predicts the Singularity will occur before the year 2030; the futurist Ray Kurzweil predicts it will occur around 2045.[11] Numerous other predictions are less specific, but typically place the event some time in the current century. According to this hypothesis, some time in this century we—or our super intelligent machines, or the hybrid organism resulting from the merging of humans with machines—will possess sufficient technological power to simulate and upload a human mind. Immortality will be there for those who want it.

Needless to say, many thinkers question whether a Singularity is likely to occur. Richard Jones, in *Against Transhumanism* (2016), offers a devastating critique of the whole Singularity notion (and related concepts such as mind uploading). The hypothesis faces numerous criticisms. To mention just one: exponential increases never continue indefinitely. Some factor always acts to moderate the increase. With computation, for example, physics itself places fundamental limits on what can be achieved and we are already seeing signs that Moore's law is slowing. Nevertheless, even if a Singularity fails to materialize, it seems certain that in the future we'll have more computing power available to us—perhaps enough to upload a mind?

If the motivation behind mind uploading is immortality, then personally I'm not convinced that the technology is appropriate. The gap between the embodied and disembodied mind seems to me to rule out a continuation of self. As I write this I'm enjoying the warmth of a beautiful sunny day, a feeling made all the more pleasant when I contrast it with the unseasonably cold days I've endured recently; I'm irritated each time my right thumb brushes against my keyboard because it sends a momentary jab, the result of a trivial but nonetheless painful cut I suffered when attempting some DIY; my eyes and nose are itchy, a result of the recent warmth launching pollen to which my body reacts by releasing histamine. As for my memories, the most vivid refer to bodily experiences: that perfect cover drive I hit when playing cricket at school; the mixture of exhaustion, elation, and pain I felt when completing a marathon; the sheer fun of playing soccer. Even if my thoughts and memories were replicated and somehow uploaded to computer could my upload, in the absence of my body, consider itself to be a continuation of me?

I guess the same level of technology capable of uploading my mind would also be able to provide the uploaded mind with a simulated reality to make it *seem* as if I had a body. We could all of us go and live in a simulation, and presumably that simulation could exist indefinitely and provide immortality. But I find myself asking, as did a character in Clarke's *The City and the Stars*, what would be the point of that?

10.2.5 Quantum Suicide

There's an interpretation of quantum mechanics which, if true, would offer you a mechanism for achieving immortality. However, I really wouldn't recommend you try it.

To begin, set up a real experiment along the lines of the famous Schrödinger's cat thought experiment: hook up a gun to a machine that measures the spin of an electron whenever the trigger is pulled. If the system measures spin-up, the gun fires a bullet; if it measures spin-down, the gun merely clicks. However, whereas Schrödinger's cat experiment puts a feline in danger this experiment involves *you*: the gun is aimed straight at your temple. And how do you go about achieving immortality? Simply keep pulling the trigger. Each time you pull the trigger you'll hear a *click*. You'll hear *click, click, click...* for all eternity.

What's happening here?

Well, the many-worlds interpretation of quantum mechanics states that whenever a quantum yes/no measurement is made the universe splits in two.

When the spin of an electron is measured, for example, the entire universe cleaves: in one universe the measurement shows spin-up and in another universe the measurement shows spin-down. In this particular experiment, it results in one universe in which you die from a bullet in the brain and another universe in which all that happens is you hear a *click*. In the universe where you die, that's the end of the matter. But in the universe where you hear a *click* you can perform the experiment again—with the same two outcomes. As time passes there will be a vast and increasing number of universes in which you are dead. But there will be one universe—the universe in which your consciousness survives, the only universe of which you can possibly have experience—which sees you live forever.

This idea goes by the name of quantum immortality, and it's based on an interpretation of quantum mechanics that several eminent physicists find plausible.

However, as I wrote above, I really, *really* wouldn't recommend that you try this.

10.3   Conclusion

At some point in the distant past the hominid brain became sophisticated enough to realize that death is inevitable. Facing up to this knowledge is not easy, and so perhaps the dream of immortality began not long after the dawn of that realization. The dream has been the spark for religious ideas, philosophical debate, great works of art—and, as we've seen, serious scientific research and numerous SF stories.

The techniques proposed for immortality (or, more accurately, for extreme longevity) are many and varied. Some of them will lead to an extension of the human healthspan—the number of years over which we can expect to enjoy good health. But will any of them significantly extend the human lifespan? I'm not sure they will. And, speaking personally, I'm increasingly able to accept that. If I stop to ponder these things at all, I find myself rejoicing in the good fortune to find myself alive at all: there are so many more ways of not existing than of existing. It is almost a miracle of luck to be alive and conscious of the fact. The cost of living, however, is the knowledge you must die. It's not such a high cost. I can live with it.

Notes

1. Ascomycin, the 'anti-agathic' drug mentioned by Blish in *They Shall Have Stars*, was developed by the pharmaceutical firm Pfitzner. When he wrote the novel, Blish was science editor for the pharmaceutical firm Pfizer.

2. *Dune* is one of the best known SF novels of all time, and won the 1966 Hugo award for best novel. Surprisingly, it did not win the award outright: another novel dealing with aspects of longevity, Roger Zelazny's *This Immortal* (1966), shared that year's Hugo. For readers interested in the theme of immortality, it's worth pointing out that many of Zelazny's novels contain a character who is effectively immortal.

3. Cyborgs—in the form of the Borg—are the most interesting creatures in the TV series *Star Trek: The Next Generation*. The Borg began as organic creatures who, in an attempt to perfect themselves, gradually became increasingly synthetic. Their cybernetic enhancements make them stronger, faster, and more resilient than imperfect carbon-based creatures such as humans.

4. Corpsicle is a portmanteau word derived from 'corpse' and 'popsicle'. Pohl coined the terms in his essay "Immortality Through Freezing" (1966) and his novel *The Age of the Pussyfoot* (1969).

5. For more about ageing, see for example Austad (1997) and Kirkwood (1999). The two authors later collaborated on a review paper (Kirkwood and Austad 2000).

6. For details of an experiment showing the effect of temperature on ageing, see for example Stroustrup et al. (2016).

7. 'Allowed a patient to wiggle individual fingers on an artificial arm—using the power of his mind': for details of this amazing medical advance, see Hotson et al. (2016).

8. A relatively recent overview of cryonics, along with a discussion of the ethical issues, is given by Moen (2015).

9. The wiring diagram of a small piece of mouse brain, referred to in the text, was published by Kasthuri et al. (2015).

10. The water lily puzzle, and many related puzzles, are described in a fascinating book by Kahneman (2011).

11. For an in-depth discussion of the Singularity, from an enthusiast's viewpoint, see Kurzweil (2005).

Bibliography

Non-fiction

Austad, S.N.: Why We Age. Wiley, New York (1997)

Ettinger, R.C.W.: The Prospect of Immortality. Doubleday, New York (1964)

Hotson, G., et al.: Individual finger control of a modular prosthetic limb using high-density electrocorticography in a human subject. J. Neural Eng. **13**(2), 026017 (2016)

Jones, R.A.L.: Against Transhumanism: The Delusion of Technological Transcendence. http://www.softmachines.org (2016)

Kahneman, D.: Thinking, Fast and Slow. Farrar, Straus and Giroux, New York (2011)

Kasthuri, N., et al.: Saturated Reconstruction of a Volume of Neocortex. Cell. **162**, 648–661 (2015)

Kirkwood, T.B.L.: Time of Our Lives: The Science of Human Ageing. Oxford University Press, Oxford (1999)

Kirkwood, T.B.L., Austad, S.N.: Why do we age? Nature. **408**, 233–238 (2000)

Kurzweil, R.: The Singularity Is Near. Penguin, London (2005)

Moen, O.M.: The case for cryonics. J. Med. Ethics. **41**, 677–681 (2015)

Pohl, F.: Immortality Through Freezing. Worlds of Tomorrow (August 1966)

Stroustrup, N., et al.: The temporal scaling of Caenorhabditis elegans ageing. Nature. **530**, 103–107 (2016)

Fiction

Adams, D.: Life, the Universe and Everything. Pan, London (1982)

Anderson, P.: World Without Stars. Ace, New York (1966)

Asimov, I.: The Last Question. Science Fiction Quarterly (November 1956)

Asimov, I.: Segregationist. Collected in: Nightfall and Other Stories (1969). Doubleday, New York (1967)

Asimov, I.: The bicentennial man. In: Stellar-2. Ballantine, New York (1976)

Benford, G.: Doing Lennon. Analog (April 1975)

Blish, J.: They Shall Have Stars. Faber and Faber, London (1956)

Blish, J.: The Seedling Stars. Gnome, New York (1957)

Borges, J.L.: The immortal. In: El Aleph. Losada, Buenos Aires (1949)

Caidin, M.: Cyborg. Arbor, Westminster, MA (1972)

Clarke, A.C.: The City and the Stars. Frederick Muller, London (1956)

Farmer, P.J.: To Your Scattered Bodies Go. Putnam's, New York (1971a)

Farmer, P.J.: The Fabulous Riverboat. Putnam's, New York (1971b)

Gibson, W.: Neuromancer. Ace, New York (1984)

Gunn, J.E.: The Immortals. Bantam, New York (1962)

Heinlein, R.A.: Beyond This Horizon. Fantasy Press, New York (1948)

Heinlein, R.A.: Methuselah's Children. Gnome, New York (1958)

Heinlein, R.A.: Time Enough for Love. Putnam's, New York (1973)

Herbert, F.: Dune. Chilcot, New York (1965)

Jones, N.R.: The Jameson satellite. Amazing (July 1931)

Kuttner, H.: Fury. Grosset and Dunlap, New York (1950)

Martin, G.R.R.: A Song for Lya. Analog (June 1974)

McCaffrey, A.: The Ship Who Sang. Fantasy & Science Fiction (April 1961)

McCaffrey, A.: Crystal Singer. Severn House, London (1982)

Miller Jr., W.M.: A Canticle for Liebowitz. Lippincott, Philadelphia (1960)

Moore, C.L.: No Woman Born. Astounding (December 1944)

Niven, L.: Becalmed in Hell. Fantasy & Science Fiction (July 1965)

Niven, L.: Wait it out. In: All the Myriad Ways (1971). Ballantine, New York (1968)

Niven, L.: The defenceless dead. In: Elwood, R. (ed.) Ten Tomorows. Fawcett, Greenwich, CT (1973)

Padgett, L.: Camouflage. Astounding (September 1945)

Pohl, F.: The Age of the Pussyfoot. Ballantine, New York (1969)

Pohl, F.: Shaffery Among the Immortals. Fantasy & Science Fiction (July 1972)

Pohl, F.: Man Plus. Random House, New York (1976)

Pohl, F.: Heechee Rendezvous. Ballantine, New York (1984)

Rucker, R.: Software. Ace, New York (1982)

Silverberg, R.: To Live Again. Doubleday, New York (1969)

Simak, C.D.: Desertion. Astounding (November 1944)

Simak, C.D.: Why Call Them Back From Heaven? Doubleday, New York (1967)

Spinrad, N.: Bug Jack Barron. Walker, New York (1969)

Stapledon, O.: Last and First Men: A Story of the Near and Far Future. Methuen, London (1930)

Swift, J.: Gulliver's Travels. Motte, London (1726)

Vance, J.: To Live Forever. Ballantine, New York (1956)

Varley, J.: The Ophiuchi Hotline. Ace, New York (1977)

Wilson, F.P.: Healer. Doubleday, New York (1976)

Zelazny, R.: This Immortal. Ace, New York (1966)

Zelazny, R.: Lord of Light. Doubleday, New York (1967)

Visual Media

Captain Scarlet and the Mysterons: [TV series]. UK, Century 21 Television (1967–1968)

Dr Who: The Girl Who Died. [TV Episode]. UK, BBC (2015)

The Bionic Woman: [TV Series]. USA, ABC (1976)

The Six Million Dollar Man: [TV Series]. USA, ABC (1973)

11

Mad Scientists

It's alive. It's alive. It's ALIVE.

Dr Frankenstein, *Frankenstein* (1931)
James Whale (Director)

On 16 June 1816, Mary Shelley had a vivid 'waking dream'[1] about the re-animation of a corpse, a dream that led to her write what, in the opinion of many,[2] is the first SF novel. *Frankenstein; or, The Modern Prometheus* (1818) is certainly science fictional: Victor Frankenstein, a man with scientific training, undertakes various laboratory experiments and succeeds in fulfilling his ambition of creating life from non-life. What could be more SF than that? Shelley established something else in her novel: the archetype of the mad scientist. Although Victor Frankenstein himself is a sympathetic character, his ambition leads him to experiment with activities traditionally forbidden by society—he decides to 'play God' (Fig. 11.1). Subsequent mad scientists might be of the certifiably insane, evil genius, or humorously eccentric variety—but the hubris of Victor Frankenstein set the pattern.

Robert Louis Stevenson followed the design some time later when he created another character who chose to 'meddle' in matters that 'only God should control', another mad scientist whose name entered the language. In *Strange Case of Dr Jekyll and Mr Hyde* (1886), Dr Jekyll is interested in the nature of human personality. He develops a chemical potion allowing him to test his theory that people have a dual nature. When he experiments upon himself he is transformed into the amoral and conscience-free Mr Hyde.

H.G. Wells wrote two novels in which a scientist's hubris leads to disaster. In *The Island of Doctor Moreau* (1896) an English gentleman finds himself on a

Fig. 11.1 A plate, entitled 'Frankenstein at work in his laboratory', from a 1922 edition of Mary Shelley's novel. Her famous story has never been out of print, and her depiction of Victor Frankenstein provided the prototype for the fictional 'mad scientist' (Credit: Public domain)

Pacific island, on which the scientist Moreau is using vivisection to reshape beasts into human form. In *The Invisible Man: A Grotesque Romance* (1897) the obsession of a scientist, Griffin, turns into full-blown insanity. Griffin's invisibility allows him to commit a series of crimes, but he exhibits no remorse or regret—he simply describes his actions as being 'necessary'.

The mad scientist also often made appearances in early films. The first film adaptation of Frankenstein took place in 1910, for example. And Fritz Lang's *Metropolis* (1927) features Rotwang, genius inventor of a machine-person. Rotwang is in many ways the epitome of the mad scientist. Shock of wild, white, Einstein-type hair? Check. Black-gloved prosthetic hand? Check. Plethora of insane schemes? Check.

These characters from page and screen—Jekyll and Moreau, Griffin and Rotwang (not to mention Frankenstein himself)—were so striking and so memorable, and the stories they inhabited were so successful and so profitable, one might have expected that *all* later science fiction would adhere to the formula. And yet. . . that didn't really happen.

11.1   Not So Mad Scientists

The character of the mad scientist holds such interest, offers so many dramatic possibilities, that of course SF writers continue to mine the trope. However, although it's the mad or bad ones who usually spring to mind when one thinks of scientists in science fiction, the truth is SF more often portrays scientists in a positive light.[3] That's hardly surprising, given that many of the best known SF writers have had at least a passing acquaintance with science and technology. Of the 'Big Three', Asimov had a PhD in biochemistry, Clarke had a degree in mathematics and physics, and Heinlein had a degree in naval engineering. Several authors have been even more qualified. If we restrict ourselves to the physical sciences then, off the top of my head, I could mention that Greg Benford is an emeritus professor of physics at UCLA, Charles Sheffield was chief scientist at EarthSat, and Robert Forward did research in the field of gravitational wave detectors; Geoffrey Landis works for NASA; Catherine Asaro has a PhD in chemical physics and David Brin a PhD in astrophysics. Similarly, several eminent scientists have written science fiction. The planetary scientist and science communicator Carl Sagan had his novel *Contact* (1985) turned into a Hollywood movie; Sir Fred Hoyle published 19 volumes of SF; and in his book *The Return of Vaman* (2015) Jayant Narlikar, who worked with Hoyle on alternative cosmological theories, has had SF stories published by Springer in this very series. The SF written by these authors, and many others I could mention, tend to treat scientists—and the scientific endeavor—positively.

For a typical example of how SF views scientists, consider two of the most enduring characters created by Asimov.

Hari Seldon, a mathematician who develops the science of psychohistory and uses it to foretell the future in terms of probability, is the leading character of the *Foundation* trilogy. Although we seldom encounter Seldon in the original *Foundation* series (he only takes center stage in the much later additions to the series), his scientific outlook on the problems of history saturate the work. Seldon's emphasis is always on resolving conflict through careful reasoning and exhaustive discussion. Indeed, throughout the *Foundation* series, as throughout all of his work, Asimov promoted the rational, scientific approach to problem-solving as humankind's best hope for survival.

Susan Calvin, the chief robopsychologist at US Robots and Mechanical Men Inc., is the leading character in several of the stories collected in *I, Robot*. In many of them Calvin comes across as a cold, analytical woman. It's true that in "Liar!" (1941) and "Lenny" (1958a) she reveals an emotional side: in the former story a robot succeeds in deceiving Calvin by telling her what she wants to hear, namely that a colleague is in love with her; in the latter she develops maternal feelings for a damaged robot that she is trying to repair. In most of the stories, however, the focus is purely on the brilliance of her mind; in none of the stories is her intellect ever in doubt. Calvin solves puzzles relating to robots with the same flare as Sherlock Holmes solves mysteries relating to people.

Asimov was often accused of creating 'cardboard' characters and he didn't try to defend himself against the accusation: it was inarguable that his characters were often less well realized than those of mainstream authors. Nevertheless, creations such as Seldon and Calvin are memorable. Through them, Asimov was able to celebrate rationality and the scientific method. Seldon and Calvin might have had their eccentricities, but they weren't 'mad scientists'; they were characters to admire.

Of course, Hari Seldon and Susan Calvin resemble real-life scientists in the same way that Hercule Poirot and Miss Marple resemble real-life crime-fighters. The portrayal of scientists in SF is as realistic as the portrayal of gunslingers in westerns or lovers in romantic novels. It's not that there are *no* realistic representations of scientists in the literature, however. Asimov himself, in *A Whiff of Death* (1958b), nicely captures the petty politics of a university science department—although that doesn't count in this discussion since it's one of his detective novels rather than one of his SF novels. A better example of the typically realistic science fictional treatment of scientists is provided, for example, in James Gunn's *The Listeners* (1972), a novel that Carl Sagan said was one of the inspirations for his own science fiction work *Contact*.

Gunn's novel examines the dedication needed to succeed in certain fields of science, and the strains such dedication can place upon those involved. Parts of the novel are set in the year 2025, and describe the 'Project'—a SETI (search for extraterrestrial intelligence) experiment. The experiment's director, Robert MacDonald, is willing to devote his career to the ongoing search even though there is no guarantee of success. The demands the 'Project' makes on him takes a heavy toll on his family life: his wife attempts suicide and his son William becomes estranged from him. The search is eventually successful—a telescope receives signals from the star Capella—but it's clear MacDonald won't be alive to hear the 'Reply' to the 'Message' humanity sends to Capella: even for a radio wave traveling at the speed of light, the round-trip time between Earth and Capella is 90 years. His son William, though, lives to become director of the 'Project'. The content of the Capellan response, when it finally arrives, is suitably downbeat and sad.

The Listeners provides an authentic account of SETI researchers doing science; Greg Benford's multiple award-winning *Timescape* (1980) is perhaps the most authentic portrayal of physicists at work. Unlike most SF, *Timescape* is fiction *about* science.

The action in Benford's novel starts in 1998 (a full 18 years after the initial publication of *Timescape* but now, at the time of this writing, 18 years in the past). The world Benford foresaw was on the brink of an environmental disaster, with ecological breakdown being matched by a breakdown in the fabric of civilized society. In England, in Cambridge, two physicists experiment with tachyons (these, as we learned in Chap. 5, are hypothetical particles that travel faster than the speed of light and that would, if they existed, travel back through time). The two physicists plan to use tachyons to send a warning signal to a previous generation that has messed up so badly. Then in 1962 (18 years before the initial publication of *Timescape*, and 36 years before the action at the start of the novel), in California, a young researcher called Gordon Bernstein discovers odd patterns of interference in his experiments. Eventually Bernstein understands that the interference contains a message, which he then struggles single-mindedly to decipher.

Timescape is a time travel story of the 'hard' variety—but, more than that, it's about how science is done (the petty politics of the lab, the struggles for grant money, the demands from management) and it's about the relationships that scientists develop (between student and professor, between academics in different disciplines, between husband and wife).

Novels such as *The Listeners* and *Timescape* are more typical of the science fictional representation of scientists than novels such as *Frankenstein* and *The Island of Doctor Moreau*. In most SF scientists aren't 'mad'; they are ordinary people who happen to be passionately curious—features which, after all, pretty much provide a reasonable definition of real-life scientists.

The examples above all related to written SF, but a similar argument could be made about fictional scientists on television. On British TV, for example, some of the most influential drama programmes ever made by the BBC were the three *Quatermass* series (1953–1958). By the time I was old enough to watch the repeats they already seemed old-fashioned—but Professor Bernard Quatermass was clearly in the heroic mold, and used his knowledge and intelligence to *save* Earth rather than try to conquer it as the classic 'mad scientist' would have done. *Dr Who*, of course, has grown to become part of British popular culture. The Doctor is a scientist (a centuries-old alien Time Lord, but a scientist nevertheless) and from the outset he always extricated himself from his many tangles by using science and argument rather than brute force and weapons. And then there's Spock, from *Star Trek*. Spock was first officer on the starship *USS Enterprise*, but he was also the ship's science officer. In many of the episodes it was the Vulcan scientist's cool rationality that saved the *Enterprise* and its crew. Spock became one of the most popular characters in the entire *Star Trek* franchise.

Even in films, scientists are as often portrayed in a positive light as a negative one. Consider some examples from that heyday of science fiction films, the mid-1950s. In *Them!* (1954) it is the FBI special agent Robert Graham who is the conventional 'hero'. But it's the gentle, mild-mannered Dr Harold Medford who understands the nature of the killings that Graham is called upon to investigate—the scientist realizes that giant ants, mutated by radiation from atom bomb tests, are to blame. (In the real world the ratio of surface area to volume places a limit on the size of insects. The giants of *Them!* would be unable to breathe. But don't let me stop you enjoying the film.) In *This Island Earth* (1955) the hero is Dr Cal Meachum, a physicist who passes a sort of interstellar intelligence test by following mysterious instructions that lead him to the construction of a device called an interocitor. The functioning interocitor in turn brings him to the attention of alien beings. It turns out that Earth's fate rests in Meacham's hands but, with the help of one of the aliens, he eventually secures the future of humanity. And in my favourite film, *Forbidden Planet* (1956), Dr Edward Morbius dabbles with alien technology and in so doing unleashes his 'evil self' as a rampaging monster; Morbius definitely qualifies as a mad scientist. But in *Forbidden Planet* it's another man with a science background, the ship's medical officer 'Doc' Ostrow, who first understands the nature of what's killing people on this strange planet. Ostrow, a scientist, is the hero.

So although the 'mad scientist' is a venerable science fictional cliché the truth is that the portrayal of scientists in modern science fiction has generally been positive. Indeed, perhaps one needs to ask whether SF has given scientists too *easy* a ride? Perhaps disaster might strike not through a scientist's hubris or overweening arrogance, but through the misguided application of increasingly powerful technology? Could it be that a real-life mad scientist might turn out to be far more dangerous than Victor Frankenstein, Henry Jekyll, and Dr Moreau combined?

11.2   A Bright Future for the Mad Scientist?

Science is our best tool for providing robust, reliable knowledge. It's the process through which we can confirm causal links, determine that one event leads to another, establish how doing X will cause Y to happen. Without science we'd be in the same situation as those who practiced human sacrifice to ensure the rising of the Sun: the practice 'works' (in the hours following a sacrifice the Sun will indeed rise) but it leaves its practitioners blinded by a confusion of 'association' with 'causality'. Science helps guide us away from such basic errors of thinking; it helps us develop technologies that really do work; it is *extremely* powerful. However, the very success of science, and the technology made possible by our ever increasing scientific understanding, raises the spectre of groups of scientists—or perhaps even an individual, a 'mad scientist'—causing havoc. Let's consider just five possibilities.

11.2.1 Mad Physicists

A few people worry about experiments taking place at particle accelerators such as the Relativistic Heavy Ion Collider (RHIC) in New York or the Large Hadron Collider (LHC) near Geneva (Fig. 11.2). It's not that these people fear a crazed individual might use the instruments for nefarious purposes—they accept that the successful operation of the RHIC or LHC is beyond any individual; it involves thousands of technicians, engineers, and physicists. Rather, the worry seems to be that scientists' blinkered, single-minded pursuit of knowledge might lead, via particle colliders, to a doomsday event.

Instruments such as the RHIC and LHC work by hurling subatomic particles at each other at high energies. The higher the collision energy, the more understanding scientists can derive about the nature of particles and the laws governing their interactions. While the RHIC and LHC were being commissioned, doom-mongers raised fears that the extreme energies involved in the experiments could trigger a catastrophe. Two particular concerns emerged. First, that the colliders might create strangelets. Second, that the colliders might create micro-black holes.

Strangelets are small chunks of 'strange matter'—a hypothetical form of quark matter which, if it exists, would be more stable than normal matter. The worry was expressed that, if a strangelet were produced, it might trigger a runaway process which would convert normal matter to strange matter. Fractions of a second after a strangelet came into existence at RHIC or LHC, we'd all be dead.

Black holes should need no further explanation. Astronomers know that a giant black hole is almost always present at the center of a galaxy, and that smaller black holes form at the end stage in the evolution of high-mass stars. The worry here was that the extremely high energies involved at the LHC might be sufficient to create micro-black holes, which then might sink to the center of the Earth and kill us all.

These catastrophic scenarios so concerned Walter Wagner, an American former radiation safety officer, that in 1999 he filed several lawsuits to try and stop the RHIC from operating. In 2008, Wagner and Otto Rössler, a German biochemist, filed a lawsuit to try and stop CERN scientists from switching on the LHC. In both cases, the lawsuits were filed despite the publication of studies from experts demonstrating clearly how the concerns were misplaced. Wagner continued to insist that the probability of the LHC creating an Earth-eating micro-black hole was 50% (based on the 'logic' it would either happen or not). There were some mad scientists involved in this sorry episode—but they weren't the scientists working on the experiments.

For billions of years, nature has been conducting particle physics experiments at far higher energies than our colliders can ever reach: cosmic rays smash into Earth's atmosphere day and night, and yet we are still here. Neither strangelets nor micro-black holes have devoured our planet. Despite the fact that LHC experiments have operated safely since 2009, RHIC experiments have operated safely since 2000, and cosmic ray 'experiments' with Earth's atmosphere have operated safely since about 4.5 billion years ago, some philosophers argue we should not operate collider experiments until physicists can *prove* the devices pose no existential risk. It's an attitude that makes one wonder how humanity ever left the caves. The risk here is surely the anti-science stance of otherwise reasonable people, rather than science itself.

Fig. 11.2 A photograph of the magnets in the Large Hadron Collider tunnel. The LHC is the most complex machine ever created by humankind. It smashes subatomic particles together at higher energies than any other collider. Some physicists have speculated that the LHC might be able to create micro-black holes or probe into parallel universes. Others, taking note of these speculations, ask the question: could the LHC spell doom for humanity? The answer: no (Credit: CERN)

11.2.2 Mad Environmental Scientists

A more realistic threat might lie in society's response to one of its greatest challenges: the problem of climate change.

There's little doubt our planet is warming. The main uncertainty is how hot things will get. If we curb carbon emissions and if we have some luck then average global temperatures might increase by less than two degrees compared to average temperatures before the Industrial Revolution. If we don't curb carbon emissions, or if we are unlucky, average global temperatures might increase by as much as six degrees—a catastrophic level. If it seems as if global warming will happen at the upper end of the scale then society might be driven to attempt geoengineering solutions to the problem—perhaps by using some Earth-spanning technique to reflect more sunlight back into space or to capture more carbon dioxide. The issue with a geoengineering approach is that if things go wrong, or even if they go right, the side-effects might be worse than the disease: it's pointless reducing temperatures if doing so worsens floods or droughts for the majority of the world's population.

The potential side-effects of geoengineering might stay the hand of governments, but it's entirely possible that sufficiently rich individuals will engage in geoengineering in an attempt to reduce global warming. Indeed, something along these lines has already been tried.

In 2012, a wealthy American businessman called Russ George dumped 100 tonnes of iron sulphate in the North Pacific. He did it as a test of a proposed geoengineering technique called ocean iron fertilization.[4] The idea is that the chemical promotes the growth of phytoplankton, which absorbs carbon dioxide during photosynthesis and then eventually carries all that carbon down to the ocean floor. Now, the idea behind ocean iron fertilization is not unreasonable; but it's an idea requiring research, analysis, debate. Russ George and his colleagues didn't ask anyone's permission, or come up with a well argued research proposal—they just went out and dumped iron sulphate in the ocean. What's to stop other individuals pumping sulphate aerosols into the atmosphere (with the intention of reducing the amount of sunlight absorbed at Earth's surface, but with possible detrimental effects on the ozone layer)? Or 'doping' trees with nitrogen fertilizer (with the intention of increasing their ability to absorb carbon dioxide and increasing their albedo, but with the possible downside that groundwater might be contaminated)? Or dumping powdered limestone into the seas (with the intention of combating increasing ocean acidity, but with who-knows-what unintended consequences)? The field is ripe for a mad scientist to come along.

11.2.3 Mad Nanotechnologists

The evolving field of nanotechnology is another in which a mad scientist (or a group of such) might, either by intention or accident, cause problems.

Nanotechnology involves the manipulation of matter on an atomic or molecular scale, and it promises applications in medicine, electronics, and many other fields. The technology will, in the future, transform human lives for the better. The worry, though, is that a nanotechnologist will one day design a self-replicating robot nanomachine, or nanobot, a mechanism that takes raw material from the environment in order to fashion a copy of itself. Why is this dangerous? Well, suppose you start off with one self-replicating nanobot capable of building a copy of itself in one hour and all this machine wants to do is to replicate. At the end of the first hour you have two nanobots; at the end of the second hour you have four nanobots; at the end of the third hour you have eight nanobots; at the end of the first day you might realize you have a problem. The power of exponential increase means the number of nanobots will soon become enormous. After a couple of days, Earth itself would consist of nothing but self-replicating robot nanomachines. This particular doomsday scenario is called the 'grey goo' problem, and was first identified by Eric Drexler in his influential book *Engines of Creation* (1986). (The 'grey goo' scenario is similar to one described by Kurt Vonnegut in his novel *Cat's Cradle* (1963). In Vonnegut's story, the Nobel award-winning physicist Felix Hoenikker develops a substance, an alternative form of water, called ice-nine. The outstanding feature of ice-nine is that it's solid at room temperature. If a crystal of ice-nine comes into contact with liquid water it acts as a 'seed' causing the molecules of liquid water to rearrange themselves into solid form: the water turns into ice-nine. When some ice-nine falls into the ocean, it's a matter of days before all life on Earth dies.)

Why would engineers or scientists create a self-replicating nanobot, if the exponential growth of such a device might cause the end of the world? Well, sane people wouldn't set out to build an Earth-killing nanobot but they might create one by accident. Suppose scientists design a self-replicating nanobot and give it the task of 'eating' hydrocarbons; they might do this for perfectly sound environmental reasons—the nanobots might be deployed, for example, to clean up an ocean oil spill. However, suppose there's a bug in the nanobot's programming: instead of 'eating' the hydrocarbons in the oil, they instead 'eat' carbon. Carbon-based life-forms, which essentially means all life on Earth, would quickly be consumed.

Our current level of technology is not so advanced that we need worry about the 'grey goo' problem happening today, and presumably engineers will have enough time to think about the development of appropriate safeguards. Drexler himself believes the dangers of the 'grey goo' scenario have been oversold. But even if safeguards are sufficient to prevent accidental events involving self-replicators, will the same safeguards be able to prevent *deliberate* acts? If scientists learn to 'program' matter at the nanoscale then that opens up the possibility of nanoterrorism: the bad guys could order swarms of nanobots to do almost anything—attack people, buildings, infrastructure. And it's unlikely that every individual would abstain from acts of nanoterror. We might expect some criminal, somewhere, to employ nanotechnology for malevolent purposes.[5]

11.2.4 Mad Computer Scientists

One might hope that the qualities needed to understand nanotechnology—a high level of education, technical proficiency, and so on—would somehow inoculate people against engaging in damaging acts. Unfortunately, the recent history of computer hacking proves such a hope to be illusory. A vast number of cases have highlighted how extremely skilled individuals are willing to try and gain unauthorized access to other people's computer systems, and in many cases they succeed. The motivations of hackers are manifold: some do it to gain information to facilitate blackmail; some want to steal goods or money; some like to test their technical prowess; some possess a warped sense of altruism and claim they are merely drawing attention to security flaws; some are simply bored and hacking gives them something to do. And it's not just individual hackers who pose a threat in our increasingly connected world: cases of state-sponsored hacking are proliferating. Nations routinely spy on the commercial and industrial sectors of their rivals, and military organizations have updated the old maxim 'He who controls the seas controls the world' to 'He who controls cyberspace controls the world'. In future conflicts we can expect to see computer scientists using their skills and knowledge to attack enemy satellites, energy infrastructure, and communications networks. Indeed, we are already seeing such attacks taking place. At best, these attacks are annoying; at worst, they hurt people. But could computer scientists, simply by carrying out their innocent and entirely legitimate research, pose a threat to the entire human species?

In Chap. 8 we considered the possibility that computer scientists might, over the next few decades or so, create an artificial intelligence (AI) that surpasses the human level of intelligence. Once this particular milestone is reached, the subsequent development of a superintelligent AI might not be far off. One can argue whether AI is likely, or even possible, but let's suppose computer scientists *can* develop such an entity and choose to do so. Should we worry?[6]

A number of respected thinkers—philosophers such as Nick Bostrom, entrepreneurs such as Elon Musk, and scientists such as Stephen Hawking—have expressed concern about the emergence of superintelligent AIs. The fundamental worry is that a superintelligent AI might possess goals that don't align with human goals and human values. Since a superintelligent AI would certainly be able to outwit us, the achievement of human goals would inevitably depend upon the whim of the AI. The future relationship of humans to superintelligent AIs would be similar to the present-day relationship of dogs to humans: our survival would no longer depend upon our own decisions and actions. Human extinction would be one possible outcome.

It's far from clear that an 'explosion' in artificial intelligence will occur. Personally, as discussed in Chap. 8, I doubt it will—at least, not on a timescale that concerns us. Nevertheless, the danger posed by superintelligent AIs falls at least into the low-probability/high-impact category. Surely now is the time for scientists to be thinking of ways to avoid these potential pitfalls—but there's little evidence that researchers are debating this topic in a rigorous way. Every day, physicists, engineers, and computer scientists are bringing us a tiny bit closer to what *might* turn out to be a runaway explosion. Scientists would indeed be mad if they didn't soon start to ponder the possible unintended consequences of such an explosion.

11.2.5 Mad Biologists

Several times in this book we've had reason to encounter Moore's law—the exponential increase in computing power over time. Although there are signs that Moore's law is finally in its end phase, it held true for about five decades: since 1965, when Gordon Moore first made his observation, the number of transistors on a microprocessor chip, and thus the amount of computation that could be done in a given volume, doubled every 18 months or so. One manifestation of this law was that new classes of device appeared every decade or so. We started out with mainframes, moved on to minicomputers, then laptops, then smartphones, and we now have embedded processors. This exponential increase in computing power has transformed our lives.

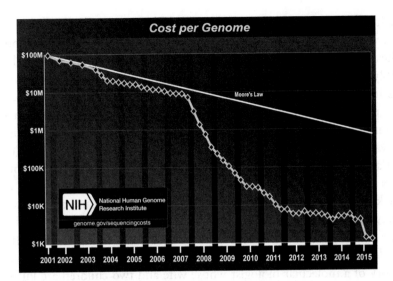

Fig. 11.3 A graph showing how the cost of sequencing a human genome dropped over the years between 2001 and 2015. Note that the vertical axis is logarithmic. Note also how Moore's law began to be vastly outpaced from the start of 2008: this was when 'next generation' sequencing technologies were implemented (Credit: NHGRI)

The growth of biotechnology began rather later than the growth of computing technology. By many measures, however, progress in biotechnology has been *faster* than the Moore's law progress of computing. For example, the cost of sequencing a human genome has dropped much more quickly than Moore's law would predict (see Fig. 11.3) and the time taken to sequence a genome has plummeted just as rapidly. The dizzying pace of biotechnological innovation will transform our lives in the future just as surely—and perhaps much more profoundly—as computers have in the past.

The potential benefits of biotechnology take the breath away. We can look forward to rapid advances in medicine, including a personalized approach to treating disease and illness. Improvements in agricultural techniques will lead to improved food security. The application of biotechnological techniques in industry might help our society reduce its reliance on fossil fuels. The promise of biotechnology is so rewarding, the possibilities so amazing, that society will surely continue to fund research and development in this area. But could this lead to disaster?

The same technology that will soon be used to transform medicine could be used to inflict untold horrors on people. What if someone—the archetypal mad scientist, perhaps—manipulates the genes of an existing virus in order to make it more deadly to humans? Or recreates the 1918 human influenza virus, an infectious agent which killed up to 5% of the world's population? Or synthesizes an entirely new microbe that kills even more effectively and horribly than Ebola? What if someone develops a process to turn off particular genes in humans, perhaps genes that are required for fertility, and uses it to attack particular groups—people with a certain skin color, maybe, or those who suffer from particular genetic complaints? What if someone uses biotechnology to target animals or crops, and thus threaten a country's (or the world's) food supply? These 'what if's' are the stuff of nightmares, but the threats need to be taken seriously. As biologists learn more about what makes life tick, it becomes increasingly possible for someone to use that knowledge for perverted ends.

The possibility has of course been considered widely in science fiction. Consider, for example, Frank Herbert's novel *The White Plague* (1982). It tells the story of a molecular biologist whose wife and two children are killed in a terrorist incident. His loss drives him insane—he becomes, literally, a mad scientist—and he devises a plan to gain revenge on the countries he believes are in some way responsible for his suffering. He designs a plague that is carried by men but kills only women: he wants others to experience his agony. It's not long before the plague gets out of hand and soon politics, society, economics, even geography are affected. The world is changed because of the actions of a single deranged man. Unfortunately, it seems entirely plausible that someone could synthesize a disease that kills selectively based on specific genetic characteristics.

One might hope no-one could so hate humanity that they'd choose to develop something similar to Herbert's 'white plague'. But think of the atrocities committed by Josef Mengele and Shirō Ishii during World War II. Both Mengele and Ishii had a medical background, and they used the technology available to them to engage in acts that are incomprehensible in their barbarity. If they'd had access to modern biotechnology they'd presumably have had no compunctions in using it. And there's little comfort to be gained by saying the actions of Mengele and Ishii were made possible because of the unprecedented situation the world found itself in during World War II. The concern is, the pool of potential wrongdoers will be so much larger in the future than it was in the period 1939–1945. The graph in Fig. 11.3 illustrates how in 2008 the cost of sequencing a human genome dropped off a cliff; it won't be long before the cost is essentially zero. *Anyone* with some basic knowledge of biology will be able to sequence a genome. It's the same with other biotechnological activities: soon *anyone* will be able to tinker with the basic building blocks of life—it will be as cheap and as easy as playing with toy wooden building blocks. So then it won't just be the likes of Mengele and Ishii we need to worry about. We'll need to worry about the blackmailers, the misguided altruists, the bored loners who just want someone to notice them.

The release of a bioweapon might well be as deadly as the dropping of a thermonuclear bomb. The fact such devices have similar death counts has lead some to conclude that the same mechanisms which have prevented nuclear armageddon might be applied to prevent biowarfare or acts of bioterror. But there's a critical difference between the two technologies: nuclear technology does not scale well but biotechnology scales extremely well. It's *difficult* to produce a nuclear bomb. The infrastructure required—uranium enrichment plant, plutonium reprocessing facilities, or whatever—is beyond the reach of individuals, while at the state level the technology is difficult to hide and so can easily be monitored. The necessarily industrial nature of nuclear weapons manufacture has made it possible to police the process. Unfortunately, the production of a bioweapon is (or at least will be) simple. The infrastructure necessary to make a bioweapon will be negligible. How will the authorities police an activity taking place in a garden shed?

In future years, the development of bioweapons will have more in common with the coding of computer viruses than the construction of nuclear weapons. And just as authorities have struggled to come to terms with the challenges posed by cyber terrorism, so they'll struggle to deal with bioterrorism. But it's something society will have to get right—for all our sakes.[7]

11.3   Conclusion

Although stories warning of the dangers of hubris have appeared throughout history (there's the Greek myth of Icarus, the German legend of Faust, Milton's portrayal of Lucifer, and so on) James Whale's 1931 film of *Frankenstein* introduced the particular science fictional concept of 'Man playing God'. In a line cut by the censor, but in a spirit which nevertheless permeates the film, the Count says "Now I know what it feels like to be God!" The concept was revisited in a number of classic 1950s SF films. In *Forbidden Planet*, for example, the Krell developed a machine that could project matter 'in any form', and could thus use it to create life. The destruction of the Krell can be interpreted as a cosmic punishment for misappropriating such power. At the end of the film, as Commander Adams watches the incineration of the Krell's home planet of Altair IV, he intones that 'we are, after all, not God'. The phrase 'playing God' has become almost meaningless through overuse, but in the not too distant future scientists will likely have the ability to manipulate matter on the atomic and molecular scale, and to alter or create life. Not quite in the league of the Krell, perhaps, but not far off. Should we be worried, as those SF films of the 1930s, 1940s, and 1950s suggested we should? Or is the scientific/technological optimism of much of my favourite written SF from that period a saner response?

The benefits of new technologies are simply too many and too profound for humankind to turn its back on ongoing research. Advances in biotechnology, nanotechnology, computer technology, and other fields will all lead to improvements in the lives of billions of people. But these advances do bring with them tremendous risks and ethical challenges. They raise questions that society as a whole must address.[8] And the best method I can imagine for bringing these questions to the attention of the public, and for highlighting the potential pitfalls as well as the rewards of future technology, is through SF. Science fiction counts as the most important literature of our day.

As well as being a science fiction writer, Asimov was famous for his popularizations of various fields of knowledge. In a magisterial overview, *Asimov's New Guide to Science* (1984), he observed that "while knowledge can create problems, it is not through ignorance that we can solve them". Although we surely need to debate the role of science, and be careful to analyze where technological advancement might be taking us, I don't think we need to fear the mad scientist. What we *do* have to fear are those who would turn us away from science.

Notes

1. The tale of how *Frankenstein* came to be written is well known. In 1815, Mount Tambora underwent the largest volcanic eruption in recorded history and ash from the explosion plunged Earth into a volcanic winter. The following year, 1816, became known as the 'year without a summer'. In June 1816 Mary Shelley, along with her lover and soon-to-be husband Percy Shelley, were visiting Lord Byron at a mansion near Lake Geneva. The weather forced the group to sit around log fires rather than enjoy the outdoors, and so Byron proposed they should amuse themselves by each writing a ghost story. A few days later Mary had her 'waking' dream. By comparing Mary's accounts of the dream with astronomical data, the physicist Donald Olson calculates the dream took place between 2 am and 3 am on 16 June 1816.

2. A number of historians of science fiction have credited Mary Shelley with writing the first SF novel. For example, Brian Aldiss (1973) provided an

extensive case for accepting *Frankenstein* as marking the start of a new genre, and Asimov (1977) wrote that: 'Mary Shelley was the first to make use of a new finding of science which she advanced further to a logical extreme, and it is that which makes *Frankenstein* the first true science fiction story.'

3. Research backs up the assertion that SF increasingly portrays scientists in a positive light. A review by Snyder (2004) of magazine SF concluded that "beyond doubt, especially when compared with the science fiction of the 1920s and 1930s, the mad scientist has been virtually banished from the pages of the best science fiction magazines". Dudo et al. (2011) showed that, on TV, scientist-type characters typically fulfill good or at least mixed roles (but, unfortunately, they are typically roles for white males). Furthermore, the authors found no significant direct relationship between TV viewing and negative attitudes toward science. It could be argued that the portrayal of scientists in films is less positive. A survey of 1000 horror films distributed in the UK between the 1930s and 1980s showed that mad scientists, or the creations of mad scientists, were the villains in 30% of the films; scientists were the heroes of only 11% of the films. See Frayling (2005). On the other hand, by their very nature, horror films are much more likely than SF films to focus on the darker aspects of science and technology. Michael Crichton, who trained in science but who became much better known for writing bestselling science- and techno-based thrillers, understood why this might be. In a speech to the American Association for the Advancement of Science (Crichton 1999) he said: 'Let's be clear: all professions look bad in the movies... Lawyers are all unscrupulous and doctors are all uncaring. Psychiatrists are all crazy and politicians are all corrupt. All cops are psychopaths and all businessmen are crooks... Why expect scientists to be treated any differently?' In a thriller or horror movie it's often the actions of a scientist that provide the mainspring for the plot.

4. To read more about ocean iron fertilization and related issues from Russ George's point of view, visit his website (George n.d.).

5. For further discussion of various aspects of the risk posed by nanotechnology, see Phoenix and Treder (2008).

6. For in-depth discussions about the potential benefits and possible dangers of artificial intelligence, see Yudkowsky (2008) and Bostrom (2014).

7. Nouri and Chyba (2008) discuss the risks surrounding biotechnology and biological weapons, but also discuss possible responses and the paths to take that might lead us to a biologically secure future.

8. I am reminded here of one of Isaac Asimov's most oft-quoted sayings: "the saddest aspect of life right now is that science gathers knowledge faster than society gathers wisdom" (Asimov and Shulman 1988).

Bibliography

Non-fiction

Aldiss, B.: Billion Year Spree: The True History of Science Fiction. Doubleday, New York (1973)

Asimov, I.: The first science fiction novel. Collected in: Asimov on Science Fiction. Doubleday, New York (1977)

Asimov, I.: Asimov's New Guide to Science. Basic Books, New York (1984)

Asimov, I., Shulman, J.A.: Isaac Asimov's Book of Science and Nature Quotations. Grove Press, New York (1988)

Bostrom, N.: Superintelligence: Paths, Dangers, Strategies. OUP, Oxford (2014)

Crichton, M.: Ritual abuse, hot air, and missed opportunities: science views media. Speech to American Association for the Advancement of Science, Anaheim, CA (1999)

Drexler, E.: Engines of Creation: The Coming Era of Nanotechnology. Doubleday, New York (1986)

Dudo, A., Brossard, D., Shanahan, J., Scheufele, D.A., Morgan, M., Signorielli, N.: Science on television in the 21st century: recent trends in portrayals and their contributions to public attitudes towards science. Commun. Res. 48(6), 754–777 (2011)

Frayling, C.: Mad, Bad and Dangerous? The Scientist and the Cinema. Reaktion, London (2005)

George, R.: Bring Back the Fish. russgeorge.net (n.d.)

Nouri, A., Chyba, C.F.: Biotechnology and biosecurity. In: Bostrom, N., Ćirković, M.M. (eds.) Global Catastrophic Risks. OUP, Oxford (2008)

Phoenix, C., Treder, M.: Nanotechnology as a global catastrophic risk. In: Bostrom, N., Ćirković, M.M. (eds.) Global Catastrophic Risks. OUP, Oxford (2008)

Snyder, L.A.: The Portrayal of Scientists in Science Fiction. Strange Horizons (May 2004)

Yudkowsky, E.: Artificial intelligence as a positive and negative factor in global risk. In: Bostrom, N., Ćirković, M.M. (eds.) Global Catastrophic Risks. OUP, Oxford (2008)

Fiction

Asimov, I.: Liar. Astounding (May 1941)

Asimov, I.: Lenny. Infinity Science Fiction (January 1958a)
Asimov, I.: A Whiff of Death. Avon, New York (1958b)
Benford, G.: Timescape. Simon and Schuster, New York (1980)
Gunn, J.: The Listeners. Scribner's, New York (1972)
Herbert, F.: The White Plague. G. P. Putnam's Sons, New York (1982)
Narlikar, J.V.: The Return of Vaman. Springer, Berlin (2015)
Sagan, C.: Contact. Simon and Schuster, New York (1985)
Shelley, M.: Frankenstein; or, The Modern Prometheus. Lackington, Hughes, Harding, Mavor and Jones, London (1818)
Stevenson, R.L.: Strange Case of Dr Jekyll and Mr Hyde. Longmans, Green, & Co, London (1886)
Vonnegut, K.: Cat's Cradle. Holt, Rinehart and Winstone, Berlin (1963)
Wells, H.G.: The Island of Doctor Moreau. Heinemann, Stone and Kimball, London (1896)
Wells, H.G.: The Invisible Man: A Grotesque Romance. Pearson, London (1897)

Visual Media

Dr Who: [Television series] UK, BBC (1963–present)
Forbidden Planet: Directed by Fred M. Wilcox. [Film] USA, MGM (1956)
Metropolis: Directed by Fritz Lang. [Film] Germany, UFA (1927)
Quatermass II: Created by Nigel Kneale. [Television series] UK, BBC (1955)
Quatermass and the Pit: Created by Nigel Kneale. [Television series] UK, BBC (1958)
Star Trek: Created by Gene Roddenberry. [Television series] USA, Desilu Productions/Paramount Television (1966–1969)
The Quatermass Experiment: Created by Nigel Kneale. [Television series] UK, BBC (1953)
Them!: Directed by Gordon Douglas. [Film] USA, Warner Bros (1954)
This Island Earth: Directed by Joseph M. Newman and Jack Arnold. [Film] USA, Universal (1955)

12

Epilogue: A New Default Future?

The future starts today, not tomorrow.

Pope John Paul II

For reasons I explained in the Introduction, the SF explored in this book was all published prior to 1985. Of course, the ten core themes I've been discussing continued to fascinate SF authors long after my self-imposed cut-off date. Asimov continued writing about robots[1] until his death in 1992; the SF community never lost interest in the possibilities of alien life[2]; and time travel stories[3] are popular to this day. Stories about antigravity[4] did lose some of their allure, but tales of immortality[5] and invisibility[6] appeared with pretty much the same frequency after 1985 as they did before, while topics such as transportation[7] and the threat of the mad scientist[8] remained obvious targets for the SF writer. Stories examining the nature of reality[9] are even more popular now than they were when Phil Dick was writing: computing advances have made immersive VR a possibility while physicists have improved their understanding of the fundamentals of quantum theory—developments to which the antennae of SF writers are attuned. The situation with that other theme, space travel, is more nuanced: many SF stories continued to be set in space,[10] but authors also began facing up to the realization that space travel is *hard*. Nevertheless, although these ten core themes remain part of the field's DNA it's clear that the 'feel' conjured up by Golden Age science fiction—the default future generated through the visionary work of writers such as Asimov, Heinlein, and Clarke—has not (yet) come to pass. The world in which we now live is not the world that SF envisaged. Why? The failure of SF to accurately foresee our world was, I believe, in large part due to the difficulty of identifying

© Springer International Publishing AG 2017
S. Webb, *All the Wonder that Would Be*, Science and Fiction,
DOI 10.1007/978-3-319-51759-9_12

technologies that scale—and in understanding what happens when technology *does* scale.

For decades, the number of components in an integrated circuit increased exponentially: every 18 months or so, the number doubled. This scaling didn't happen by accident: chip manufacturers had to work together in order to make components steadily smaller, had to cooperate in order to develop new technologies to maintain the scaling. Competitors were willing to coordinate their research and development activities because it was profitable for them to do so: each generation of computer was faster than its predecessors and so customers would buy the new devices. The era of exponentially increasing component numbers on chips appears to be at an end, but computer manufacturers hope that a different approach will continue to deliver exponential improvements in our digital world. Just as a decades-long period of computer scaling changed our world in ways Golden Age SF writers did not predict, a new period of scaling—building on the proliferation of smartphone use, the Internet of Things, and one day perhaps even quantum computing—will change the world in ways we ourselves might struggle to comprehend.

Technology improvements in DNA sequencing have in recent years followed a better-than-exponential path. The implications of such rapid developments in biotechnology are yet to be fully understood by society; the technological horizons are vast. And if we ourselves flounder when trying to predict where this technological scaling will take us, what chance did the authors of Golden Age SF have to foresee our brave new world?

On the other hand, those same SF authors hoped certain technologies *would* scale when it's now clear they won't. For example, many authors assumed the conquest of space by humans would follow a similar pattern to the conquest of the American west. But that central element of the erstwhile default future— the belief that people such as you and me might choose to live their lives away from Earth—hasn't come to pass. The reason is clear: the technology needed to get people into space cheaply, and keep them there safely, hasn't been developed. The first rocket capable of reaching space was launched by Germany in 1942, five years *before* Bardeen, Brattain, and Shockley built the first transistor. Compared to the seven decades of exponential improvement in transistor technology, advances which enabled computing technology to conquer the world, the development of rocket technology over the same period has been glacially slow. The modern rocket is better in all respects than the primitive V2 rocket of 1942; future rockets will be better still, and the appearance of private companies investing in space travel research will speed progress. But it's hard to see how conventional rocketry can deliver the orders of magnitude improvement necessary to turn a niche technology into the mundane.

To the eyes of a modern reader, old-time SF can appear over-optimistic to the point of naivety: the writers trusted science to deliver wonders that haven't happened. But they were *right* to be optimistic—it's just that the wonders science delivered weren't the ones those writers were expecting.

***

SF writers of the past developed a default future—a vision of times to come capable of inspiring generations of scientists and permeating popular culture—so an obvious question arises: have modern SF writers, in response to our deepened understanding of science, settled on a new shared vision of the future?

I believe they have.

One element of a new default future is formed by questions around the continued development of computers. If computing power continues to scale exponentially how might our political and economic systems be affected? What will happen if, through some catastrophe, self-inflicted or natural, we *lose* a technology we are already so reliant upon? Can the concept of privacy survive in a world in which information 'wants to be free'? SF authors, by writing so many stories that address these sorts of issues, have cemented the role of digital technology as a key element of the new default future.

The continued development of biotechnology forms a second element of the new default future. This seems reasonable, given our ever-improving knowledge of how life works. Francis Crick and James Watson identified the molecular structure of DNA in 1953; work on mapping the human genome began in 1990 and was complete by 2003; in 2016, scientists announced a project, HGP-Write,[11] with the aim of *synthesizing* the human genome—a project which, if it's successful, would in principle allow someone to build a human from scratch, with no genetic parents. And, as the technology has improved, costs have dropped. Furthermore, if we consider DNA to be a store of biological information then the advances in digital technology mentioned above can be brought to bear on the problem of life. This is another field that seems likely to continue to scale exponentially, raising possibilities that are exciting yet profoundly troubling—an ideal playground for SF writers.

A third element of the new default future arises through a different example of scaling. Averaged across the globe, the rate at which human civilization is using energy has grown exponentially. Our greed for using energy has been met mainly by the burning of fossil fuels. Over the past century, humans have dug up untold megatons of fossil fuels—in essence, millions of years of frozen sunlight—and burned it all (on geological timescales) in an eyeblink. Our industrial civilization, and all the conveniences it provides, has been made possible only through the venting of hundreds of billions of tons of carbon dioxide into Earth's atmosphere. It is inconceivable that such a massive release of a greenhouse gas will fail to warm our planet. Not since the dawn of civilization have humans experienced a hotter global climate than we're now encountering, and no serious scientist doubts that in future decades Earth will become even hotter. This, then, is another element of the new default future: a world in which climate change has caused disruption.

Computing, biotechnology, climate change—these elements are central to the new shared vision of what lies ahead, to the new default future. My monthly reading of *Asimov's* gives me stories exploring the consequences of having information flow as freely as water does now; stories set on a hothouse Earth, where water flows less freely than information; stories where the form of a human body is a mere lifestyle choice. And, suffusing many of these stories, is a sense of the persistence of inequality. If you are rich, you'll be able to afford the treatment that fixes your Alzheimer's or cancer or infection; if you are poor, you won't. Some things never change.

The new default future has a less optimistic feel than the one I read as a youth. But perhaps it's more realistic. It's certainly just as interesting.

Notes

1. *Asimov's* later stories robot were collected in *Robot Dreams* (1986), *Robot Visions* (1990), and *Gold* (1995). Asimov, as we saw in Chap. 8, was keenly pro-robot. However, non-Asimovian killer robots have become popular in the literature, perhaps because stories in which they feature are more likely to be turned into movies than stories in which robots work peacefully alongside humanity: the success of the *Terminator* franchise, and in particular *Terminator 2: Judgment Day* (1991), led to a spate of 'robots vs people' films. There are signs, though, that directors and producers are starting to explore some of the more subtle aspects of robot technology. For example, the Disney animation *WALL-E* (2008) had a sympathetic robot as its star, while films such as *The Machine* (2013) and *Ex Machina* (2015) featured robots that are complex entities, not necessarily bent on killing their creators.

2. Two of the most interesting aliens of recent years appeared in Ted Chiang's Nebula award-winning novella "Story of Your Life" (1998)

and Peter Watts's novel *Blindsight* (2006). Chiang highlighted the problems inherent in trying to communicate with extraterrestrial intelligence: the alien heptapods in his story take such a different view of the world that human linguists struggle to understand their language. The alien in *Blindsight* has a high level of intelligence and functions extremely effectively, but over the course of the novel it becomes clear that it does not possess consciousness. How effective would communication be with creatures that are intelligent but not conscious? SF excels at posing this sort of question!

3. Inventive authors have continued to ring the changes on the theme of time travel. (An anthology by Ann and Jeff VanderMeer called *The Time Traveller's Almanac* (2013) weighs in at just under 1000 pages, but even at that length it fails to capture the full range of time travel fiction.) For example, Joe Haldeman's Hugo- and Nebula-award winning novella "The Hemingway Hoax" (1990) begins with a scholar being persuaded to write a fake Hemingway manuscript; the academic soon becomes involved with entities who control the destiny of multiple parallel timelines involving our world. Stephen Baxter's award-winning *The Time Ships* (1995), an authorized sequel that appeared a century after the publication of Wells's *The Time Machine*, continued the adventures of the Time Traveller. Connie Willis has imagined a near future in which Oxford University historians conduct field research by travelling into the past as observers, a setting that forms the springboard for her award-winning novels *To Say Nothing of the Dog* (1997), *Blackout* (2010a), and *All Clear* (2010b). Ted Chiang's Hugo- and Nebula-award winning novelette "The Merchant and the Alchemist's Gate" (2007) was a beautiful meditation on our natural desire to move through time and fix mistakes we think we might have made. And the hero of Claire North's novel *The First 15 Lives of Harry August* (2014) was a so-called 'kalachakra'—a person who, at the moment of death, is reborn in the same body, to the same parents, and at the same time as before, but who retains all memories of previous existences. North cleverly demonstrated how, by dripping messages from child to dying adult, kalachakras can alter the past and thus change the future. Five different authors, five different types of work. What a rich variety of storytelling the time travel paradigm permits! The temporal paradoxes explored by SF have even made inroads into general culture. For example, Ray Bradbury's famous story "A Sound of Thunder" is parodied in "Time and Punishment", one of the segments in The Simpsons's *Treehouse of Horror V* (1994): while trying to fix a broken toaster, Homer turns it into a contraption that takes him back in

time. He swats a fly, then finds himself returning to a dystopian horror in which his nerdy neighbor Ned Flanders is world dictator. And the key elements of Lupoff's story "12:01 P.M.", referred to in Chap. 5, were echoed in Grimwood's novel *Replay* (1986) and the wonderful film *Groundhog Day* (1993). These mainstream stories are fun, but science fiction remains the best place to ponder the paradoxes inherent in time travel.

4. Nowadays, anti-gravity does not often feature in stories. There are some exceptions. Duncan Long's novel *Anti-grav Unlimited* (1988), for example, was the tale of an engineer who develops an anti-gravity rod—and thus the key to space travel, perpetual motion, and free energy. Naturally, when big business gets wind of the discovery his life is under threat. It's an old-fashioned story that could have been written decades earlier, when anti-gravity stories were in their pomp. However, while anti-gravity itself might have slipped from fashion, there have been some interesting fictional explorations of the properties of general relativity, our best theory of gravity. The film *Interstellar* (2014), for example, was not only a suspenseful tale—it treated the science seriously. Indeed, the relativist Kip Thorne was a science consultant on the film and he has written about the scientific background to the movie (Thorne, 2014). His book is a great way to learn about some of the counterintuitive possibilities inherent in general relativity, while the movie itself is probably unique in that it formed the basis for a publication in a peer-reviewed physics journal (see James et al. 2015a, b): the scientifically accurate animations used in the film revealed details that were previously unknown.

5. Science fiction excels at highlighting the downsides of immortality. Poul Anderson's *The Boat of a Million Years* (1989) was the story of immortals such as Hanno, a Phoenician sailor born about 1000 BC, who must hop from identity to identity as those around them age. The immortals are stuck forever at a biological age of 25, but they are wise enough to understand that normal humans will be jealous of them—and even an immortal can't withstand a knife through the heart. Dan Simmons's *Hyperion* (1989) posited the existence of an infectious organism called a cruciform. When an infected human dies, the cruciform resurrects its host. The drawback? Well, time and again, for seven years, a priest finds himself resurrected from a death caused by crucifixion to a tree that continuously electrocutes him. Pain unrelenting. Fred Pohl's novella "Outnumbering the Dead" (1992), which appeared in the special tribute issue of *Asimov's Science Fiction* following Asimov's death, posed some interesting questions regarding immortality. In the future sketched by

Pohl, unborn babies in the second trimester are given a treatment to make them immortal. The treatment usually works—but not always. What happens to a mortal in a world of immortals? And, more intriguingly, if you have all the time in the world to do something why bother doing anything? In *Altered Carbon* (2002), Richard Morgan described a future in which memory and personality can be stored digitally and downloaded into a 'sleeve'—either your own cloned body or the body of someone else. Immortality is achieved, therefore, through the implantation of a cortical 'stack' into a person's spinal column: the stack stores the person's memory and then, when the body dies, the stack can be put in store. In principle, a person could live indefinitely. Morgan cleverly examined the implications of all this, and its impact on a society in which such technology is common but not shared equally between the rich and the poor. And of course the repercussions of immortality have always been deemed worthy of exploration by the writers of *Dr Who*. The ninth series of the revamped franchise, which featured the Hugo-nominated episode "Heaven Sent" (2015), played with numerous variations on the theme of longevity. Immortality remains one of the central science fictional themes.

6. SF writers have continued to examine the concept of invisibility from a variety of angles—from straight thriller-type updates of Wells through to more literary meditations. In H.F. Saint's *Memoirs of an Invisible Man* (1987), for example, an accident in a nuclear laboratory causes everything within a 15 m radius to become invisible. The protagonist, who was within the blast radius, wakes to find himself hunted by government agencies—his invisibility would, after all, make him the perfect spy. Saint's novel was heavily influenced by the Wells classic. China Miéville's *The City and the City* (2009), on the other hand, explored the notion of psychological invisibility. The two cities of the title are Besźel and Ul Qoma, the precise locations of which are not mentioned but would seem to be somewhere near the Black Sea coastal regions of Bulgaria and Romania. In fact, there's only *one* location for these cities: Besźel and Ul Qoma share the same geographical space, but their citizens are taught from childhood to 'unsee' anything to do with the other city. A Besź resident will recognize an Ul Qoman resident, will apprehend the existence of Ul Qoman buildings and of events taking place in the twin city, but will consciously erase from memory all those sights. And vice versa: an Ul Qoman will 'unsee' anything to do with Besźel. Against this bizarre background the novel tells the story of a Besź detective who, charged with investigating the murder of a foreign student, finds the case takes him into

Ul Qoma. Christopher Priest, in his haunting novel *The Glamour* (2005), dissected the notion that invisibility could be a failure of memory. A man wakes up after being caught by the blast from a car bomb to find he's lost entire sections of his memory. He tries to recover his recollections with the help of a woman who claims to have been his girlfriend. Both characters share the narration and, as the story progresses, we meet a character who is invisible—although our view of whether that invisibility is metaphoric or literal depends upon who is narrating. And in his novel *Blindsight* (2006), already mentioned above, Peter Watts suggested an extremely ingenious approach to obtaining invisibility. Psychologists have discovered how, in the human visual system, the eye undergoes rapid movements and the brain assembles a mental image of a scene from the data it receives. Watts suggests that a sufficiently intelligent creature would be able to read the state of a person's nervous system in real time and so compute the moments during which movement would not be noticed. If it could do that then the creature would be concealed in the inevitable blind spots of a person's visual system.

7. One of the most realistic depictions of our near-term automotive future, and of how our transport requirements will fit in with our increasingly urban existence, appeared in Kim Stanley Robinson's novel *The Gold Coast* (1988). Robinson paints a dystopian picture of Southern California in 2027, a place where autonomous cars run on tracks stacked high in the air—an engineering approach made necessary because developers have run out of space on the ground. Drug taking is rife; people are rich, but unhappy; the 'autopia' Robinson describes is horribly plausible. More recently, Rich Larson's humorous short story "Bidding War" (2015) gave a 'day-after-tomorrow' look at a world in which people combine smartphone apps with car ownership to offer Uber-style journeys. Larson's future is probably already happening. More science fictional transport possibilities remain popular; SF writers continue to make use of matter transmission, shortcuts through space, and projects requiring vast feats of engineering. In *Pandora's Star* (2004), for example, Peter Hamilton imagined an interstellar transport system based on wormhole-based transmitter gates through which run a system of railways. In Kim Stanley Robinson's Nebula-award-winning novel *2312* (2012) humans on the planet Mercury live in a city called Terminator. The city has been constructed on vast railtracks, so that it's able to trundle along and stay in the shade of the planet's nightside. (Perhaps Hamilton and Robinson were telling us trains and track-based transportation have a future!)

8. The mad scientist cliché limps on in horror films (the genre contains any number of psychotic medical practitioners), but for more interesting treatments of the idea it remains necessary to look to SF. Michael Crichton's *Jurassic Park* (1990), for example, contains the character Dr Henry Wu—a biotechnologist who heads a team that brings dinosaurs back to life through cloning technology. (Steven Spielberg, of course, turned the novel into the blockbuster film *Jurassic Park* (1993).) And Greg Bear's novel / (1997)—yes, / is the title, although for obvious marketing and categorizing purposes it is also called *Slant*—contains one of the more fascinating mad scientists in the genre. Seefa Schnee, a scientist who studies AI, induced in herself a form of Tourette's in order to enhance her creativity. Schnee is obsessed with the idea of creating an organically based artificial intelligence; she is also a tragic victim of unrequited love. She succeeds in creating a 'recombinant optimized DNA device' whose mind/body is made of mud, but the object of her love finds nothing but disgust in her brainchild.

9. Of all SF writers it's Phil Dick, of course, who is most closely associated with stories that raise questions about the nature of reality. His novels have been the inspiration, either directly or indirectly, for a number of movies on this theme. For example, the creators of *The Truman Show* (1998) must surely have been influenced by *Time Out of Joint*—Dick's novel in which the protagonist inhabits a simulated reality. More recent examples of Dick-influenced films include *The Adjustment Bureau* (2011), which was based on the story "Adjustment Team" (1954), and *Total Recall* (2012), based on the story "We Can Remember It For You Wholesale" (1966). However, stories involving computer-mediated virtual reality—a concept that's nowadays more a matter of engineering than fantasy—only began to appear with increasing frequency after 1985, which was three years after Dick's death. Gibson's seminal novel *Neuromancer* (1984) was the trigger for the avalanche of stories involving virtual reality (and also for the numerous films on the theme, including of course *The Matrix* (1999)). To give just one example of the sort of story that began to appear, consider the novelette "The News from D Street" (1986) by Andrew Weiner. The protagonist is Joseph Kay, an 'inquiry agent'—some type of hard-boiled private detective—who is asked to investigate a missing person case. The case widens, and Kay learns of the disappearance of several more people from his city. The story starts out in a style typical of film noir, but Kay can't figure out some key elements—such as why he doesn't know the name of his city and why people's memories are so restricted. Eventually he understands that 'the news is not good; the news is very bad': it turns

out (and here comes the inevitable spoiler) that the city and everything in it, including the detective, are simulations developed for the purposes of sociological study. The missing people were just data; their disappearance on a bus that takes them out of town is, in reality, the result of data transfer on a computer bus. Since then, a number of SF writers have written novels in which virtual reality plays a key role—from Neal Stephenson's *Snow Crash* (1992) and Greg Egan's *Permutation City* (1994) to Charles Stross's *Accelerando* (2005) and Hannu Rajaniemi's *The Quantum Thief* (2010)—and the theme continues to grow in influence as, in the real world, VR devices become consumer items.

10. The solar system continues to offer a backdrop for SF stories, and writers still develop backgrounds in which space travel is integral to the plot. The Moon remains a preferred location. *Asimov's Science Fiction* alone has published enough stories to merit a lunar-themed anthology—*Isaac Asimov's Moons* (1997)—and the magazine continues to publish stories with a lunar setting. The March 2015 issue, for example, contained "Inhuman Garbage" by Kristine Kathryn Rusch; it's a detective story set in Armstrong City, Luna, where both police and organized crime struggle to understand the motive behind a young woman's murder. Many novels have a lunar setting—Varley's *Steel Beach* (1993), Bova's *Moonwar* (1997), and McDonald's *Luna: New Moon* (2015) spring immediately to mind. Astronauts might have abandoned the Moon after Apollo, but SF writers never left our twin planet. Mars, too, remains popular. One of the most successful Martian writers is Kim Stanley Robinson. His story "Green Mars" (1985) appeared in *Asimov's Science Fiction* and later he published three books—the award-winning *Red Mars* (1993), *Green Mars* (1994), and *Blue Mars* (1996)—that describe the colonization, terraforming, and long-term future of the planet. *The Martians* (1999) is a collection of stories set in the same fictional universe. Robinson explores not only the scientific aspects of future human life on Mars, but also the political and ecological debates that surround it. Allen Steele is another who has published stories with Mars as the setting. His Hugo award-winning novelette "The Emperor of Mars" (2010), for example, was about a roughneck stationed on a Mars colony who seeks mental refuge from a personal tragedy by reading old-time SF about the Red Planet. Geoffrey Landis, a NASA scientist who engages in real-world research on robotic Mars missions, set his novel *Mars Crossing* (2000) just a few years from now, in 2028. It tells of the difficulties faced by the third international expedition to the Red Planet. And Andy Weir's tale of interplanetary rescue, *The Martian* (2014), formed the basis of the 2015 Ridley Scott

film of the same name. So Mars remains part of the heritage of SF—and it's also now in the public consciousness. The wider solar system has also continued to provide a canvas for SF writers. Alastair Reynolds's novel *Blue Remembered Earth* (2012), for example, contains scenes set on the Moon, Phobos, Mars... but also in the Kuiper Belt, in the reaches far beyond Neptune. Closer to home, Robinson, in his novel *2312* (2012), imagined a Mercury in which the great city Terminator rolls continuously on elevated tracks so as to stay out of the Sun's glare. And in his 'Grand Tour' series of novels, of which *Moonwar* mentioned above is part, Bova explored settings on all the planets of the solar system. But I believe stories set *beyond* the solar system appear less often than they did in the Golden Age—and when authors *do* venture further afield they tend to accept it's going to be a long, slow trudge to the nearest stars. In Steele's novel *Coyote* (2002), for example, a starship travelling at $0.2c$ carries a crew of 104 in suspended animation and delivers them to a new world, Coyote, which is 46 light years distant. The novel and its sequels describe the tedious journey and the hardships involved in colonizing Coyote. Robinson's novel *Aurora* (2015) describes an attempt, based on generation ship technology, to reach and colonize the moon of a planet in the Tau Ceti system, 12 light years away. The fundamental human urge to explore is set squarely against the huge technical difficulties involved in operating a generation ship. And beyond the nearest stars? SF authors continue to write space opera. Galaxy-spanning empires, constructed upon some form of faster than light spacecraft, remain a popular theme. But the technology that might deliver such a future is as plausible now as it was when Asimov wrote his famous *Foundation* series—in other words, it won't come to pass any time soon.

11. For details of the HGP-Write proposal, see Boeke et al. (2016).

Bibliography

Non-fiction

Boeke, J.D., et al.: The Genome Project – Write. Science. (2016). doi:10.1126/science.aaf6850

James, O., von Tunzelmann, E., Franklin, P., Thorne, K.S.: Gravitational lensing by spinning black holes in astrophysics, and in the movie Interstellar. Class. Quant. Grav. **32**(6), 065001 (2015a)

James, O., von Tunzelmann, E., Franklin, P., Thorne, K.S.: Visualizing Interstellar's wormhole. Am. J. Phys. **83**, 486 (2015b)

Thorne, K.S.: The Science of Interstellar. Norton, New York (2014)

Fiction

Anderson, P.: The Boat of a Million Years. Tor, New York (1989)

Asimov, I.: Robot Dreams. Berkley, New York (1986)

Asimov, I.: Robot Visions. Roc, New York (1990)

Asimov, I.: Gold. Harper Prism, New York (1995)

Baxter, S.: The Time Ships. HarperCollins, London (1995)

Bova, B.: Moonwar. Hodder and Stoughton, London (1997)

Chiang, T.: Story of your life. In: Hayden, P.N. (ed.) Starlight 2. Tor, New York (1998)

Chiang, T.: The Merchant and the Alchemist's Gate. Fantasy & Science Fiction (September 2007)

Dick, P.K.: Adjustment Team. Orbit (September–October 1954)

Dick, P.K.: We Can Remember It For You Wholesale. Fantasy & Science Fiction (April 1966)

Dozois, G., Williams, S.: Isaac Asimov's Moons. Ace, New York (1997)

Egan, G.: Permutation City. Orion, London (1994)

Gibson, W.: Neuromancer. Ace, New York (1984)

Grimwood, K.: Replay. Arbor House, Westminster MA (1986)

Haldeman, J.: The Hemingway hoax. Asimov's (April 1990)

Hamilton, P.F.: Pandora's Star. Pan Macmillan, London (2004)

Landis, G.A.: Mars Crossing. Tor, New York (2000)

Larson, R.: Bidding war. Asimov's (December 2015)

Long, D.: Anti-grav Unlimited. Avon, New York (1988)

McDonald, I.: Luna: New Moon. Gollancz, London (2015)

Miéville, C.: The City and the City. Macmillan, London (2009)

Morgan, R.K.: Altered Carbon. Gollancz, London (2002)

North, C.: The First 15 Lives of Harry August. Orbit, London (2014)

Pohl, F.: Outnumbering the dead. Asimov's (November 1992)

Priest, C.: The Glamour. Gollancz, London (2005)

Rajaniemi, H.: The Quantum Thief. Gollancz, London (2010)

Reynolds, A.: Blue Remembered Earth. Gollancz, London (2012)

Robinson, K.S.: The Gold Coast. Tor, New York (1988)

Robinson, K.S.: Red Mars. Random House, New York (1993)

Robinson, K.S.: Green Mars. Random House, New York (1994)

Robinson, K.S.: Blue Mars. Random House, New York (1996)

Robinson, K.S.: The Martians. Random House, New York (1999)

Robinson, K.S.: 2312. Orbit, New York (2012)

Robinson, K.S.: Aurora. Orbit, New York (2015)
Rusch, K.K.: Inhuman garbage. Asimov's (March 2015)
Simmons, D.: Hyperion. Doubleday, New York (1989)
Steele, A.M.: Coyote. Ace, New York (2002)
Steele, A.M.: The Emperor of Mars. Asimov's (June 2010)
Stephenson, N.: Snow Crash. Bantam, New York (1992)
Stross, C.: Accelerando. Orbit, London (2005)
VanderMeer, A., VanderMeer, J.: The Time Traveller's Almanac. Head of Zeus, London (2013)
Varley, J.: Steel Beach. New York, Berkeley (1993)
Watts, P.: Blindsight. Tor, New York (2006)
Weiner, A.: The news from D Street. Asimov's (September 1986)
Weir, A.: The Martian. Crown, New York (2014)
Willis, C.: To Say Nothing of the Dog. Bantam Spectra, New York (1997)
Willis, C.: Blackout. Bantam Spectra, New York (2010a)
Willis, C.: All Clear. Bantam Spectra, New York (2010b)

Multimedia

Dr Who (Heaven Sent): Written by Steven Moffat. [TV episode] UK, BBC (2015)
Ex Machina: Directed by Alex Garland. [Film] UK, Universal (2015)
Groundhog Day: Directed by Harold Ramis. [Film] USA, Columbia (1993)
Interstellar: Directed by Christopher Nolan. [Film] USA, Paramount (2014)
Terminator 2: Directed by James Cameron. [Film] USA, TriStar (1991)
The Adjustment Bureau: Directed by Gorge Nolfi. [Film] USA, Universal (2011)
The Machine: Directed by Caradog James. [Film] UK, Content Media (2013)
The Martian: Directed by Ridley Scott. [Film] USA, 20th Century Fox (2015)
The Matrix: Directed by Wachowski Bros. [Film] USA, Warner Bros (1999)
The Truman Show: Directed by Peter Weir. [Film] USA, Paramount (1998)
Total Recall: Directed by Len Wiseman. [Film] USA, Columbia (2012)
Treehouse of Horror V (The Simpsons) Time and punishment: Directed by Jim Reardon. [TV episode] USA, Gracie Films (1994)
WALL-E: Directed by Andrew Stanton. [Film] USA, Walt Disney (2008)

Index

© Springer International Publishing AG 2017
S. Webb, *All the Wonder that Would Be*, Science and Fiction,
DOI 10.1007/978-3-319-51759-9